RLC CIRCUITS

Time Constants

$$t = \frac{L}{R} \qquad t = RC$$

Reactance

$$X_L = 2\pi f L \qquad X_C = \frac{1}{2\pi f C}$$

Impedance

$$Z = \sqrt{R^2 + X_T^2}$$

Phase Angle

$$\text{cosine } \theta = \frac{R}{Z}$$

RESONANT CIRCUITS

Resonant Frequency

Q

Bandwidth

$$f_R = \frac{1}{2\pi \sqrt{LC}} \qquad Q = \frac{X_L}{R} \qquad BW = \frac{f_R}{Q}$$

TRANSFORMER CIRCUITS

Voltage

$$\frac{V_s}{V_p} = \frac{n_s}{n_p}$$

Impedance

$$\frac{Z_s}{Z_p} = \left(\frac{n_s}{n_p}\right)^2$$

F V

 St. Louis Community College

Forest Park
Florissant Valley
Meramec

Insructional Resources
St. Louis, Missouri

SECOND EDITION

INTRODUCTION TO ELECTRONICS

PATRICK CROZIER

Quincy Vocational–Technical School
and Quincy Junior College

Breton Publishers
Boston, Massachusetts

PWS PUBLISHERS

Prindle, Weber & Schmidt ·✺· Duxbury Press · ♠ · PWS Engineering ·△· Breton Publishers ·✲·
20 Park Plaza · Boston, Massachusetts 02116

PWS Publishers is a division of Wadsworth, Inc.

Library of Congress Cataloging-in-Publication Data

Crozier, Patrick, 1934–
    Introduction to electronics.

    Includes index.
    1. Electronics. I. Title.
TK7816.C74 1987   621.381   86-25522
ISBN 0-534-07644-0

Printed in the United States of America
1 2 3 4 5 6 7 8 9 — 91 90 89 88 87

Sponsoring editor: *George J. Horesta*
Editorial assistant: *Susan M.C. Caffey*
Production: *Technical Texts, Inc./Sylvia Dovner*
Art coordinator: *Elizabeth Slinger*
Interior design: *Technical Texts staff with Jean T. Peck*
Cover design: *Sylvia Dovner*
Composition: *Compset, Inc.*
Cover photo: *Steve Grohe/The Picture Cube*
Cover printing: *New England Book Components, Inc.*
Printing and binding: *The Maple-Vail Book Manufacturing Group*

Acknowledgments

   Chapter 1: Figure 1.1, Wang Labs; Figure 1.2, U.S. Army.
   Chapter 2: Figure 2.2, General Electric Company; Figure 2.4, Texas Instruments.
   Chapter 3: Figure 3.1, GC Electronics, Rockford, Ill.; Figure 3.3, Static Control Systems/3M.
   Chapter 5: Figure 5.10, Photo-Fiber Optic Products–Amphenol Products.
   Chapter 6: Figure 6.3 and 6.9, Murata Erie North America, Inc.
   Chapter 7: Figure 7.1, Simpson Electric Company; Figure 7.5 and 7.6, Triplett Corporation; Figure 7.7, Simpson Electric Company.
   Chapter 10: Figure 10.1, 10.2, and 10.3, Klein Tools, Inc.; Figure 10.5, Hexacon Electric Company; Figure 10.6, GC Electronics, Rockford, Ill.
   Chapter 11: Figure 11.1, Dynapert; Figure 11.2, J&J Industries, Inc.; Figure 11.11, Panduit Corporation; Figure 11.12, Hexacon Electric Company; Figure 11.13, Ungar, Division of Eldon Industries, Inc.; Figure 11.15, NASA.
   Chapter 13: Figure 13.6, Murata Erie North America, Inc.
   Chapter 19: Figure 19.1, Tektronix, Inc.
   Chapter 20: Figure 20.8, Simpson Electric Company.
   Chapter 21: Figure 21.3, Hewlett Packard; Figure 21.4 and 21.5, Hickok Electrical Instruments Co., Inc., Cleveland, Ohio; Figure 21.7, B-K Precision Products/Dynascan; Figure 21.17, Triad-Utrad.
   Chapter 22: Figure 22.2, Triplett Corporation; Figure 22.7, Simpson Electric Company.
   Chapter 29: Figure 29.16, © Martin Marietta Corporation, 1984; Figure 29.21B, Motorola, Inc.
   Chapter 32: Figure 32.1, 32.2, and 32.3, Instron Corporation.

To my parents, Patrick and Mary Crozier

PREFACE

Introduction to Electronics is intended to assist each student in taking that first step toward a technical-level career in electronics. Some electronics technicians build, test, and troubleshoot electronic products, while others install, operate, and service large electronic systems. There is a wide range of opportunities for qualified technicians in the electronics industry. The key word here is *qualified*.

The primary qualification for employment in electronics is an understanding of how electronic products and systems operate. For that understanding, a technician needs to know how active devices such as diodes and transistors behave, both individually and when combined with other components. But first, a technician must understand resistance, capacitance, and inductance and their relationships with DC and AC voltages.

The text begins with DC theory, and topics range from Ohm's law and series circuits to series-parallel circuits and networks. Next, the text considers AC theory; topics extend from generation and transformers to *RLC* circuits and resonance. Finally, the text discusses diodes, transistors, and integrated circuits as an introduction to future courses.

Knowledge alone will not produce success, for a technician must also perform. To assist the student with future performance, the book introduces various skills. Thus, tools, safety, and assembly techniques are introduced in the early chapters. Theory and operation of multimeters and oscilloscopes follow. Test, evaluation, and troubleshooting techniques are then discussed. The need for and methods of lab report writing are also covered.

Underlying both theory and application is a message about attitude. A technician's attitude determines work habits. Safe work habits are expected, as is respect for co-workers and equipment. Thus, technicians should strive for quality, since many people may depend on the results of their work. In summary, this book focuses on the factors that lead to success as an electronics technician: knowledge, skill, and attitude.

This second edition incorporates the industry changes in component descriptions, symbols, and codes. Also, more digital and less analog instrumentation is included. Integrated devices are emphasized, since vacuum tubes maintain only a limited presence today. However, troubleshooting and diagnosis are still considered essential skills, as they were in the first edition. Finally, instructor comments have resulted in more examples, questions, and problems in this second edition.

Many people and organizations have assisted in the preparation of this book, and to them, I express my gratitude. I thank those who provided photographs and other materials. I also thank Frank Mann, Southern Technical College, Little Rock, Arkansas; Gerald R. Sevigny, Southern Maine Vocational-Technical Institute, South Portland, Maine; Edward H. Waller, Sr., Delgado College, New Orleans, Louisiana; John Fitzen, Idaho State University, Pocatello, Idaho; E. F. Marsoobian, California State University, Los Angeles, California; and Gordon Eddins,

Mid-Florida Technical Institute, Orlando, Florida, for reviewing and commenting on the manuscript. Finally, special thanks to my editor, George J. Horesta of PWS-Kent Publishing Company and to Sylvia Dovner and the production staff of Technical Texts, Inc., for producing a book from my manuscript and getting it into your hands. With all of this assistance, however, I accept full responsibility for the accuracy of content.

TO THE STUDENT

The electronics industry employs millions of people in a wide range of occupations. Some work for the government, and others work for universities. Most, however, are involved in the manufacture, installation, operation, or servicing of electronic products and systems. Job levels in electronics are based on the amount of technical knowledge and skill required to perform the tasks. Common levels are semiskilled, skilled, technical, and professional. Semiskilled and skilled workers generally need little theoretical preparation; those at the professional level usually have four or more years of college. This book is written for the person preparing for an electronics career at the technical level.

Technical-level workers build experimental circuits, test circuits for proper operation, locate defects, and make necessary repairs. Technicians install products such as mobile telephones and systems such as mainframe computers. In broadcasting, technicians operate cameras, mixers, switchers, and transmitters. They also service cable and satellite TV and personal VCRs. Medical electronics technicians install, align, calibrate, and repair critical instruments and systems. In sum, the opportunities for an electronics technician are many and varied.

To be employable as an electronics technician, you must meet certain requirements. First, you need knowledge, an understanding of the principles of electronics theory. These principles begin with DC and AC theory, the primary focus of this book. You will learn about components such as resistors, inductors, and capacitors. You will learn how they behave, alone and combined with others, when subjected to DC and AC voltages. More complex components such as diodes, transistors, and integrated circuits, which will be extensively detailed in later courses, are introduced in this book.

The second qualification you need is skill. You must be able to perform. Knowing is not enough; in electronics, you get paid for doing. Therefore, this book introduces tools, assembly skills, and the handling of sensitive components. For example, once a circuit is built, it must be checked; if defective, it must be repaired. Thus, this book shows you how to use multimeters and oscilloscopes for the basic measurements needed in such maintenance work. And since you must be able to communicate your activities and results to others, lab report writing is included.

Attitude is the third qualification. Employers are very concerned about how you perform; and your behavior is greatly influenced by your attitude. Do you work safely? Are your results neat and accurate? Can you get along with others? Do you accept your employer's way of doing a job, even though you know that there are many ways to do it? These factors affect production costs, marketing success, company reputation, and, in turn, whether or not your employer remains in business and you keep your job. Therefore, attitude is as important as knowledge and skill. You will be reminded of this issue throughout the book.

This book is called *Introduction to Elec-*

tronics because it is the first in a series you will use in your study of electronics. When you finish this book, you will understand the electrical part of electronics. You will know how passive components react to DC and AC voltages, and you will be ready for the next level. That next level is a more detailed look into active devices, such as diodes, transistors, and integrated circuits. At the end of this course, you will also know some basic facts about circuits. And again, succeeding courses will teach you more. Thus, the study of electronics is progressive, with each subject building on the previous one and leading to another. So, whether your goal is computers, communications, instrumentation, or some other specialty, it all begins here.

CONTENTS

CAREERS IN ELECTRONICS

State definitions of electricity and electronics.

Name the levels of occupations in electronics and give an example of each.

Name and describe the technical-level occupations in electronics.

Describe the tasks performed by individuals employed in the various technical-level occupations in electronics.

Describe the preparation for a technical career in electronics.

ELECTRONICS AND ELECTRICITY

The purpose of this chapter is to provide a general introduction to the ever-expanding field of electronics. The chapter first considers electronics and electricity both as science and as industry. Then, it focuses on the range of occupational levels and the requirements for success in a career in electronics.

As an Industry

Electronics is among the most widespread and rapidly expanding fields in the industrial world. Every month, new products enter the market. Most of them do not eliminate old products but add to the variety available. For example, television went from black and white to color, to videocassettes, to videodisks, and to video games. Electronic calculators were followed by home computers. Automobiles have become increasingly electronic, with AM and FM radios, CB radios, tape players, alarms, and even microprocessor systems.

Millions of people in the United States work in the design, manufacturing, sales, installation, and servicing of electronic devices. They may work for a large company, a school, or a government agency, or they may be self-employed. The range of jobs is extensive and the opportunity for growth almost unlimited.

While the electronics industry covers a wide range of occupations, these occupations have much in common. All electronics workers share a language of terms and symbols, a knowledge of certain theories, and some essential skills. Many of these terms, symbols, theories, and skills also apply to the electrical industry. The main differences between the two industries are that the electrical industry involves work with home and industrial wiring, motors, and generators, and that its main concern is the generation and transmission of electric power. This book focuses on the electronics industry.

As a Science

A typical dictionary definition of electronics is a scientific one. The definition usually describes physical events that occur in space or in a gas, events that involve ions, electrons, and other invisible elements. To advance in the field of electronics, you must understand this scientific view. While this book is simply an *introduction* to the science of electronics, an overview can be presented. This overview involves electricity.

Electricity involves the movement of electrons through wires. The source of energy to move these electrons may be a battery, a power station, or some other generator. This electricity is available at different voltages and can be used to operate lights, motors, stoves, and a wide range of devices. Therefore, products that operate from this source of energy are called electrical products.

Televisions are electrical products, but they are also electronic. Electronic devices determine the way electricity is used by the television in the process of receiving a signal and producing sound and a picture. Electronic devices also control the electricity going from the rear of the picture tube to the front to produce a picture. The method through which the transistors process this weak signal picked up by the antenna and amplify or strengthen it to produce sound at the speaker is also electronic.

While you probably do not know how a television works, you do know the result. A television uses electricity to receive a signal with its antenna and reproduce sound and a picture. A scientific description of what happens inside includes oscillation, amplification, and rectification performed with the aid of parts and processes to be described later. These processes are all part of the world of electronics.

Electronics Defined

As indicated earlier, a precise definition of electronics is difficult since there are many ways to view electronics. What is more valuable is a general understanding of the scope of electronics and how it differs from electricity.

Recall that electricity can be defined as the flow of electrons through circuits. Motors, lights, air conditioners, and refrigerators are electrical. Stereos, radios, televisions, and pocket calculators are electronic.

While transistors or other semiconductors provide the best indication that a device is electronic, there are other indicators. Vacuum tubes were the first electronic components, and semiconductors replaced them to form a second generation. Now, integrated circuits have become the third generation. Therefore, **electronics** may be defined as that field of science and industry that involves the use of vacuum tubes, semiconductors, and integrated circuits for information processing. It is the physics of electrons flowing in these active devices.

RANGE OF OCCUPATIONAL LEVELS

People are employed in electronics over a wide range of occupational levels. Semiskilled jobs require the least amount of preparation, while professional-level jobs require four years of college. Duties, responsibilities, and pay differ at the occupational levels. Semiskilled positions require the least preparation and, therefore, generally provide the lowest salary.

Semiskilled Level

Many people are employed at semiskilled occupational levels in electronics. Stockroom workers sort, inventory, and distribute a multitude of parts. Other workers pack, transport, and install equipment. Some manufacture individual parts. Still others lace wires into cables and attach connectors.

Semiskilled positions require little training and involve sorting, packing, and distribution of parts. They provide an excellent way for people to begin a career in electronics. The opportunities for advancement are very good. For those who may choose not to advance to a higher level, the advantage of job security remains. Parts will always be manufactured, stored, and shipped. While the types of parts needed may change, many basic skills will not. When one part is no longer needed, these people will produce, sort, inventory, and distribute another. Entry into electronics at this level requires a thorough knowledge of the names of parts, tools, and equipment, as well as the ability to work safely and efficiently.

Skilled Level

At the **skilled level,** electronics work requires some training and is primarily assembly work. Electronic assemblers produce a wide range of items from small printed circuit boards to hand-wired cabinets. The quality of their workmanship is evaluated by quality control inspectors, while proper operation is ensured by production testers. These skilled workers, along with many others like them, are primarily responsible for the manufacture of most of our electronic products.

The qualifications of skilled workers are somewhat more advanced than those of semiskilled workers. More theoretical knowledge is necessary. Many of the people in skilled positions began their career in a semiskilled job. The chance for advancement often inspired them to take courses or attend company training programs in order to qualify for higher-level, higher-paying jobs.

Skilled workers in one segment of electronics can often find work in other segments because of their broad electronics knowledge and skills. With some preparation beyond the high school level, the average person can develop a solid background and qualify for entrance into a skilled electronics occupation. Transfer into other positions at the skilled level becomes easier as experience is gained.

Technical Level

The skilled worker needs further preparation to enter a technical-level position. While workers at the **technical level** often build and test devices, just as skilled workers do, they also have other, more complex tasks to perform. People in technical jobs operate test equipment and repair electronic devices. They also install and maintain equipment like computers or operate equipment such as that found in the broadcasting industry. Technical-level workers are expected to review test results, diagnose the cause of equipment malfunctions, make repairs, and check equipment for proper operation.

Diagnosing requires an understanding of how devices function. The need for this theoretical understanding leads to another major difference between skilled and technical-level workers, namely, preparation. A minimum of two years of technical schooling and a commitment to continued study is typical of those who work at this level.

Professional Level

Products that are assembled at the skilled level and evaluated at the technical level are usually designed at the **professional level.** Most professionals are engineers who have a minimum of four years of college.

In the progression from the skilled level to the professional level, the technician works less with equipment or parts and more with design manuals or drawings. For example, someone at the professional level may hardly ever touch a soldering iron, a meter, or an electronic circuit. The work location also changes, from a workbench to a desk. Mental abilities become more important than manual abilities, and projects often take months rather than minutes. The greater theoretical understanding of electronics that is needed requires, in turn, a greater understanding of other subjects, such as mathematics and physics.

Many engineers were once technicians who began with some training and returned to school to learn more. Such advancement is typical in electronics. Continual development of new principles and devices and the opportunity for growth give people in the industry the desire to learn more.

How you prepare for an electronics job depends mostly on which of the four levels you wish to enter. Since this text is intended for use in the first year of a technical school or a community college, it is assumed that you wish to enter at the technical level, and therefore, the discussion will continue accordingly.

TECHNICAL-LEVEL JOBS

Community college and technical school students usually find employment at the technical level. On the job, they are usually called technicians. While duties vary from one employer to the next and from one type of product to another, some job characteristics are similar. In general, technicians build, test, evaluate, and repair electronic products.

Manufacturing

Think of all of the electronic devices you can. For example, for the home, there are televi-

FIGURE 1-1
COMPUTER MANUFACTURING

sions, radios, stereos, and computers, to name a few. Then, there are electronic devices for offices, banks, schools, cars, planes, and boats. The companies that manufacture these products employ many people at the technical level.

Every new product is built as a prototype, or model. Technical workers assemble, test, evaluate, and modify every section of the prototype to ensure the quality of the design. This process can take months and sometimes years to accomplish.

Once a new product is ready for mass production, the technical worker's job is done, and skilled workers take over. However, although the technical worker may have finished with' the prototype, improved models must be developed. Because the time from design to production is so long, new designs are usually well under way once production on the initial model has begun. In addition, a customer may want a standard product with slight modifications, and technical people are usually assigned the task of producing the modified device. Other technical workers get new products into production by helping skilled and semiskilled workers learn efficient ways to perform their tasks. Of course, complex systems, such as computers, require qualified technicians as the product approaches completion. Figure 1–1, for example, shows a technician working with microcomputers on the production line.

Another group of technical people is employed in companies that manufacture non-electronic products. These technicians maintain electronic devices used in the manufacturing processes, such as the computer-controlled robots used in the manufacture of automobiles. There are few manufacturers who do not use electronic equipment to make their products. Thus, in manufacturing alone, opportunities for the technical worker are great.

Research

Some technical people work with the professional staff in research departments of manufacturing concerns, assisting in experiments that will lead to the development of new products. But others are involved in research efforts not intended to produce new products. Most universities conduct research in agriculture, space, and a wide range of other areas, and they usually use electronic equipment in their experiments. Hospitals use electronic equipment in medical research. The most powerful telescopes are electronic, as are the most powerful microscopes. While the location and purpose of research may vary from place to place, the basic electronic equipment used works the same way. And technical people are needed to operate and maintain it.

There are, however, some major differences between technical work in manufacturing and technical work in research. The manufacturing scene includes the pressure of time, for time is money. In research, the pressure of time is not as intense. Manufacturing also provides the reward of seeing the finished product. In research, the product is often simply a report or a prototype. You should carefully consider these differences between the two types of technical jobs before selecting a specific direction.

Operation

The manufacturing and research fields just described do have one thing in common. They generally involve the assembly and evaluation of experimental or prototypal circuits. The operation of equipment, however, involves little assembly and essentially no experimentation.

Radio and television broadcasting as well as TV cablecasting provide employment for many technically qualified people. While it is true that some of the workers maintain equipment, most of them operate it. For example, in television, positions such as audio technician, camera operator, and technical director are open to well-trained people. Since high-quality cameras and videotape recorders can cost more than $100,000, it is not hard to understand why their operators need extensive technical training.

Many manufacturers use electronic process control systems, and technical people are needed to operate these systems. While some duties may include maintenance and repair, most of the technician's time is spent monitoring the operation of equipment, making adjustments when necessary, and keeping a record of meter readings according to a schedule.

An operator generally works with high-quality and expensive equipment that is well beyond the design or prototypal stage. The equipment has been tested at the factory, shipped to the site, installed, and put in working order. Other technical people at the factory did the troubleshooting; now, the operator's job is to keep the equipment functioning. While operators do not generally do troubleshooting, they must have the ability to do so on those occasions when equipment fails. Then, operators must quickly identify and resolve the problem.

Installation

At one time, a new television was installed by a person from the store where it was purchased. Installation consisted of mounting, aiming, and connecting the antenna, followed by positioning, adjusting, and demonstrating the set. Today, TVs are easier to install and more reliable than the early versions, making customers more confident. Most people buy a television, plug it in, and adjust it themselves.

Installation covers a wide range of complexity. While many people install their own car radios, telephones used in cars and trucks must be installed and adjusted by licensed technicians. Aviation and shipboard radios and radar systems also require a licensed installer, even though these are relatively simple devices to install.

The installation of a large mainframe computer requires a factory-trained team and can take a number of days. Every step is carefully spelled out in installation manuals, and system tests are made along the way. Because of the size and complexity of these systems, bugs arise and must be removed as the installation progresses. Debugging a system like the one shown in Figure 1–2 can only be accomplished by someone who has a working knowledge of the complete system and who is highly skilled in troubleshooting.

Some common and familiar institutions illustrate the range of electronics installation jobs. Inside a bank, for instance, are fire, smoke, motion, and intrusion alarms; video-

FIGURE 1-2
SYSTEM CHECKOUT

cameras and monitors; electronic copiers; and computers and computer terminals. Supermarkets use music and paging systems as well as electronic checkout devices. Wherever you go — school, office, store, or vehicle — you can almost always find equipment that had to be installed by someone with technical electronics training.

Service

All the equipment that is designed, manufactured, and installed must be serviced. Service involves two areas. The first and the most well known area is repair for equipment that fails. For example, the television technician comes to your house when the television stops working properly. You return your computer to the store when it fails.

The second area of service is field service. Most suppliers of complex electronic equipment offer a field service plan whereby a technician comes to the site on a regular basis, not to repair a failure but to reduce the possibility of failure. The process varies from one type of equipment to another, but it generally involves cleaning, lubricating, inspecting, replacing wearing or worn parts, and making any necessary adjustments with the aid of test equipment.

A good analogy to field service is car maintenance. An occasional check of the front-end alignment can reduce tire wear. Likewise, a regular chassis lubrication and oil change can extend the normal life of the car. The avoidance or early detection of possible failure can mean considerable savings to a customer, both in dollars and in the inconvenience of downtime.

Service work has many challenges. First, service workers must know the product better than most other people because they see almost every possible thing that can go wrong at one time or another and must be able to put the equipment back in working order. Second, service workers have to find the cause of new and different problems each day. Finally, service workers must deal with the pressure of a customer who wants the unit repaired at once.

NATURE OF THE WORK

A technician's duties as an employee depend on the product of the employer and the technician's specific role in its production. There are, however, some similarities in technicians' jobs from one company to another. The tools and test equipment are generally the same. Symbols and components are often the same. Differences can depend on the size of the company. In a large company, a technician may have a very specific assignment; in a small company, a technician may have a wide range of tasks to perform.

Common Practices

The basic knowledge received in a technical school provides a solid base that allows for entry or future transfer into many positions. When considering new employees, for example, companies look for a solid background in the basic practices common to all electronics industries. They then train the new employees to work with their specific products. Consider the analogy to an auto mechanic. While engines vary from one manufacturer and model to another, all have parts with similar names, and all procedures are generally the same. If a mechanic knows how to overhaul one model, he or she can easily learn to overhaul another.

What are the practices and features common to all electronics work? First, all manufacturing, research, service, and other technical-level workers share a common electronics language. Tools, parts, equipment, and theories all have names; and the basic elements are the same in all areas of electronics. For example, a meter is the same instrument no matter what area of electronics it is used in. A technician can read a wiring diagram whether it is for a radio, a television, or a computer.

Other common features include the ability to read technical manuals and diagrams, the ability to use basic test equipment for the determination of proper operation of a circuit, the ability to troubleshoot, and the ability to remove and replace defective parts in a variety of electronic devices. Most technical people must maintain records of their work, such as a lab report or an itemized bill for a customer. Thus, keeping complete and accurate records is an important and common practice. Technical workers are on their own quite often, so the ability to work independently is another common feature. While seeking help is unavoidable sometimes, technical people must be able to solve problems and work on their own.

Company Size

Organizations can range in size from an international corporation to a one-person business. The size of the company quite often determines the range and the nature of the tasks performed. Generally, the larger the organization, the more specialized a technical worker must become. For example, a broadcast technician at a small radio station may operate the transmitter, install new equipment, do all the repair work, and have an afternoon music program. In a large station, a person at a similar level may just perform maintenance or record commercials.

While the small companies offer variety, the large companies offer opportunity for advancement. Also, large companies quite often provide better salaries and more benefits than small companies do. Yet, employees in small companies more often feel noticed and satisfied that they are an important part of the organization.

While most people want to operate their own business at some time, most never start one. Electronics, though, offers excellent opportunities for people to go out on their own. For example, you could start your own home entertainment equipment repair service in your home, in a shop, or from a truck. Some people install master antennas and video, intercom, or alarm systems on their own. Many of today's medium-sized companies were started by one person in a basement, subcontracting the assembly of small electronic devices for a large company. Electronics is such a wide field that the opportunities for starting your own business are limited only by your imagination.

The choice of being self-employed, working for a small company, or working for a large company requires some careful thought. How much security do you want? What risks are you willing to take? Is salary most important? Are opportunities for advancement most important? What about freedom on the job? Your answers to these questions will determine the company size you should consider.

People versus Equipment

Television repair provides an excellent example of two extremes in the orientation of work. Field service persons work mostly with people. Consequently, they are expected to be nicely groomed and attired, polite, and well mannered. Their work is in another person's home. While attempting to diagnose the problem, they must be careful with the customer's property, be understanding, and be reassuring. If the problem is complex, they remove the television from the home for shop repair.

When they return with the repaired equipment, their professional behavior helps in the presentation of the bill.

The same is true of computer service persons. In the office of a customer, they are representatives of the company. For the customer, they *are* the company. In people-oriented work like the examples given here, customer goodwill rests on the technician's performance. Interpersonal skills are as important as technical ability.

When a television is brought back to the shop, a different type of person gets involved. Bench work is equipment-oriented. Shop technicians work with test equipment and repair defective items. The only people they interact with are their supervisor and possibly another bench worker. The freedom of getting around the community is sacrificed for the freedom to dress and behave with fewer restrictions. Technically, shop work is more difficult than field work, because only the complex problems are brought back to the shop.

As you can see in Figure 1–3, the selection of an occupational goal involves many decisions. For example, how much preparation are you ready to undertake? What type of duties are you interested in performing? Do you like to work with people or with equipment? Some who enter electronics and like working with people may find their way into sales positions in a company. Others who like working with equipment may find their way into testing and troubleshooting positions in the same company. So you have many choices. Technical-level jobs can range from working totally with people as a sales representative for an electronics manufacturer to working totally with equipment as a radio transmitter operator.

Job Location

There are many factors to consider when deciding who to work for and where to work. The job you select will be influenced by your

FIGURE 1-3
SELECTING A CAREER GOAL

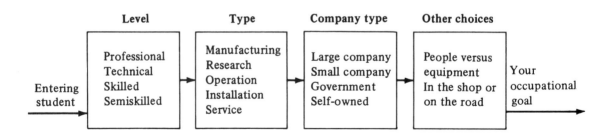

attitudes toward travel, relocation, and the area where you want to live. For instance, some people may choose to work in the general area where they are now in order to be close to friends and family, while others may wish to travel around the country. A person hired by a large corporation may be sent to school in another part of the country. When that training is over, the person may be relocated again. Advancement in a company is often accomplished through relocation. Thus, your attitude toward relocation is a factor you must consider when you are seeking a job.

To enter certain fields of electronics, you may have to travel to specific areas. Opportunities in broadcasting increase with the community size, while television repair shops are almost everywhere. The large electronics manufacturers were once clustered on the coasts in the Northeast and Southwest. But new areas are opening up almost everywhere. As a general rule, most manufacturing, research, and service opportunities are found around the large industrial areas.

Another possibility for job location is available to some sales and service representatives. Sometimes, a person may represent a company in several states, traveling from one place to another by car. A job that requires the sales representative to be away from home and office 60% to 80% of the time is difficult but to the liking of many people.

So as you begin your preparation for a technical career and learn about the types of jobs available, keep these questions in mind. What is the nature of the job? What is the environment like? What will the duties be? Is the job typically found in large or small companies? Does the job involve working with people or with products? Does the work involve travel? Finding out the answers to these questions will help you make a wise choice about a career.

PREPARATION FOR A TECHNICAL CAREER

Your career preparation involves three dimensions, as shown in Figure 1–4. You need specific knowledge about electronics theory and product operation. You need skill in the performance of various tasks and in the operation of electronic test equipment. In addition, the attitude you develop toward your work is a major factor. In fact, some employers consider attitude the most important factor. Your strength in these three dimensions, plus your experience, will contribute to your opportunities for success.

Knowledge

To qualify for technical-level employment, you must know many things about electronics. The content of this book represents a small part of that electronics knowledge. Basically, you need a good working knowledge of DC and AC theory, circuit theory, and systems theory. You need to know how semiconductor devices and integrated circuits work individually and how they are grouped into circuits. To determine whether you have acquired this understanding, many employers will require you to take a competency exam, while others will require a technical school certificate. For some jobs, you will have to possess federal or state licenses.

Learning about electronics is a step-by-step process; you cannot understand step C unless you understand step B. Likewise, you cannot understand semiconductors unless you understand resistors. In the same manner, knowledge of electricity precedes knowledge of electronics. Proficiency with mathematical processes, including algebra and trigonometry, is also necessary. A good physics course along the way will be a benefit. And you need to know more than theory; you must also know the vocabulary. You must know the names for parts, tools, equipment, procedures, symbols, and the like.

In addition to theory and vocabulary, you must know rules. There are standard procedures and practices that you must follow to ensure your safety, the safety of those working with you, and the safety of the equipment you are working with. Although each company has its own safety guidelines, you must know the basic ones that are universal to electronics.

You must also be able to apply your knowledge. An employer wants to know what

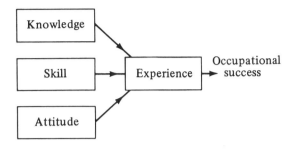

FIGURE 1-4
REQUIREMENTS FOR SUCCESS

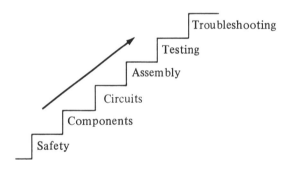

FIGURE 1-5
STEPS TOWARD OCCUPATIONAL COMPETENCE

you can do. You are paid for what you do rather than what you know, so the effective application of this knowledge is essential.

As shown in Figure 1–5, the steps in acquiring knowledge are similar to climbing a ladder. You begin by learning the language and the basic safety practices. Parts are introduced, and you learn how each works. Next, the single parts are combined into circuits, and these circuits are assembled to produce products. Finally, you will understand how all the parts work together and will recognize the symptoms that indicate they are not working properly. Troubleshooting to find problem causes is the highest level in the climb to occupational competence.

Skill

In addition to having a knowledge of theory, you must be able to perform many tasks. Some of the tasks you must be skilled in are the following:

1. Stripping the insulation from a conductor without cutting any of the strands;

2. Attaching a wire to a terminal and producing a good-quality solder joint;

3. Reading a meter indication and interpreting the correct value, and doing both in a prompt and efficient manner.

These skills take practice and are excellent indicators of the quality of a worker.

Communication skills are also very important. If you are demonstrating how a product works, you must be able to clearly communicate instructions to the customer. If you are doing television field service work, you must understand the customer's complaint and effectively communicate it to the person back at the shop. When the work is complete, you must prepare the bill clearly and accurately or the customer may question your honesty.

Persons doing research or experimental work are usually required to prepare lab reports. Therefore, writing skills are essential. Written lab reports contain an accurate record of your work, your results, and your conclusions. They must be complete enough so that someone at a later date will have no question about what you did or what you concluded. If the notes from an experiment are not well made, even the person who wrote them may not understand them a year later. Thus, good lab reports are mandatory.

Attitude

A solid theoretical background and the ability to skillfully apply your knowledge are necessary but are still not enough. Your attitude is also very important. At one level, attitude means the ability to get along with co-workers, supervisors, and customers, to be both cooperative and willing to accept suggestions. These traits are essential if you are to maintain a position. At another level, attitude means the desire to produce a quality product and to advance to a higher-level position through experience and additional education. These characteristics are recognized by employers and, in a little time, should help you get promoted.

Experience

The "little time" just mentioned usually means experience. Many companies like to hire experienced workers, assuming they have made their mistakes elsewhere. Experience also provides other advantages, both to you and to your employer. The more products you repair, the faster you can locate trouble. The more parts you install, the more efficient you become. These skills make you valuable.

Experience also helps if you are not sure of the direction you want to take. When you try one job and find you do not like it, you have narrowed your choices. Finding what you do not like helps you find what you do like.

Getting a job in a large company is another experience that helps you decide your future. In a large company, you are exposed to many other kinds of jobs. You can select jobs that you like and those you do not like. In addition, managers in the company can observe your work; and if you are valuable, they will want you to stay, Many companies first look within when searching to fill higher-level positions. In this way, they know the kind of person they are getting. They also know that person's experience applies to their specific products. Thus, your experience in a large company is of value to both you and your employer.

SUMMARY

Electronics is both a science and an industry. By definition, it involves the theory and application of information-processing devices built with semiconductors and integrated circuits. Millions of people are employed worldwide in electronics occupations at different competency levels ranging from semiskilled to skilled, to technical, and to professional. This text is intended for individuals preparing for entrance at the technical level.

Employment can be found in a number of areas, including manufacturing, research, operation, installation, and service. The organization may involve only one person, or it may employ thousands of people. Opportunities, salary, and conditions vary with company size.

Some positions can be found in your own community, while others will require relocation. You may choose a people-oriented job or one that is equipment-oriented. You may also choose an on-the-road job or one at a fixed location.

Whatever your choice, you will need a solid theoretical background as well as the skill to apply your knowledge before you are employable. Once you get a job, you will only keep it if you have a satisfactory attitude. Finally, your attitude and experience will help you advance in your career.

CHAPTER 1

REVIEW TERMS

electricity: flow of electrons through a wire (conductor)

electronics: physics of electrons in active devices such as diodes, transistors, and integrated circuits

professional level: requires four years of postsecondary education and involves circuit design

semiskilled level: requires little training and involves sorting, packing, and some assembly activities

skilled level: requires some training and involves assembly work

technical level: requires two years of postsecondary schooling and involves circuit fabrication, testing, troubleshooting, and repair

REVIEW QUESTIONS

1. Describe electronics as a science.

2. Describe electronics as an industry.

3. Describe the skilled level.

4. Describe the technical level.

5. Describe the professional level.

6. Describe two tasks a skilled-level worker may perform.

7. Describe two tasks a technical-level worker may perform.

8. Discuss the preparation for a technical-level job.

9. Discuss the preparation for a professional-level job.

10. Describe two jobs available for technical-level workers in electronics manufacturing.

11. Discuss why knowledge and skill are necessary qualifications for jobs in electronics.

12. Explain why your attitude is important to an employer.

13. Compare the characteristics of manufacturing jobs with those of research jobs.

14. Compare the characteristics of installation and service jobs with those of an operator's job.

15. Compare the aspects of working for a large company with those of working for a small one.

16. Compare working with people and working with equipment.

17. Compare the characteristics of a shop job with those of an on-the-road job.

18. Review Help Wanted ads in your regional newspaper and collect those for electronics jobs. Make a list of the jobs, requirements, duties, and benefits.

19. Visit a local electronics manufacturer and write a report describing electronics job opportunities and requirements.

20. Talk with someone you know who works in electronics, and prepare a report about this person's job.

ELECTRONIC PRODUCTS AND EQUIPMENT

Define products, systems, and equipment.

Name at least three electronic products.

Name at least three electronic systems.

Identify the common equipment used in electronic testing.

KNOWING SPECIFIC NAMES

The objective of this chapter is to introduce terms used in electronics. In addition, some items you will work with will be described. This introduction is still only a beginning, for a great many new names will appear in later chapters. In those chapters, the names of tools, small parts or components, wire types, and symbols will also be introduced.

With the introduction of new names, you must realize that opinions vary. We noted that point in the definition of electronics given in Chapter 1. This text will present many definitions in that context, not as absolute and universal definitions but as common, usual, or typical names or descriptions.

When you enter a job in electronics, you are expected to be able to speak the language. Devices have names, and you should know them. If you merely say, "this thing over here" or "that thing over there," those who work with you will not know exactly what you mean.

There is another reason for knowing names. Your first job quite likely could be as an assistant to a senior technician or an engineer. Assistants are often asked to get one device or another. If you are asked to get a spectrum analyzer or a heat sink, you must know what the device is in order to bring back the correct one.

The ability to speak the language is an indicator to others of your overall familiarity with the field of electronics. A person would be judged unfamiliar with the game of hockey if he or she said, "The player just shot the little black thing into the net." You would be judged in a similar way in an electronics shop if you called an item "that machine over there." The names of components, products, and equipment are important in the language of electronics.

By Use

A good way to begin learning names is to know what an item is used for. What does it do? Is it used in the home, the office, or the shop? Does it work by itself, or is it connected to other items in some sort of combination? Is it used to produce something or to test something else? Categorizing items by use is one good way to remember what they are named.

To illustrate the enormous number of electronic products, let's consider one familiar category by use: home entertainment. There are stereos, televisions, radios, and games, and new items are added each day. Furthermore, for example, radios can be AM or FM, portable or plug-in, receivers or transceivers. Other variables include price, quality, and manufacturer. And there are just as many variations in all other electronic products.

By Type and Brand Name

There is another way to describe types of products: by the brand name. From your knowledge of automobiles, you know that the brand name describes many things. It is an indication of cost, capabilities, and quality. Similarly, in the electronics industry, brand names do become important, and the people you work with will have likes and dislikes. Preference is usually based on quality. Therefore, it is important that you investigate different brands and determine which are most common and why.

By Price and Quality

We just noted that preference is usually based on quality. Therefore, why doesn't everyone have the best of everything? The obvious answer is that they cannot afford it. The product purchased is a result of a trade-off between quality and cost. How good does it have to be,

FIGURE 2-1
ELECTRONIC SYSTEMS

Components   Products   Systems

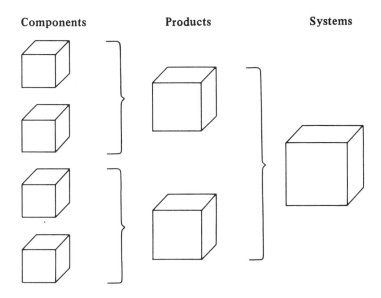

and how much can you afford? The decision is a cost-versus-value compromise.

Television cameras used in banks, for example, only have to be good enough to see what the customers look like. Color, high resolution, and a smooth range from black to white are not critical. However, broadcast cameras do need these qualities and others. And the quality requirements for broadcast equipment increase its cost to the point where one broadcast camera would pay for many simple cameras of the type used in banks. In other words, high quality means high cost. Therefore, businesses buy only the qualities that they need and can afford.

ELECTRONIC PRODUCTS

In the course of the day, you work with many items, some electrical and others electronic.

While all electronic items are electrical, all electrical ones are not electronic. We indirectly described this characteristic in the discussion of the field of electronics in Chapter 1. Now, we offer another definition. Items that operate on the principle of electronics and are assembled from many parts to form one physical package or unit are called **electronic products.** Some examples of electronic products are televisions and calculators. A stereo is also an electronic product. Note that while transistors make a stereo electronic, the transistors themselves are not products. They are only individual electronic parts, or **components,** which, when assembled with other components, produce a product.

A stereo amplifier is generally assembled out of many individual components. The amplifier is a product that can then be connected to other products to produce a system, as indicated in Figure 2–1. Systems will be described later in the chapter.

As we noted, a radio is an electronic product. A fan, however, is an *electrical product*. A fan is electrical because it is an independent unit that operates on electricity but does not have any transistors or integrated circuits.

Reason a Product Is Electronic

We have just considered why a product is *called* electronic. Why products are *made* electronic is another issue. To understand this issue, we must first consider what the other options are. A good example to begin with is automobile ignition systems.

The primary purposes of an automobile distributor are to make and then interrupt the electric current of a coil and to direct that current to the spark plugs in a predescribed order. The current interruption can be produced by metallic contacts on distributor points, which are opened and closed by a moving cam. Those familiar with automobile tune-ups know the problem that results from misoperation of the distributor. The current interruption causes sparking, which, in turn, causes pitting and other deterioration of the contacts, eventually making them useless. In addition, the moving cam tends to wear out the part it moves against.

Electronic ignition systems replace the mechanical contacts with a transistor. Since a transistor has no moving parts, there is nothing to wear out. Since it does not physically open and close, there is no arcing and, therefore, no pitting. The simple transistor has greatly extended the effective life of automobile distributors and reduced the frequency and extent of ignition tune-ups. Thus, we have one reason why products are made electronic.

Another good example is the electronic computer used in banks and other businesses. The original version of the computer, designed decades ago, was a mechanical device with many moving parts. In fact, the early versions were called machines. Since they were mechanical, they were large, slow, expensive, and very limited in their data-processing capacity.

The incorporation of vacuum tubes converted the computer from a mechanical machine to an electronic device and reduced the size and cost while increasing speed and capacity. Newer generations of computers used transistors and then integrated circuits. With each step, the computer became faster, smaller, and more powerful. Microcomputers clearly demonstrate the size reduction produced by integrated circuits. If they were made with tubes instead of integrated circuits, they would occupy the larger part of a room. Thus, we see another example in which the quality of task performance was improved by changing the approach from mechanical to electronic.

Products for the Home

Several electronic home entertainment devices have already been described, including radios, stereos, and televisions. Many other electronic devices are also found in the home today, and the number continues to increase. The electronic microwave oven such as the one shown in Figure 2–2 is one example. Electronics has brought about a revolutionary method of cooking in the home. While the theory will not be described here, it is enough to say that this oven cooks in seconds or minutes rather than in minutes or hours.

Many homeowners today also install complex electronic alarm systems. Early alarm systems were usually installed in business establishments and consisted of units that detected the breaking of glass and the opening of doors and windows. New alarms for homes and businesses sense these conditions as well as heat, smoke, water, motion, and a variety of other indicators of trouble. The imagination of the electronics industry is sure to develop new products for the public, which you can help design, build, sell, install, or service.

Products for the Office

The electronics industry has created new ways for business to spend money to save money. Small computers, for instance, have become commonplace in almost every business office. Electronic copiers have provided businesses with an inexpensive method of copying documents, which has, in turn, created the habit of copying almost everything that comes along. By now, you have probably found it easier to machine-copy pages from texts rather than make handwritten notes of desired information.

Electronic paging devices are now quite common. At one time, only a few professionals, like doctors, carried electronic paging devices (also called "beepers"). Now, many people whose job takes them away from their desk, office, or workbench have devices.

Another popular device for businesspeople is the pocket cassette recorder. It allows a person to make notes quickly without a pencil and note pad. Businesspeople can dictate letters on the cassette while driving from one meeting to another. Meetings can also be recorded for future reference without the need for a stenographer and typist. The recorded cassette can be labeled and filed as a complete and accurate record of the meeting discussions.

Products for the Automotive Industry

Automotive electronics is a very rapidly expanding field from both a sales and a servicing point of view. For many years, AM and FM radios were the only electronic products in au-

FIGURE 2-2
MICROWAVE OVEN

tomobiles. Then, tape players, CB radios, and alarms were introduced. Now, we are beginning to see electronic equipment taking the place of many mechanical devices, as it has in other areas. For instance, the electronic ignition, which has already been described, is popular. Digital electronic clocks and speedometers are being used, as are other digital displays. Many people have electronic mobile telephones for their cars.

Compact and complex electronic microprocessors are also used in automobiles. They monitor and adjust many of the engine's operating functions. With the increased concern for the efficient operation of engines in recent years, these products have become quite popular. With them, such functions as air/fuel mixture, water flow, and timing can be monitored and adjusted electronically according to the needs of the engine.

Most automotive technicians do not have the electronics background to service these new automotive electronic devices. Therefore, there is excellent opportunity for a person with electronics training to find employment in the service phase of the automotive

FIGURE 2-3
STEREO SYSTEM

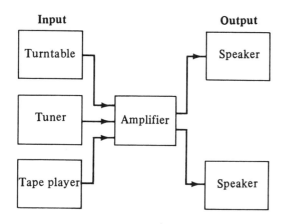

industry. One requirement certain to be a necessity in this area is the need to keep up to date. If the electronics industry continues to perform as it has in the past, there will be new and different devices developed for use in automobiles. In five years, no doubt there will be electronic devices in automobiles that you cannot even imagine today.

ELECTRONIC SYSTEMS

You have probably heard the word *system* used often and in a number of ways. For example, you attended a school system, and your automobile has a cooling system. While there is little similarity between these two uses of the word *system,* they do have one thing in common. Each term describes a group of parts that work together to perform some function. Therefore, we define a **system** as a group of parts interconnected to perform some common function.

The function of an automobile is to transport an individual from one location to another. The engine, transmission, and rear axle

join to form the drive system of the car. These parts must all work together to perform the overall function of the drive system. This system, in turn, must work with other systems, such as the electric and cooling systems, in order to keep the larger system, the automobile, operating properly.

Inputs and Outputs

The stereo shown in Figure 2–3 is an example of a system. The tape player, tuner, turntable, amplifier, and speakers are all connected together to provide music. Each unit has a specific task and is somewhat useless alone. However, the system needs each unit in order to operate properly.

Notice in Figure 2–3 that there are two key sections of a system: input and output. The term **input** has two meanings. The tuner is an input to the amplifier, since it is the device that provides information that the amplifier will process. The second meaning of the term relates to the signal that the tuner must process. That signal is also the tuner's input.

The term **output** has two meanings as well. The output is the place where information comes out of the tuner. The output is also the information that comes out of the tuner. The output of the tuner is a signal for the amplifier, while the output of the system is the sound from the speaker.

Notice that Figure 2–3 shows how a stereo system is connected and how the information moves. One characteristic to note is that the output of one stage is the input of the next. Overall, the inputs are signals at the antenna and the grooves on a record, while the output is the speaker sound.

Systems Concept

A good understanding of the systems concept is important in electronics. Your future employer will have an organizational system describing how jobs interrelate with other jobs. There will be a paperwork system describing how orders, repairs, and other forms of information must be processed. The company product will be manufactured, tested, and shipped through a complex system, and you will be part of that system.

Furthermore, understanding systems in general and the systems you work with specifically is important if you are to locate problems. You must know the input to, the process within, and the output of each device or stage in a system. If there is a problem, it will be indicated by an unacceptable output — for example, poor sound coming from the stereo speakers.

A systematic approach to solving a problem requires that you understand the desired output of the system. For the stereo system, for example, the desired output is a good quality of sound. By checking each preceding stage, you will probably find a stage, like the amplifier or turntable, that has a good signal going in and a poor one coming out. This checking is beyond the level of our discussion here, but a good understanding of electronic systems and the use of test equipment is necessary.

Thus, understanding the concept of systems as it applies to the many areas you will work in is necessary because successful troubleshooters use a systematic approach. For example, troubleshooters consider first the symptoms of the faulty system. Then, they determine what parts of the system are working and which ones are not. To correct the problem, they must know how those parts interrelate. Therefore, much of the problem-solving task can be accomplished with a good understanding of the systems approach.

Systems within Systems

The human body's subsystems, such as digestive, nervous, circulatory, and skeletal, all add

FIGURE 2-4
PRINTED-CIRCUIT SUBSYSTEM

up to form one large system, the body. Likewise, any electronic system is composed of subsystems. For example, in the stereo system described earlier, one subsystem is the tuner. The subsystems within the stereo add up to the one large system. Another example of an electronic subsystem is the printed circuit board shown in Figure 2–4.

This point about systems and subsystems is important because it makes you think about problems without first applying any electronics theory. If you understand the overall function of a stereo, then you can localize problems to a subsystem or part of a subsystem. All it takes is some logical thought about the parts of a system and what they are supposed to do.

For example, if you had a stereo with poor sound coming from one speaker, where would

FIGURE 2-5
FREQUENTLY USED TEST EQUIPMENT

you look first to identify the problem? If you switched the wires of the two speakers, and the poor sound changed sides, then the problem would not be the speaker subsystem but in a subsystem before the speaker. If the poor sound did not change sides, the problem would be in the speaker subsystem. Thus, you can take the first step in troubleshooting a system without applying any theory of electronics, only logic.

ELECTRONIC EQUIPMENT

The term *equipment* may mean many things. For instance, baseball or hockey equipment refers to those items the player uses or works with. So far in this book, many items a technician will work with have been mentioned, including radios, televisions, and stereos. In the electronics industry, though, those devices are not equipment. While you may work *on* those devices, you do not work *with* them.

The things you work with are the tools of the trade, the items used to accomplish your assigned responsibilities. They consist of a variety of hand tools as well as power supplies, meters, and other testing devices. While hand tools like cutters and soldering irons are not equipment, the electric and electronic devices are. Therefore, **equipment** is defined as those electric and electronic devices used in the performance of electronics testing and troubleshooting tasks. Figure 2–5 shows some frequently used test equipment.

Equipment Categories

Equipment can be categorized in many ways. An initial grouping distinguishes fixed versus portable equipment. Fixed equipment is primarily intended to be kept on a workbench. Portable equipment is generally small and battery-operated, and it is intended for use in the field. Portable equipment is usually much more rugged than bench equipment.

Another method of grouping is by function. Some equipment is used for measurement of signals while some is used for a signal source. There are many other groupings, such as general-purpose versus specific-purpose equipment. Shops and labs usually have a variety of general-purpose instruments. Instruments designed for one specific function are often more accurate but also more expensive.

Common Test Equipment

Voltage, resistance, and current are probably the most frequently measured elements in electronic circuits. Therefore, a portable multipurpose meter, or **multimeter,** has been developed to measure all three factors. It is called a **volt-ohm-milliammeter,** or **VOM,** after the units being measured: voltage, in volts; resistance, in ohms; and current, in milliamperes. While VOMs may vary in brand, quality, and cost, they are often similar in appearance. A VOM is an analog instrument with a pointer, a scale, one or more switches, connections for the two test leads, and a plastic case. **Digital multimeters,** called **DMMs,** have a digital display in place of the VOM pointer and scale. VOMs and DMMs, shown in Figure 2–6, are the most common test instruments used in electronics.

An **oscilloscope** is an instrument that displays voltage and time waveshapes. These instruments are easy to identify by their display screens, which use a cathode ray tube, and their many knobs and switches. Oscilloscopes are used everywhere, and they range in cost from a few hundred dollars to many thousands of dollars.

Diode, transistor, and integrated circuit testers are other common test equipment items used in electronics. These items vary greatly in cost and complexity according to what they measure and how well they do it. Other frequently used test devices are **function generators,** which provide test signals when circuit performance is being evaluated.

One last category of common test equipment is the **power supply.** It produces a variety of voltages in order to operate the devices that are tested. For example, automotive equipment requires power supplies that produce 12 volts. Aircraft equipment requires 24-volt power supplies. Research and experimental work requires a wide range of voltages and, thus, many types of power supplies. As in other equipment, cost is determined by quality, which, in turn, is determined by the needs of the situation.

The test equipment just described includes the most common devices you are likely to encounter. There are, of course, many other devices used in testing. One way to become aware of the different types of test equipment is to look in a catalog. There you will find frequency meters, impedance bridges, and many more items. For every parameter found in electronics, there is test equipment to measure it.

FIGURE 2-6
ANALOG AND DIGITAL MULTIMETERS

SUMMARY

Some of the first things you must learn in any new endeavor are names. For example, you need to know the names of the people you work with and work for. Similarly, you must know the names of the tools, parts, equipment, and processes you work with in electronics.

Electricity and electronics are defined in this chapter. Many of the tools used in electronics are common to other occupations, but most of them will be described in a later chapter. Products and equipment are introduced in this chapter and will be described in more detail as the book progresses.

Products can be described in a variety of ways. In electronics, it is common to describe a product first by its application, what it does and what it is used for. Thus, a product can be a stereo, a computer, or a VCR. Products are also described by brand name, price, and quality. Finally, the product may be an individual part or component, a complete product, or part of a large system.

Products are also categorized by where they are used: in manufacturing, offices, homes, or automobiles. While these products may not be manufactured and serviced by the same company and may be sold in different markets, they are typically assembled, tested, and serviced with similar techniques.

The word *system* is a common electronics term and it means a group of items that work together to perform some function. It is a relative term, since a number of systems can work together to produce a much larger system. For instance, a tuner, turntable, amplifier, and speakers can be assembled to produce a home music system. This system can then be combined with a VCR, television, laser disk system, and other products to produce what is called a home entertainment system.

Each part of a system has an input and an output. In traditional usage, the word *input* is a noun and it means the location where signals or information enters a product. It also can mean the information itself. Likewise, output means the location where information leaves and that information itself. In a system, the output of one part connects to the input of the next. For example, in a home music system, the output of a tuner connects to the input of the amplifier.

The equipment used in electronics is many and varied. Much of your study will involve learning about equipment, how it works, what it is used for, and how it is properly used. The most common test equipment item is the multimeter, which can be analog (the VOM) or digital (the DMM). In either case, the multimeter is used in basic testing to measure voltage, current, and resistance; it is described in detail in later chapters.

CHAPTER 2

REVIEW TERMS

component: small individual electronic parts

DMM (digital multimeter): instrument for measuring voltage, current, and resistance and having a digital display

electronic product: complete unit such as a radio, TV, or computer

equipment: items such as measuring instruments, power supplies, and generators used in product evaluation

function generator: device used to produce test signals for the evaluation of products

input: location where signals or information enters a product or system and that information itself

multimeter: instrument for measuring current, voltage, and resistance

oscilloscope: instrument that displays voltage and time waveshapes on a cathode ray tube

output: location where signals or information leaves a product or system and that information itself

power supply: a device that produces a variety of voltages to operate other devices

system: a group of parts interconnected to perform some common function

VOM (volt-ohm-milliammeter): instrument for measuring voltage, current, and resistance and having an analog pointer and scale display

REVIEW QUESTIONS

1. Define electronic products.

2. Define electronic system.

3. Define electronic equipment.

4. Describe the input and output of a system.

5. List three ways of categorizing electronic items.

6. Discuss the relationship between quality and price.

7. Describe some electronic products used in an office.

8. Describe some automotive electronic products.

9. Explain why it is important to know the names of items used in electronics.

10. Name five electronic products used in the home. Name five electrical products.

11. Compare the characteristics of the items in the two lists from Question 10.

12. Compare inputs with outputs.

13. Identify four common test equipment items used in electronics.

REVIEW PROBLEMS

1. Write a report comparing the characteristics of an electronic watch with those of an electric watch. Consider characteristics such as accuracy, moving parts, cost, and features.

2. Review the test equipment section in an electronic equipment catalog and prepare a report on the cost range for an item such as a multimeter or oscilloscope.

3. Make a list of the various items of test equipment in the lab at your school.

OCCUPATIONAL SAFETY

OBJECTIVES

Describe environments and activities where safety is a special concern.

List the steps to follow to improve worker safety.

List the steps to follow in case of an accident.

List several ways for maintaining tools.

Name two safety concerns of electronic components.

Describe the areas and scope of common industrial safety regulations.

PERSONAL SAFETY

The objective of this chapter is to introduce you to many concerns about your safety and the safety of the people, equipment, and materials you will be working with. The chapter focuses on accident prevention but also outlines the steps to be taken in case of accident.

Kinds of Accidents

Safety is a primary concern in every occupation, because the possibility of injury exists in most jobs. It is important to know what to do in case of an accident; however, it is more important to know how to avoid accidents. In many accidents, it doesn't matter what you do afterwards; by then, it's too late. The point to be stressed is *accident prevention.*

Electric shock can kill you. The amount of electricity required for electrocution depends on many conditions, including how you come in contact with the circuit, the path the electricity takes through your body, how well you are insulated from the floor, and your general health.

Consider the electricity in your home. You may have already had the unpleasant experience of being shocked by a 120-volt electric circuit. Consider yourself lucky, for although you survived, you might not have. Many people die from this type of shock.

Another hazard facing electronics workers is the possibility of accidental burns from soldering irons. You might touch a hot tip, touch parts that have just been soldered, or splatter hot solder onto your face or hands. Thus, you learn to avoid burns.

As an electronics worker, you can get shocks, burns, cuts, or foreign matter in your eyes while just sitting and working at your bench. If you get up and move around, the dangers increase. Co-workers drop items on the floor, and you may trip over them. You must learn to look carefully. If you drop a television tube or pick it up by the neck, it will break, and you can be cut by flying glass. If you pick up a heavy chassis improperly, you can strain muscles. In other words, there are a great many ways in which you can be hurt. Most can be avoided.

Accident Prevention

Calling people accident-prone is a nice way of saying that they are careless. Consider almost any two-car crash. Would it have happened if both drivers were paying attention, driving with regard for the road conditions, and driving properly maintained cars? Not likely. Most accidents are a result of negligence and are not accidental or a result of bad luck.

Key point: Most accidents are avoidable.

When wiring circuits at a bench, you must wear safety glasses to prevent solder or cut wire — from your work or from that of persons working near you — from getting in your eyes. You must learn how to cut wire and make solder connections without causing material to fly. You must learn not to drop things. While safety glasses and safety shoes may reduce damage, they do not eliminate the cause of accidents. You must learn to do that.

The ways to avoid accidents vary from one work situation to another. Here are a few general considerations:

1. Pay attention to what you are doing.
2. Do not rush.
3. Keep your bench and work area clean.
4. Wear safety glasses when appropriate.
5. Work in a well-ventilated area.
6. Know the characteristics of the circuits you are working on.

7. Know the characteristics of the chemicals you are working with.

8. Do not work with faulty tools or equipment.

Employers are safety-conscious for various reasons. Government regulations demand safe work areas. Industrial insurance rates depend upon the number and cost of employee injuries. Lost time increases labor costs and reduces productivity. Therefore, employers are concerned about their safety record and expect their workers to be equally concerned. Workers at all levels are expected to demonstrate work habits that reduce the probability of an accident.

What to Do in Case of Accident

Regardless of how careful we are, accidents do occur. When they do, there is one point to keep in mind.

> **Key point:** Do not make the situation worse.

Many times, when a person is involved in an accident, more harm is done after the fact than is done by the accident itself. A cut that is left untreated can become infected. A person left unconscious after a fall can be injured further if he or she is improperly moved. If you touch someone who is being electrically shocked, you also can be shocked.

Injured persons do need professional medical care. Before the professionals arrive, try to contact an emergency medical technician or someone trained in first aid. While waiting for this help, follow these rules:

1. Do not touch anyone undergoing electric shock. If you cannot turn off or discon-

nect the circuit, move the person with a board or other nonconducting material.

2. Do not move a person unless his or her life is threatened by remaining in that position. Let trained persons do the moving.

3. Have the injured person lay down and relax until help comes so that he or she does not faint or fall.

After the injured person is attended to, you should take some additional steps:

4. Report the accident to your supervisor, and fill out all of the appropriate accident report forms.

5. Reconsider your work habits to avoid a repetition of the accident.

SAFETY OF OTHERS

The safety of those around you is greatly affected by your work habits. While you must protect yourself, of course, you also should not create hazards for those around you. For instance, you should avoid leaving things around for others to trip over. While you may feel that others should watch where they step, you can reduce their risks by your concern for their safety.

Taking Precautions

"Look before you leap" is a saying that has been around for generations. Small children are warned to look both ways before crossing the street. You must learn to take the same attitude whenever you are working with electricity.

> **Safety tip:** Never plug in or otherwise activate a circuit without knowing everything that may be connected to it.

You can shock someone who is working on another device operated by the same power source.

Electricians have a practice of attaching warning tags to main circuit breakers if the circuit is being worked on elsewhere. This practice should be followed in the shop and in the home. Suppose you turn off a circuit breaker in the basement to disconnect the power in your room in order to safely replace a socket. In the meantime, a member of your family, not realizing what you are doing, notices that the power is off in the room and turns the circuit breaker back on. The wires you are working on then become activated and may cause injury to you. To avoid injury, always attach warning tags when turning off a circuit breaker.

Cleaning the Work Area

Develop the habit of cleaning up your work area when you are finished with a project. A clean work area makes a good appearance. A clean work area also keeps you from losing parts and tools. Most important, though, is that a clean work area is safe. People can slip on liquids that you spill on the floor. They can trip on loose wire since much of the wire you will work with is small and difficult to see on the floor. They can lose their footing when they step on screws or other small parts you may have dropped. Some of the chemicals that you will be working with can irritate the skin, ruin clothing, or damage circuits. So do not leave chemicals or equipment lying around. Loose parts or spilled chemicals can cause injury to yourself and to others.

WHAT YOU WORK WITH

The chance of an accident occurring depends on how you approach a task. Being careful in the procedures that you select is one way to avoid accidents. Choosing the proper tool is also a critical step since the wrong tool can break or cause damage. The proper tool used according to the manufacturer's intentions and with a little patience can greatly reduce the risk of accident.

Proper Tools and Procedures

Electronics workers use a variety of tools and test instruments. Each item has been designed for a specific purpose. Thus, it is important to use the right tool for the right job. In other words, screwdrivers are for turning screws, not for chiseling. Wire cutters should not be used for pulling or cutting nails. Tools used for the wrong job are often damaged in the process and will no longer work for their intended purpose.

In addition to using the right tool, electronics technicians must also use the right procedure. Inexperienced workers sometimes attempt to drill a hole in sheet metal on a drill press without first clamping the metal. This careless method can lead to serious injury, for most people simply cannot hold the metal once the drill begins operation. The metal gets caught on the drill bit and spins around at 500 or more revolutions per minute, causing serious cuts to the hand of the operator.

Determining and following the proper working procedures are the best lessons of experience. Most older workers can attest to that fact. It is part of human nature to try to find an easy way to do things, to not take the time to get the proper tool or to be careful. But after ruining several tools or projects, a worker finally begins to get the point: Use the right tool and follow the proper procedures. Your safety and that of those working around you are at stake. For instance, the addition of a simple soldering iron holder, as shown in Figure 3–1, can make a significant difference in the safety and efficiency of your proce-

FIGURE 3-1
SOLDERING IRON HOLDER

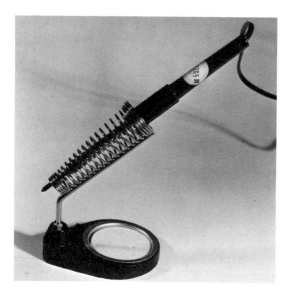

dures. An iron placed on the bench could easily fall onto your lap. With a holder, you will always know where to reach for the iron, and most important, you will be protecting the iron, yourself, the bench, and other items around the area. Thus, one small addition to your tools can make a big difference in safety.

Adequate Maintenance

The tools and equipment that you work with will not remain new forever. However, there are steps that you can take to prolong their life. It is not enough to just be careful in their use; you must also perform preventive maintenance. For example, tools, like automobiles, must be lubricated on a regular basis in order to reduce damage from excessive wear.

Some electronic products have fans for cooling and filters to keep out dust. If the fil-

ters are not clean, air does not flow well, and the resulting excessive heat can cause damage. Thus, you should always check the filters on a product to make sure that they are clean.

Lubrication of equipment is a special concern. Most equipment needs lubrication to ensure long life, and it also needs correct lubrication. The wrong lubrication can be more harmful than none at all.

The preceding examples lead to an important point: You must follow the instructions that come with equipment. Manufacturers tell you not only how to operate their product but also how to maintain it to make it last. Tools and equipment will be safer to work with and will last longer if they are well maintained.

Many electronic devices are battery-operated. Leaving the batteries in too long can do more than just make the device inoperative. Dead batteries can corrode and destroy an electronic product. While you can easily detect dead batteries by inoperation, you cannot detect corrosion unless you look inside the device and inspect the batteries. You should check batteries on a regular basis.

Another major problem with electric equipment is frayed or worn wires and cables. You may have seen worn wires in your own home. Frayed wires on a lamp cord can lead to someone getting a shock — sometimes fatal. They can cause a burned-out fuse or even sparks and a fire in the house. With electrical hand tools, frayed wires lead to electrocution. In test instruments, frayed wires cause the instruments to give incorrect information. The pulling and bending that occurs when connectors are plugged in and removed can loosen the wire connections inside. If you are too rough, if you disconnect plugs by pulling the wire, or if you otherwise abuse connectors, you can cause damage inside the equipment. Often, this damage is not noticed until it causes other problems. Therefore, always handle wires and connectors properly to protect the equipment and yourself.

In summary, successful workers know that they must use good-quality tools, take care of them, use the correct tool for each task, be careful in their procedures, and be patient. Remember: Poor or worn tools break and cause damage. Good tools last for a long time when properly maintained.

WHAT YOU WORK ON

Another safety responsibility is to be careful of the things that you work on. You should not cause damage or make existing problems worse than they already are. To avoid damaging parts or products, you should know what you are going to do and the correct way to do it before you begin.

There are two important concerns when you work with many electronic components: heat and static electricity. The heat from a soldering iron can easily damage diodes, transistors, and similar components. This concern is discussed in more detail later in the text.

Static electricity — the electricity caused by friction — can also damage parts. You have no doubt experienced the shock produced by walking across certain carpets. The shock results from voltage produced by friction, voltage that can reach levels up to 5000 volts. Yet, minor sparks that you might not even feel have enough voltage to instantly destroy the inner workings of some types of transistors and integrated circuits.

One of the methods used to reduce the sparks caused by static electricity, called **electrostatic discharge (ESD)**, is to store spark-sensitive components on conductive foam. Figure 3–2 shows an integrated circuit for a computer mounted on such foam. Another common practice is for each person working with these parts to be grounded with a wrist strap and to use a grounded bench cover, as shown in Figure 3–3. **Grounding** establishes a common

FIGURE 3-2
INTEGRATED CIRCUIT ON CONDUCTIVE FOAM

FIGURE 3-3
STATIC REDUCTION THROUGH GROUNDING

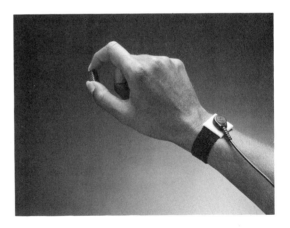

electrical connection to the earth or ground. These common safety practices are followed for the protection of electronic components.

Products

If someone brings a product to you for repair, you must not make the problem worse. That may seem like an unlikely possibility, but it is not. Improper test procedures can damage sensitive parts. Rough handling of a television can damage the picture tube. And you cannot charge a customer for a new picture tube in a television that was brought in for audio problems. Furthermore, customers are usually aware of scratches on their cabinets and will notice any you might accidentally add. The point is: Do not add any problems to those that already exist.

Systems

Systems such as stereos have special safety concerns. For instance, the instructions that come with stereos advise you not to turn the system on without first connecting the speakers. Care must be taken when you are making the connections to ensure that there are no shorted wires. One strand can cause an output transistor to short, requiring its replacement. Thus, always check the connections before turning a unit on. Always read the manual that comes with a system.

All CB radios have a similar requirement: An antenna must be connected before the transmitter is turned on. A CB radio and its antenna are parts of a system.

Most system parts are intended to be operated only when they are connected to each other; they should not be operated alone. When you become more familiar with electronic systems, you will know what parts of a system can operate alone. Until then, assume that a system will safely operate only when properly connected with all other sections of the system.

A final point about systems involves putting parts from one system into another. While the battery and tires from one car model may fit another model, the transmission and radiator probably will not. Similarly, while the speakers from one system may work with an-

other, they also may not. If you do not know whether system parts can be interchanged, ask someone who understands the system before you try to exchange subsystems. Otherwise, assume that the parts cannot be interchanged.

REGULATIONS

A brief review of almost any history book will convince you that, for generations, working conditions in many factories were both unsafe and unhealthy. Polluted air, toxic chemicals, unsafe facilities, and dangerous tools were a part of many industries. The U.S. Congress passed the Occupational Health and Safety Act in the 1970s as an attempt to ensure safe and healthy working conditions for men and women. The Labor Department's **Occupational Safety and Health Administration,** called **OSHA,** has the responsibility for establishing working standards, conducting inspections, and enforcing rules.

One initial concern was the working environment. Were shops well illuminated? Did they provide fresh air and comfortable temperatures? Was there excessive noise? Other concerns were tools and machinery. Were the tools safe to work with? Was there enough room between machines? Workers' responsibilities were considered, too. Good-quality safety glasses had to be worn for some jobs and hard hats for others. Certain clothing could not be worn, for it could easily catch fire or be caught in machinery.

More recently, the federal government has established Right to Know regulations. These regulations obligate manufacturers and employers to inform people who work with chemicals of the precautions and hazards of the materials that they are likely to come in contact with.

What Regulations Mean to You

The regulations are extensive, complex, and difficult to totally understand. But there is no need to know them all. What you need is a good understanding of your rights and responsibilities.

Many tasks performed by an electronics worker require safety glasses. Cutting wire leads requires them because a cut lead can fly into your eye. Soldering also warrants safety glasses because solder and/or rosin can splash into your eye. Furthermore, proper glasses must be used. Glasses appropriate for the job of cutting leads may not be good enough for the job of soldering. Glasses must cover the area to be protected and be able to withstand the forces and temperatures they may be subjected to. Your instructor or employer should select appropriate glasses for you after consideration of the task, the environment, and the safety glass manufacturer's recommendations.

Chemicals are also a safety concern. They can be harmful to breathe and may require the use of masks and proper ventilation. Chemicals can irritate the skin and may thus require the use of gloves and aprons. Most chemicals can be disposed of only by following specific procedures described by the manufacturer and the government.

> **Safety tip:** When you work with chemicals, know what you are working with and follow all safety procedures and precautions.

Because of the widespread use of dangerous chemicals in the workplace, the federal government enacted the **Hazard Communication Standard,** which requires that employees be given information and training about hazardous chemicals that may be in their work area. Chemical manufacturers and distributors are required to provide their customers with **material safety data sheets (MSDS)** describing hazardous chemicals and mixtures. This information, in turn, must be provided to employees. It includes descriptions of ingredients, hazards, and safety procedures, precautions, and protections.

What Regulations Mean to Your Employer

One of your employer's responsibilities is to ensure safe working conditions and procedures. Another is to ensure that you know the safety procedures that you must follow. While employers do not have to stand behind workers to make sure that they keep their safety glasses on, employers can penalize workers who violate the rules.

One thing that an employer cannot do is penalize you for reporting unsafe conditions. As mentioned earlier, when there is an accident, an employer faces many complications, such as lost employee time, medical expenses, low morale, and higher insurance costs. These results are usually enough to make an employer try to operate a safe area.

SUMMARY

The words "safety first" best summarize this chapter. Before you start any job, before you perform any task, determine the safety precautions you should follow. Then, observe them.

An electronics worker can be shocked, burned, cut, or injured in some way. Injury can result from the worker's own actions or from the actions of someone near the worker. The degree of injury can range from a cut requiring a small bandage to death by electrocution.

The best approach to accidents is prevention. Most accidents are not accidents at all but are the results of negligence. Accidents happen because people are careless. Some people work with unsafe and defective tools. Others are irresponsible in the ways they work. In both cases, injuries can result.

When an accident occurs, a few steps must be taken. Injured persons need emergency care at once and professional medical care later. Accidents must be reported to the employee's supervisor. Then, all concerned should decide how to avoid a recurrence of the accident.

Wherever you work, you must know what to do in an emergency. You must know where the fire extinguishers and alarms are and what to do if someone gets shocked. You should consider this knowledge part of your job responsibilities.

Your work habits can prevent — or cause — accidents. Think before you turn on a switch. Determine whether someone else is working on another part of a circuit before you energize it. Don't leave things around for other people to fall over.

The proper care and use of tools are also important safety considerations. Check to be sure that power cords are not frayed, exposing the bare wire. Always use the proper tool for a job. Proper tool care means to use the correct tool and to use it properly. You can damage tools if you use them incorrectly. In addition, tools should be maintained if they are expected to last. They should be stored in a toolbox or tool cabinet. Power tools should be unplugged by disconnecting the plug, not by pulling the cord. The life expectancy of tools depends on the care of the user.

You are responsible for the care of the products and systems that you may be working on. Do not make matters worse than they already may be. Handle items with care so that you do not produce scratches or other damage. Operate equipment according to the instructions. If you do not know the proper operating procedure, read the manual. Treat the product as if it were your own.

In an attempt to ensure safe working conditions for all employees, Congress passed the Occupational Health and Safety Act in the 1970s. As a consequence of the act, many companies have spent large sums of money to make necessary changes. Lighting, ventilation, sound, and the size of work areas all have been improved in order to make the environment safer and more pleasant for the workers. The new regulations are not just for employers, though. Workers also have their responsibilities. Safety glasses are required for many tasks. Safety procedures for doing other tasks are also carefully spelled out.

By following the guidelines and safety precautions outlined in this chapter, you can reduce the chance of accidental injury to yourself, to others, and to the devices you work with and service.

CHAPTER 3

REVIEW TERMS

ESD (electrostatic discharge): electric current (sparks) caused by static electricity

grounding: establishment of common electrical connection to earth or ground

Hazard Communication Standard: federal regulations requiring that employees be given information about hazardous chemicals

MSDS (material safety data sheets): documentation from manufacturers and distributors describing hazardous chemicals and mixtures

OSHA (Occupational Health and Safety Administration): federal agency responsible for establishing safe working conditions, conducting inspections, and enforcing rules

static electricity: electricity caused by friction

REVIEW QUESTIONS

1. List the steps to be taken in case of accident.

2. Describe preventive maintenance.

3. What do the letters OSHA stand for?

4. Describe the purpose and scope of OSHA.

5. Describe ways to protect yourself from danger while doing electronics work.

6. Describe what is meant by "look before you leap" in electronics work.

7. Describe ways to protect those you work with in an electronics shop.

8. Describe what you would do if someone near you got an electric shock.

9. Explain how you can protect items or equipment that you work with.

10. Name some safety precautions you can take with equipment you work on.

11. Compare improper procedures with improper maintenance.

12. Name four possible accidents in an electronics shop, and explain how they can be avoided.

13. Select five chemicals, such as a cleaner, solvent, flux, or finish used in your shop or home, and list the manufacturer's use and disposal precautions.

14. Review an electronics supply catalog and select safety glasses recommended for the following jobs: soldering, drilling, and working with chemicals. Describe how the glasses differ.

15. Survey your electronics shop and note the safety rules that are being observed.

16. Prepare a report about ways your personal shop practices can be improved.

17. Prepare a report about ways to improve the safety of your school shop or lab.

18. Prepare a safety poster for use as a reminder in your electronics shop or lab.

ELECTRONIC COMPONENTS AND SYMBOLS

Identify common electronic components, including resistors, inductors, capacitors, transformers, and switches.

Identify common electronic symbols, and describe their function.

Define schematic, and describe its use.

Define pictorial, and describe its use.

INTRODUCTION TO ELECTRONIC COMPONENTS

The objective of this chapter is to introduce the names and the symbols of the common components used in electronic products. We will examine some of the products and see what they are made of.

The parts described in this chapter are those items that make an electrical or electronic product. Therefore, an **electronic component** can be defined as an individual part that is involved in the electrical or electronic processes of a product. Figure 4–1 shows a few components. Transistors and resistors are components, while screws and washers are not. Screws and washers are hardware, and they will be discussed in Chapter 10.

The variety of component types and sizes is almost beyond imagination. A quick look at an electronics parts catalog will convince you of that. This chapter and later chapters provide a brief introduction to some components and their symbols. However, the discussions in this book are only introductions. You will learn more about other components as you continue your studies of electronics and as you work in the field.

Resistors: The Most Common Component

Resistors provide resistance to current flow. They are suitable as an introduction to electronic components since they are the most commonly used component in electronic products and systems. For instance, if you were to look inside a radio or television, you would see more resistors than any other component. Most resistors, such as those in Figure 4–2, can be identified by their stripes. More will be said about stripes in Chapter 6.

FIGURE 4-1
ASSORTED RESISTORS, CAPACITORS, AND INDUCTORS

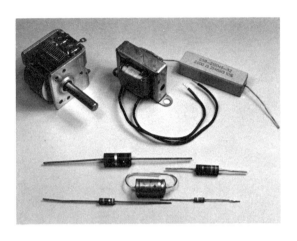

FIGURE 4-2
FIXED RESISTORS

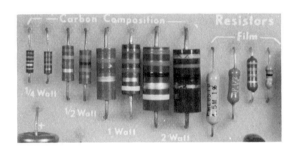

A resistor is about ½ inch long and has two wires, or leads, one coming out from each end. At this point, you only need to know how to recognize resistors; you do not need to know what they do. Their theory of operation comes later. In this section, we will take a brief look at resistors as an introduction.

Resistors come in a variety of brands, types, sizes, shapes, and prices. The most common resistor is the carbon resistor. You should also try to locate resistors in an electronic product, such as a small battery-operated radio.

Safety tip: Do not open a television to look for resistors, because you may get shocked.

Instead, your instructor can provide you with a safely disabled unit to look at. You can also look for other components while you're inspecting the disabled unit.

Types of Resistors

The resistors shown in Figure 4–2 are called fixed resistors. Each has a fixed specific value. There are many types of fixed resistors. Each type is named according to its composition — to what it is made of.

Resistors may also be adjustable. Adjustable resistors have a fixed total resistance value, but they also have controls that allow only a portion of that resistance to be used, if desired. Some adjustable resistors, called potentiometers, are used as radio volume controls.

SYMBOLS OF FAMILIAR ELECTRONIC COMPONENTS

Traffic signals use colors rather than words because colors are easy to recognize. They are also clearly seen from a distance. So you can decide quickly from a distance whether you must stop or go. Colors as symbols, such as red for stop, are understood by everyone. The advantages of the use of symbols are simplicity, clarity, and universal understanding.

Drawings of electronic circuits use sym-

FIGURE 4-3
FIXED RESISTOR

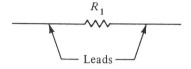

A. Fixed-resistor symbol

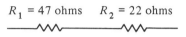

R_1 = 47 ohms R_2 = 22 ohms

B. Resistors in a series

bols for similar reasons. Symbols are easy to draw and recognize. There is little chance of mistaking one for another. They are simple to learn and easy to remember. Some common symbols are introduced in this chapter, and others will be introduced in later chapters. A more complete set of symbols appears in the inside covers of the book.

Resistors

As mentioned earlier, fixed resistors have one lead coming out of each end. The purpose of resistors is to oppose, or resist, electrons as they flow from one lead to the other. The resistor symbol in Figure 4–3A gives that impression with its jagged line. In the figure, the letter R stands for resistor, and the subscript 1 means it is the first resistor in the circuit.

Since all fixed resistors are drawn in the same way, more information must be added to the symbol. For example, a circuit may have two resistors of different size or value. The numbers and units added to the symbols in Figure 4–3B indicate the difference in the size of the resistors.

Adjustable resistors, or potentiometers,

FIGURE 4-4
ADJUSTABLE RESISTORS

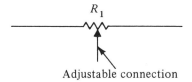

R_1

Adjustable connection

A. Potentiometer

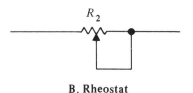

R_2

B. Rheostat

FIGURE 4-5
TAPPED RESISTOR

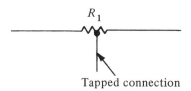

R_1

Tapped connection

FIGURE 4-6
FIXED CAPACITORS

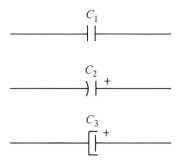

C_1

C_2

C_3

have three connections available. The symbol in Figure 4–4A shows the two end connections and the adjustable center connection. The arrow indicates that the middle connection can be moved from one end to the other, which is done by turning a shaft or moving a slider. The rheostat in Figure 4–4B is similar to the potentiometer, but one end is connected to the moving contact, as shown.

As we see in Figure 4–4B, arrows are often used to indicate that devices are adjustable or variable. However, there will be other uses for arrows also.

A tapped resistor is one that has a tap or connection part way from one end to the other. Although it has a third connection between the ends, the connection cannot be moved around. Its symbol looks like the one in Figure 4–5.

Capacitors

Another very common electronic component is the capacitor. **Capacitors** store electric energy, and like resistors, they come in a variety of shapes, sizes, and styles. Sometimes, the only way to identify a capacitor is by the information printed on it. Usually, this information includes the capacitance value and voltage rating. The capacitor in Figure 4–1 is centered between the resistors in the bottom row.

The symbol for a fixed capacitor has a few different forms. Figure 4–6 shows the three most common symbols. The letter C stands for capacitor, and the subscripts 1 and 2, as usual, identify one capacitor from the other. The symbol shows two leads that are connected to plates; the plates are separated by an insulator.

While capacitor theory and applications will be described in detail in Chapter 13, one point must be made here. In the upper symbol, the two leads are connected to plates or elements represented by parallel lines. Since these lines are similar, the elements that they represent are the same. There is no preference in position when you connect these two leads to two points in a circuit. In contrast, the elements connected to the leads in the other two symbols have different shapes and a plus (+) sign. These symbols indicate that there is a difference between the two leads and that there are right and wrong ways to install these components. Other components to be described later also have leads that are not interchangeable.

Adjustable, or variable, capacitors also come in a few basic styles. One familiar adjustable capacitor is the capacitor used as a station selector in radios. As shown in Figure 4–7A, the symbol for adjustable capacitors utilizes an arrow drawn through the two plates. Figure 4–7B shows another practice with symbols — that is, to use dashed lines for components that work together. When you adjust one capacitor, the other also changes because they are mechanically connected. However, their capacitances are separate since they are not electrically connected.

Inductors and Transformers

Basically, an **inductor** is a coil of wire. It can induce (produce) electric current. Some inductors consist of a few turns around a plastic form, and others have many turns of wire around an iron or air core. Inductors like the one shown at the top center of Figure 4–1 can be found in radios and televisions. Inductor theory will be discussed in Chapter 12.

The symbol for an inductor indicates the coil of wire (in an air core) and shows the two leads at the ends, as seen in Figure 4–8A. The Letter L is used to identify inductors. This symbol can be modified to indicate that the coil has a fixed iron core, as shown by the parallel lines in Figure 4–8B. An arrow added, as in Figure 4–8C, indicates an adjustable inductor.

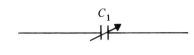

FIGURE 4-7
VARIABLE CAPACITORS

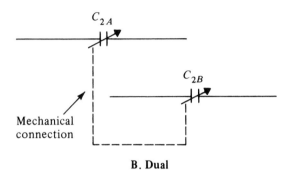

C_1

A. Single

C_{2A}

C_{2B}

Mechanical connection

B. Dual

FIGURE 4-8
INDUCTORS

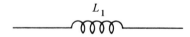

L_1

A. Air core

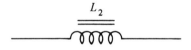

L_2

B. Iron core

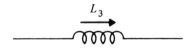

L_3

C. Adjustable core

FIGURE 4-9
TRANSFORMERS

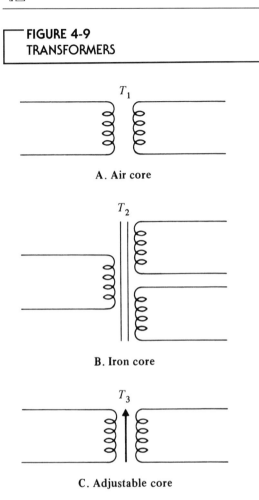

T_1

A. Air core

T_2

B. Iron core

T_3

C. Adjustable core

Transformers are similar to inductors in that they also operate on a coil-and-core principle. While inductors have only one coil, transformers have more than one. Transformers are used to change the level of an AC voltage. A more detailed explanation of transformers is given in Chapter 21.

Because there are many variations in transformer design, there are also many transformer symbols. The symbol for a transformer with an air core is shown in Figure 4–9A. It indicates two or more coils near each other but not touching. The coil on the left is called the primary, and the coil on the right is called the secondary. The letter T identifies a transformer.

Like inductors, transformers also can have cores that are iron or adjustable. Figure 4–9B shows the symbol for an iron-core transformer that also has two secondaries. The arrow in Figure 4–9C means that the transformer has an adjustable core.

Tubes, Transistors, and Integrated Circuits

The components described to this point — resistors, inductors, and capacitors — are categorized as **passive components.** Their characteristics — resistance, inductance, and capacitance — tend to remain constant during normal circuit operation. The components about to be introduced are called **active** because they are dynamic. Some of their characteristics change during normal circuit operation. For example, both current and temperature can change the resistance of a transistor. It is because of this behavior that transistors are used in electronic products.

Tubes, transistors, and integrated circuits, in that order, indicate the development of devices in electronics. Tubes, for decades, dominated the field. Figure 4–10 shows various styles of tubes, which are easily recognized by their glass envelope and their connector pins. Tubes can still be found in some electronic equipment today.

The symbol for a tube shows the outer glass envelope and shows which pins the internal parts, or elements, are connected to. Figure 4–11 shows a very common tube, a triode, along with the names of its three elements. The letter V stands for vacuum tube. More tube symbols are shown in the inside covers of the book.

The next step in the development of elec-

FIGURE 4-10
VACUUM TUBES

FIGURE 4-11
VACUUM TRIODE

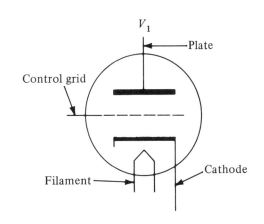

tronic components was the solid-state or semiconductor generation. Some common semiconductors are shown in Figure 4–12. This development began with the **diode,** a small component that allows electrons to flow in one direction but not the other. Once semiconductor diodes were introduced, they found their way into almost every product. The bottom and left components in Figure 4–12 are diodes.

The symbol for a semiconductor diode is shown in Figure 4–13. The two leads and the arrow pointing into the device identify the component as a diode. Diode theory, including the significance of the arrow, will be discussed in Chapter 28.

A **transistor** is a device that can amplify current or can change alternating current to direct current (called rectification). Transistors gradually replaced tubes in a number of circuits because they were smaller, used less power, lasted longer, and used no filament power while performing the same job tubes performed. Transistors appear at the top left and lower right in Figure 4–12.

FIGURE 4-12
ASSORTED SEMICONDUCTORS

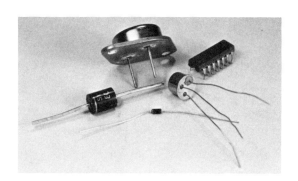

FIGURE 4-13
SEMICONDUCTOR DIODE

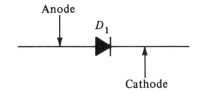

FIGURE 4-14
TRANSISTORS

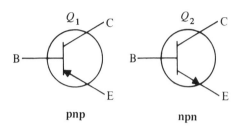

pnp   npn

FIGURE 4-15
DIGITAL-LOGIC INTEGRATED CIRCUIT

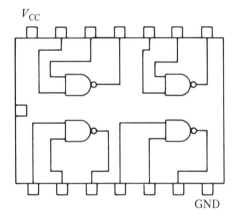

FIGURE 4-16
BATTERY

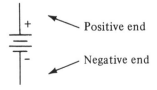

The most common transistor symbol shows three leads and an arrow, as presented in Figure 4–14. The letter Q identifies a transistor. The leads are called emitter (E), base (B), and collector (C), as shown, with the emitter lead arrow indicating the transistor type. An arrow pointing into the device identifies a pnp transistor; an arrow pointing out of the device designates npn. These terms, along with transistor theory, are discussed in Chapter 29.

As the field of electronics continues to develop, **integrated circuits (ICs)** are becoming increasingly popular. One of the main advantages of ICs is that they fit many resistors, capacitors, transistors, and other components into a small package. In fact, over 100,000 single components might now make up a single integrated circuit. One appears in the upper right of Figure 4–12.

Integrated circuit symbols and drawings — called schematics — are very complicated because they must show many parts. Therefore, they are often represented by a diagram that indicates function and pin connections. Figure 4–15 shows one example.

SYMBOLS OF OTHER PARTS OF ELECTRONIC PRODUCTS

There are many additional parts in electronic products besides the resistors, inductors, capacitors, and semiconductors already mentioned. For instance, there are connectors, cables, switches, antennas, and a wide range of other items too numerous to mention here. However, we will consider some of the more common ones.

Batteries and Lamps

One item that you are probably already familiar with is a **battery.** It is an electrochemical device used as a source of electricity and can

FIGURE 4-17
LAMPS

FIGURE 4-18
CONNECTOR

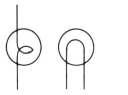

A. Incandescent   B. Neon

be found in flashlights, cars, and radios. The symbol for a battery is a series of long and short parallel lines with a line coming out of each end, as shown in Figure 4–16.

Symbols tend to represent what actually exists. If you looked inside a car battery, you would see a group of evenly separated plates. The symbol for a battery shows these plates as well as the two places where connections are made to the battery. These connections are marked + and −. The + (plus symbol) represents the positive terminal of a battery, and the − (minus symbol) represents the negative terminal. While this symbol was designed to represent a battery, it is generally used to represent any DC voltage source.

Another familiar item is the incandescent **lamp,** like the lamp in a flashlight. The symbol in Figure 4–17A shows two connections, or leads, a glass enclosure, and a filament (wire) inside.

Since all incandescent lamps are drawn the same way, more information is needed. You need to know the voltage and wattage of the lamp to accurately describe the component. The letter L stands for lamp in most electronics diagrams. Figure 4–17B shows another form of lamp, the neon lamp.

Cables and Connectors

Cables and connectors combine a group of products into a system. The obvious advan-

FIGURE 4-19
CONNECTORS

Output connections   Input connections

A. Plug-in connectors

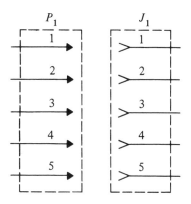

B. Five-pin connectors

FIGURE 4-20
GROUNDS

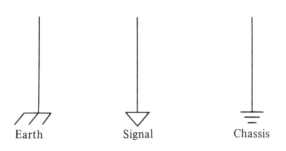

Earth   Signal   Chassis

FIGURE 4-21
ASSORTED SWITCHES

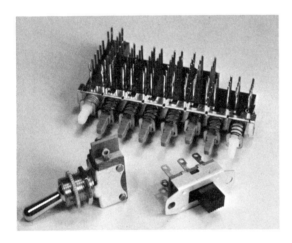

tage of using **connectors** is the ease with which the items in the system can be separated. Figure 4–18 shows the power cord connector on the back of an electronic product.

The symbol in Figure 4–19A represents a simple two-circuit connector. The arrows (output connections) indicate the load — what is going to be connected to the power source. A television, for example, is a load, and as you know, its connector has prongs. The indented symbol represents the source of power. For a television, the power source is a house outlet.

Figure 4–19B shows a five-pin, or five-circuit, connector pair. The plug is marked P and the jack is marked J. The plug is inserted into the jack.

The number of variations for connector symbols is beyond count. Generally, they all distinguish one pin from another and one wire from another, and they show which connector plugs into the other. For instance, the symbol in Figure 4–19B shows five circuits of plug 1 (P_1) connecting to the five circuits of jack 1 (J_1).

Grounds

An electric **ground** generally refers to some part of a circuit that is electrically connected to the earth. A ground is used for two reasons. First, the earth or ground is a common return point for voltages and signals. Second, a ground connection is made for safety reasons. Ungrounded circuits can cause shock when a bystander touches the cabinet and becomes the ground path. Metallic water pipes or rods driven into the earth make good grounds. Part of the wiring in your house is connected in this way.

Ground can also refer to a connection made to the metal chassis of a device. Thus, a wire connected to a metal chassis is said to be grounded. Figure 4–20 shows the various ground symbols.

Switches

Switches are used to quickly and conveniently connect and disconnect electric circuits. The simplest switch is the on-off switch. Laboratory devices, for example, often use a simple toggle switch, which moves one way for on and another way for off. Figure 4–21 shows various switches.

FIGURE 4-22
TOGGLE SWITCHES

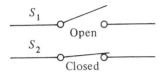

A. Single-pole–single-throw (SPST) switch

B. Single-pole–double-throw (SPDT) switch

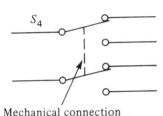

C. Double-pole–double-throw (DPDT) switch

The symbol for a switch shows two places to make connections, and it shows that the switch can be closed to connect a circuit to the power source or opened to disconnect it. Figure 4–22A shows a single-pole–single-throw (SPST) switch in the open and closed positions. The letter S represents a switch. Figure 4–22B shows a single-pole–double-throw (SPDT) switch, which is used to transfer a circuit to one of two possible other circuits. The double-pole–double-throw (DPDT) switch in Figure 4–22C can transfer two circuits to two different circuits each.

Switches can also come in many other types. For instance, a doorbell uses a push-button switch. There are also slide switches

FIGURE 4-23
SWITCHES

A. Normally open (NO) push button

B. Normally closed (NC) push button

C. 5-position rotary

and rotary switches. The name of a switch generally refers to the manner in which it operates.

A doorbell switch is normally opened since you must push it in to close, or energize, the circuit. See Figure 4–23A. It is also said to have momentary contact since it shuts off when you release the button. The door switch for a car or refrigerator light is normally closed; when you release the switch, by opening the door, the light goes on. This switch's symbol is shown in Figure 4–23B.

One important characteristic of switches is their ability to be multisectional — that is, to be able to switch several different circuits at one time. For instance, the rotary switch on a stereo, which selects AM or FM, tape, phono, or off, changes many circuits as it moves from one position to another. This switch's symbol is shown in Figure 4–23C.

There are so many possibilities for switch

**FIGURE 4-24
CIRCUIT DIAGRAMS**

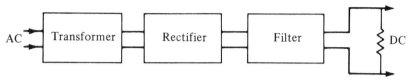

A. Block diagram

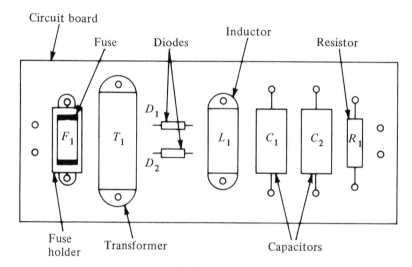

B. Pictorial

symbols that we cannot print them all. More are presented in the inside covers of the book. If you understand the basic concept of switch symbols, you can learn to interpret all of them.

COMBINING SYMBOLS

Symbols are primarily used to show how a complete circuit is connected. Pictures of complete circuits have two basic forms. **Circuit diagrams** show the circuit's interconnections. A **schematic** uses symbols to show how each specific component is connected.

Circuit Diagrams

The simplest form of diagram is the block diagram, like the one shown in Figure 4–24A. It does not necessarily show single components; rather, it shows the interconnection of sections that perform the basic functions. In other words, the **block diagram** gives a general idea of how a product functions. The process that occurs usually flows from left to right in the diagram. More complex block diagrams will appear later in this text.

Another form of diagram is called a pictorial. Like a picture, a **pictorial** is a drawing that shows the exact physical placement of components in a circuit. The pictorial in Fig-

FIGURE 4-25
SCHEMATIC DIAGRAM

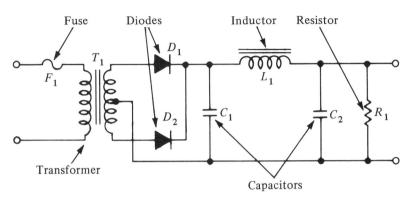

A. Complete schematic

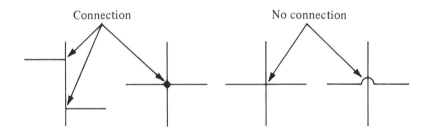

B. Wire-crossing symbols

ure 4–24B shows the top view of a power supply. It can be used by an assembler to find the exact place to install one component, or it can be used by a troubleshooter to find a particular component's physical location in a circuit.

Schematics

The second basic type of circuit drawing is the schematic. It is a map that shows the exact electrical location of all components and which ones are connected to each other. After you work with schematics like the one in Figure 4–25A, you will be able to read schematics quite easily. As in block diagrams, the electronics information in a schematic tends to flow from left to right. The troubleshooting process, however, may go from right to left, top to bottom, or any other direction, depending on the circuit symptoms. Note that parts electrically connected to one another may not be physically next to one another. That's why both pictorials and schematics are used.

The schematic in Figure 4–25A uses symbols that have been introduced in this chapter. The only new symbols are the symbol for the fuse (F_1) and the indications for wires that cross without being connected and wires that do connect. These symbols are highlighted in Figure 4–25B.

Parts Lists

The complete description of a resistor may be "1000 ohm 10% ½ watt carbon resistor." A schematic would quickly become cluttered if it included all information for all the components. Therefore, on a schematic, it is common practice to just label one resistor R_1, another R_2, and so forth. Capacitors are then labeled C_1, C_2, and so on. A parts list is prepared to accompany the schematic. The **parts list** names and describes all items on the schematic, such as "R_1 1000 ohms 10% ½ watt carbon resistor."

The parts list, then, serves a few purposes. It keeps the schematic uncluttered, and it provides a complete, up-to-date parts list for ordering. The parts list also provides a convenient way to change the description of a component without revising the schematic. Schematics are difficult to prepare and revise, while parts lists are not.

SUMMARY

Electronic components are items that make a product electrical or electronic. Resistors and wires are components, while screws and washers are not.

Symbols are a common and convenient way to represent a component on paper. The symbol may resemble the physical shape of the component, or it may characterize the way the component functions. The symbol also has the same number of leads as the component it represents.

This chapter discussed some common electronic components and their symbols, including resistors, capacitors, inductors, transformers, tubes, transistors, and integrated circuits. At this point, it is unlikely that you can completely define and describe these components, but you should be able to identify them and recognize their symbols. Brief definitions of these components are included in this chapter; detailed explanations of theory and application will be given in later chapters.

Pictorial diagrams, schematics, and parts lists are ways of describing a complete product. Pictorials are drawings that show where the components are actually located. They are used by assemblers to put parts in their proper place, and they are used by troubleshooters to find the physical location of parts. Schematics use symbols connected with lines representing wires or electric connections to illustrate a circuit. Schematics can be considered a map of the circuit. Parts lists are just that, a printed list of all parts with a complete description of each. The separate parts list keeps the schematic or pictorial from being cluttered and provides an easy way to change descriptions of parts without revising the schematic.

CHAPTER 4

REVIEW TERMS

active component: device whose characteristics change during normal circuit operation

battery: electrochemical device used to produce DC voltage

block diagram: drawing that shows the flow of information or functions of a circuit

capacitor: component that stores electric energy

circuit diagram: drawing showing a circuit's interconnections

connector: device that allows convenient electrical separation of one part of a system from another

diode: two-terminal component that allows electrons to flow in one direction but not the other

electronic component: individual part involved in the electrical or electronics processes of a product

ground: electrical connection to earth

inductor: component, such as a coil of wire, that induces (produces) current

integrated circuit: small multiterminal complex of electronic components and their connections

parts list: list of components used in a circuit, by type, identification number, and value

passive component: device whose characteristics remain constant during normal circuit operation

pictorial: drawing that shows the physical location of components in a circuit

resistor: component providing resistance to electron flow

schematic: drawing that uses symbols to show the interconnections between components in a circuit

switch: device for quickly connecting or disconnecting electric circuits

transformer: component that can change the level of AC voltage

transistor: three-terminal component used to amplify or rectify current

REVIEW QUESTIONS

1. Define electronic component.

2. Describe the purpose of using symbols.

3. Define schematic, and describe its purpose.

4. What is a pictorial, and what is it used for?

5. Describe the difference in appearance between a fixed resistor and a fixed capacitor.

6. What information is usually placed next to a resistor symbol on a schematic?

7. What information is usually placed next to a capacitor symbol?

8. Explain why complete schematic diagrams are not usually drawn for integrated circuits. What is usually drawn?

9. What do dashed lines mean on a schematic?

10. Why is there more than one symbol for ground?

11. What does an arrow usually mean?

12. Name the three leads on an npn transistor.

13. What is the significance of the parallel lines on inductor and transformer symbols?

14. The two leads of a fixed resistor are interchangeable, but the two leads of an electrolytic capacitor are not. Name four other components whose connections are not interchangeable.

REVIEW PROBLEMS

1. Draw three common resistor symbols and name them.

2. Draw three capacitor symbols and name them.

3. Draw the symbols for two types of inductors and name them.

4. Draw two transistor symbols. Name them.

5. Draw the symbols of five different switches. Name them.

6. Draw the symbols of two transformers.

7. Draw symbols for a ground, a connector, a wire crossing showing connection, and a wire crossing showing no connection.

8. Obtain a schematic of a simple electronic circuit and see whether you can identify all component symbols.

9. Make a list of the symbols that you do not recognize in the circuit of Problem 8. Find out what these items are.

10. Draw a pictorial and then a schematic of a circuit provided by your instructor. Label all symbols. Make a parts list.

DIRECT CURRENT, CONDUCTORS, AND INSULATORS

OBJECTIVES

Define electron, and describe electron current.

Define voltage, and name several voltage sources.

Define conductor, and name some electric conductors.

Define insulator, and name some electric insulators.

Describe various types of wire and cable.

ELECTRIC CURRENT

The objective of this chapter is to provide you with a basic understanding of how electrons flow from one place to another. You will also learn how their flow can be encouraged or resisted.

Atoms and Electrons

Let us suppose that we could take a drop of water and divide it into smaller sections. If we could do that, the smallest section we could produce would be a section called a molecule. While it is too small to see, the smallest section we can produce of any material is called a **molecule.**

Molecules are made up of atoms. For example, a water molecule is made of 2 atoms of hydrogen and 1 atom of oxygen. This composition gives water its formula name, H_2O. Basic substances like oxygen and hydrogen are called **elements,** and there are more than a hundred elements. Copper, lead, and zinc are others.

Atoms, the smallest particles of elements, are made up of protons, neutrons, and electrons. The protons and neutrons are in the center of the atom and form a **nucleus,** or center. Electrons orbit this nucleus, just as the planets orbit the sun. Figure 5–1 shows that the number of electrons in a normal atom equals the number of protons. For example, the hydrogen atom has one electron and one proton. Also, the number of protons, neutrons, and electrons in an atom determine its type — that is, whether it is iron, copper, oxygen, or some other element.

Protons have a positive charge, **electrons** have a negative charge, and **neutrons** have no charge. Since atoms have an equal number of positive and negative charges, they are balanced. Thus, the normal atom is neutral or stable.

There are two important forces that work within each atom. As the electrons speed around the nucleus in their orbits, centrifugal force makes them try to leave the atom. But they do not leave because another force, the attraction of the positive protons, holds them in their orbit. These forces are similar to the forces acting on a ball attached to a string. If you spin the ball around your head, you can feel the force of the ball trying to leave, but the string holds it in. The forces balance as they do in an atom.

Laws of Charges

We begin here with two basic points. Electrons have negative charges, and atoms have balanced charges. If, by force, we remove an electron from an atom, the atom will no longer be balanced. Instead, the atom will appear to be positively charged. Such an atom is called a **positive ion.**

In another situation, a free electron may join an atom, causing that atom to have an excess of electrons. The atom is then negatively charged and is called a **negative ion.**

If a material has a number of atoms with an excess of electrons, the material is negatively charged and will attempt to get rid of some electrons. If the atoms in the material have a deficiency of electrons, the material is positively charged and will attempt to get more electrons.

One example of this effect occurs when you comb your hair. The combing produces heat, which causes the electrons in your hair to move more rapidly than normal. As they go faster, they break away from the normal bond they have with the nucleus, and they leave your hair for the comb. You can often see and hear the electric spark as many electrons jump across to your comb. What happens next is another important electrical concept.

If you hold the comb near your hair, your hair will stand up on end. In this situation, two charged bodies are being placed near each other, and they are interacting. Your hair has a deficiency of electrons and the comb has an excess. So there is a force of attraction between your hair and the comb. This force follows a basic law of charges.

First law of charges: Unlike charges attract.

When the comb is allowed to touch your hair again, your hair will seem to stick to the comb for a moment and then fall away. While the comb is touching your hair, the electrons move from the comb, where there is an excess of electrons, to your hair, where there is a deficiency of electrons. This transfer continues until both objects, hair and comb, have similar charges. At this time, the attraction will stop, and your hair will fall away from the comb. In other words, the transfer of charges stops when both objects are equally charged. Note that although the atoms may be equally charged, they may not be normal.

The electroscope in Figure 5–2A is uncharged, and so its two metal plates hang side by side. But if you touch the charged comb to the top of the electroscope, the comb charge is transferred to the electroscope and down to the plates. Figure 5–2B shows what happens: Both plates become equally charged and separate. That feature shows another law of charges.

Second law of charges: Like charges repel.

Direction of Current Flow

When unlike charges are placed near one another, an attraction occurs, as was described with the hair-and-comb example. If the charged bodies are placed near enough to each other, sparks can be seen and heard. The sparks are the result of an electric current

FIGURE 5-1
HYDROGEN AND OXYGEN ATOMS

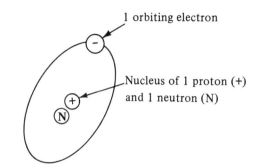

Hydrogen atom

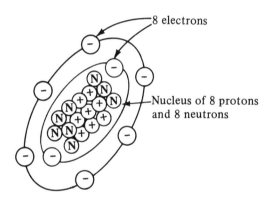

Oxygen atom

(moving electrons) that flows between the two bodies as they balance their charges. The flow stops when the two bodies have equal charges.

Which way does the current flow? Since electrons travel at the speed of light, which is 186,000 miles per second, you cannot tell the direction of flow by watching. Lightning, for example, results from charges caused by friction in the clouds. Does the lightning bolt go up or down? It occurs too fast to tell. Thus, the direction of current flow has been the subject of dispute for decades. You will find both

FIGURE 5-2
ELECTROSCOPE

books and people who disagree. The direction current travels depends on how you define current.

At one time, scientists knew that electricity — sparks — could be produced and that an electric current could be made to flow. But sparks were seen before they were understood. In order to have a common view, scientists agreed to define current as a flow of charges that went from a place of many charges, which they called positive, to a place of few, which they called negative. This view was agreed upon by convention — that is, by a meeting of scientists. Their definition for

conventional current, then, described a flow from positive to negative.

As time went on, research techniques improved, and scientists began to get a better understanding of what was occurring in electricity. They realized that *electrons* were moving from one place to another. Bodies became charged by having an excess or deficiency of electrons. And current flow was a result of a balancing of these charges. Therefore, another definition was presented, a definition for electron current. This new definition states that **current** is a flow of electrons from a place where there is a deficiency of electrons.

Key point: Electron current flows from negative to positive.

The debate over the direction of current flow goes on, and although both views have just been described here, this text will use the negative-to-positive approach. This approach is better for an understanding of concepts to be considered later.

The current described in this section travels in one direction only. Therefore, it is called **direct current (DC).** Current that changes direction back and forth is called **alternating current (AC),** and it will be described later in this text.

Definition of Terms

In the preceding sections of this chapter, some terms were defined that you must remember if you are to understand future subjects in electricity and electronics. While these definitions are not the only way to define the terms, they are commonly accepted definitions. The following list summarizes the definitions you should understand and remember.

— Electrons are negatively charged particles.
— Negatively charged materials are materials with an excess of electrons.
— Positively charged materials are materials with a deficiency of electrons.
— Current is a flow of electrons from a place where there is an excess to a place where there is a deficiency.
— Current flows from negative to positive.

Also, remember that current flows as long as there is a difference of charges and a reasonable route to travel. And direct current flows in one direction only.

VOLTAGE

You probably know from experience that if you put your finger into a wall or lamp socket, you will be shocked. You will get a shock because there is voltage at the socket. While you may think that nothing is occurring at a socket, the possibility, or potential, is there. This potential energy (available voltage) is described as having a value of about 115 volts. What exists at the socket is a force that can move free electrons if they are made available under the right circumstances. The right circumstances will be described later, in Chapter 8.

The force that moves electrons is called the **electromotive force (emf).** This force is also called **voltage,** and it is measured in units called **volts** (abbreviated V). Sometimes, you will hear a circuit described as having an emf of 150 volts. However, keep in mind that voltage basically describes what is available to cause current to flow. Voltage describes the potential and does not necessarily mean that current is flowing. Thus, sometimes we say that a circuit has a potential of 150 volts.

Chemically Produced Voltage

A very common source of DC voltage — that is, voltage associated with direct current — is a battery. Batteries come in a wide variety of sizes and shapes, as Figure 5–3 shows. Voltage is produced in a battery by chemical means. The simple demonstration in Figure 5–4 shows zinc and copper strips in a glass of salt water. These dissimilar metals undergo a chemical reaction in the liquid and produce a small voltage. The voltmeter in Figure 5–4 shows that this battery produces a voltage across its two terminals (the zinc and copper plates). Since this battery uses liquid, it is called a **wet cell.**

A **dry-cell battery** is shown in Figure 5–5.

FIGURE 5-3
ASSORTED BATTERIES

FIGURE 5-5
DRY-CELL BATTERY

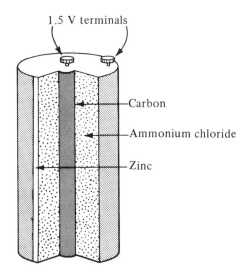

FIGURE 5-4
SIMPLE WET-CELL BATTERY

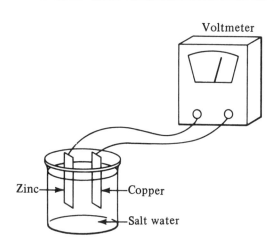

It operates on an interaction between zinc and carbon immersed in a salt (ammonium chloride) mixture. This basic dry cell has a voltage of 1.5 volts and can be connected with others to produce higher voltage levels. The main advantage of dry cells has been their portability.

Batteries have also been defined as primary and secondary cells. Primary cells use an irreversible chemical process; they cannot be recharged. Some dry cells used in radios and flashlights are in this category. Secondary cells, in contrast, can be restored by reversing the chemical process. The lead-acid wet cells used in automobiles are examples. Figure 5–6 shows how six 2-volt cells of a car battery are connected to produce 12 volts.

Many changes have occurred since these original batteries were developed. Chemicals such as nickel, cadmium, and mercury are now used in batteries, and their life and stability have thus been improved. As in all other aspects of electricity and electronics, new approaches are introduced each year to chemically produce voltage in smaller units and in better and less expensive ways.

Alkaline primary cells for use in radios, cassette players, and the like, have a longer life than the older dry cells. Nickel-cadmium (Ni-Cd) secondary cells, also called Ni-cad batteries, are used in portable VCRs because they are relatively compact, portable, and rechargeable.

Other Ways of Producing Voltage

Electricity can also be produced mechanically with generators. The theory of generators will be explored in Chapter 20, but a brief introduction is given here. Basically, in a **generator,** large coils of wire are moved through magnetic fields. These large generators are turned in a number of ways. They can be turned by moving water in a river or at a dam. They also can be placed with an impeller (the blade of a rotor) on a tower and turned by the wind. The most common way of producing voltage mechanically is by steam-driven generators. The heat to produce the steam comes from the burning of coal, oil, or gas or from a nuclear reaction.

Solar cells also produce electricity. They are called photovoltaic devices because they produce electricity with light. Solar cells are commonly used on spacecraft as a source of electric energy.

Two dissimilar metals such as iron and constantan (an alloy of copper and nickel) can be bonded together to form what is called a thermocouple. A thermocouple produces a small voltage that varies with temperature. This voltage can be measured and, with the aid of a chart, will indicate the temperature of the thermocouple. Since the metals can withstand temperatures that glass thermometers cannot, they can be used to measure very high and very low temperatures.

**FIGURE 5-6
AUTOMOBILE WET-CELL BATTERY**

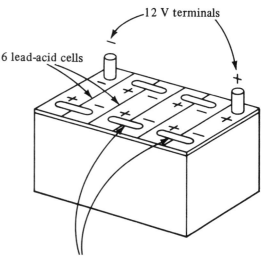

Connections between cells

Some materials like quartz change shape when voltage is applied to them. They also produce a voltage when pressure is applied to them. This method of producing voltage is called the **piezoelectric effect.**

Voltage Sources and Levels

Voltage level or amount is described in volts (V). A voltage description also includes terms like direct current (DC) or alternating current (AC) and frequency; these terms are important but will not be discussed here. They will be covered in Chapters 20 and 21. As mentioned earlier, the level of voltage indicates the amount of force available to move electrons. Here are some examples of voltage:

— Flashlight and radio batteries, 1.5 V, 9 V, DC;

— Car batteries, 12 V DC;

— Most household products, 115–120 V AC;

— Most household stoves and dryers, 220 V AC;

— Subways, 400–600 V AC;

— High-tension lines, 4400–220,000 V AC;

— Lightning, 1,000,000 V DC or more.

The level of voltage is an obvious indicator of danger, and it should be a guideline in your observance of safety rules. While you can assume that 9 volts are safe to work with and 220 volts are not, there is no voltage level that represents the line between safety and danger. Treat all voltage as potentially harmful, and as the level increases, make sure that your methods and materials increase in protection. While a 12-volt battery cannot shock you, shorted battery cables can cause heat that can burn you.

CONDUCTORS

A conductor of an orchestra directs or leads a group. A conductor pipe on a house carries the rain from the roof gutter to the ground. While it is true that the rain would have found its way to the ground anyway, the conductor directed it to a specific location. Similarly, an electric conductor serves two purposes: It allows the current to flow, and it directs it to a specific location.

While this fact may seem odd, electricity does not move easily through air. Light, sound, heat, and people move through air easily, but electricity does not. If it did, electrons would be shooting out of the electric sockets in your house all the time. But they do not. Instead, the force is there, ready to move electrons as soon as you attach something that allows them to move and provides a direction. Therefore, we have the following definition: An **electric conductor** is a material that carries or directs electric current.

Current flow through a conductor is very quick, although an electron may find its way through the conductor much more slowly. Current flow is caused by a strong negative voltage, or pressure, forcing an electron away from one atom and toward another atom. This action, in turn, forces an electron away from that second atom. In the meantime, the negative ion, with its excess of electrons, is replacing those that leave. The positive ion, with its deficiency of electrons, is drawing in those that are displaced. So an electron entering a conductor at the negative end begins a rapid bumping effect that results in another electron leaving the positive end almost immediately. The process shown in Figure 5–7 just continues to repeat itself as long as there is an excess of electrons at one end of a conductor and a deficiency at the other end.

Conducting Materials

A material can be a conductor of some things and not of others. Metal is a good conductor of electricity. Salt water is a reasonably good electric conductor, while distilled water is a poor conductor. Your chances of being hurt by an electric shock increase if your hands or feet are wet, because the moisture makes your skin a better conductor.

Table 5–1 compares the quality of many conductors. As the table shows, silver is the best conductor, and copper is a close second. While the conductive quality of carbon is far from that of copper, it is still very good.

Factors Affecting Quality

The primary factor affecting the quality of a conductor — that is, the **conductivity** — is how easily a given number of electrons can be freed to move from one atom to another. If the atoms of a given material release electrons easily, the material is a good conductor. If the

FIGURE 5-7
ELECTRON FLOW THROUGH A CONDUCTOR

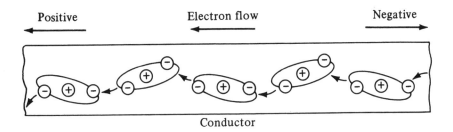

Positive   Electron flow   Negative

Conductor

atoms release electrons only with extreme difficulty, then the material is a poor conductor.

Since heat causes electrons to move faster, temperature affects the conductivity of most materials. Some materials have a positive **temperature coefficient;** that is, the conductivity increases with temperature. Other materials have a negative temperature coefficient. The rate of change also varies among materials. For some materials, it is slight; and for others, it is significant.

Another factor affecting conductivity is the cross-sectional area of a conductor — that is, the width of the path the electrons can follow. When a group of people walking down a school corridor come to double doors, the flow through the doors is greater when both doors are open. Closing one door reduces the conductivity, even if the corridor is very wide, since fewer people can pass through the door. Thus, the important consideration is the area of the opening at its smallest part.

Popular Electric Conductors

Although gold is an excellent conductor, it is not used very often because of its cost. It is used only in those situations where its excellent opposition to corrosion is necessary.

Many switches use silver contacts when

TABLE 5-1
CONDUCTOR QUALITY OF SOME COMMON MATERIALS

Material	Conductivity (%)
Silver	100
Copper	98
Gold	78
Aluminum	61
Tungsten	32
Zinc	30
Platinum	17
Iron	16
Lead	15
Tin	9
Nickel	7
Mercury	1
Carbon	0.05

good quality and long life are required. While silver is the best conductor, it is also expensive and will tarnish. Silver, like gold, is used only when its special qualities are required and can be afforded.

Aluminum is used when light weight is important. Mobile homes are often wired with aluminum because of its low weight and low cost. Aluminum is also used as an alternative when copper is not available.

FIGURE 5-8
RESISTIVITY OF MATERIALS (IN OHM-METERS)

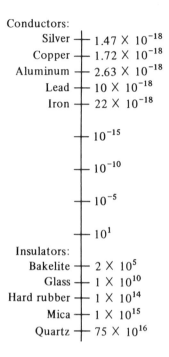

Conductors:
Silver — 1.47×10^{-18}
Copper — 1.72×10^{-18}
Aluminum — 2.63×10^{-18}
Lead — 10×10^{-18}
Iron — 22×10^{-18}

— 10^{-15}

— 10^{-10}

— 10^{-5}

— 10^{1}

Insulators:
Bakelite — 2×10^{5}
Glass — 1×10^{10}
Hard rubber — 1×10^{14}
Mica — 1×10^{15}
Quartz — 75×10^{16}

Copper is the most commonly used conductor. Copper is preferred over other metals because of its availability, low cost, strength, flexibility, and durability. Note that copper should never be connected to aluminum. The combination of these two metals may cause corrosion and sparking.

INSULATORS

An insulator is a very poor conductor. While a conductor allows and directs current flow, an insulator opposes it. The outer electrons of an insulator atom can be freed only with extreme force. Quite often, this force is enough to damage the insulator material. While it is true that electrons can be moved in an insulator, the amount of movement is often very small. But given the right conditions, such as a very strong force or a small distance, all insulators can break down. The scale of materials in Figure 5–8 shows the two extremes — excellent conductors on one end and excellent insulators on the other end.

The scale in Figure 5–8 is a resistivity scale. **Resistivity** is the opposite of conductivity, and it describes the extent to which a material resists, or opposes, the movement of free electrons. As we can see by comparing Figure 5–8 and Table 5–1, there is an inverse relationship between conductivity and resistivity. Moving down the conductivity scale in Table 5–1 is the same as moving down the resistivity scale in Figure 5–8. If all the known elements were listed on the table, we would eventually arrive at the good insulators: glass, ceramic, and so forth.

It is not enough to describe an insulator by what it is not, though. An insulator is a material that has very few free electrons. So we have the following definition: An **insulator** is a material that opposes the free movement of electrons.

Insulating Materials

A material may be a good conductor or insulator for light but not for heat. It may be a good insulator for sound but not for electricity. Therefore, note that this section discusses only electric insulators.

The best insulators are quartz and mica. Plastics and rubber are also good insulators. Distilled water is a good insulator, but impure water with salt or other minerals is not. Glass and ceramic are also relatively good insulators.

You should keep in mind what insulators are and what they are not while you are working. If some wires or a metal chassis are on a wood or plastic bench top, you need not be as concerned about safety as you would have to be if the bench had a metal top or edge. Tools with insulated handles are much safer to work with than those without insulation. The more insulation there is in your work area, the less likely it is that you will be affected by contacting circuit parts.

Factors Affecting Quality

The most significant factor affecting the quality of an insulator is its resistivity. The resistivity of some common materials is listed in the Appendix. The thickness of an insulator does not increase its basic resistivity factor, but it does increase the insulation it provides to a circuit. An analogy is the thickness of clothing: When it gets cold outside, you put a coat on over your sweater. Thickness in an insulator works in the same way.

Since temperature affects the movement of electrons, it also has some effect on the characteristics of insulators. For example, copper has a positive temperature coefficient; thus, its resistance increases as the temperature rises. This effect can be seen in the resistance columns of the copper wire table in the Appendix. Extremes of temperature can also affect the quality of insulators. If an insulator becomes too cold, it can crack. If it becomes too hot, it can melt or burn.

Moisture reduces the quality of some insulators by getting into small cracks and providing paths for free electrons to travel. While the insulator appears to be good, it may not be doing its job. For example, sometimes on a quiet night, you may hear soft snapping at the top of power lines. Small cracks in the glass insulator have become moist and are allowing current to leak through.

So to summarize, the main factors that affect the quality of an insulator are resistivity and thickness; time and the environment may also cause deterioration of insulators.

Popular Electric Insulators

As with conductors, many factors must be considered in the selection of an insulator. These factors include environment, use, quality required, and cost. At one time, paper was used to insulate the wiring in homes. Paper is a reasonably good insulator, and it could be used in homes because there it was protected from the weather and would not be moved around. However, as it happened, mice would get into the walls, take the paper for nests, expose the wires, start electrical fires, and get electrocuted. Therefore, the paper was later encased in metal.

PVC (polyvinyl chloride) insulation is used under the hood of automobiles. It must withstand winter and summer temperatures along with engine heat and oil, gas, and other materials that may be spilled. Toasters use mica to hold the heating wires because it is firm and can withstand the heat. Glass is used in fixed positions where high voltages exist.

In the past, rubber was a commonly used insulator for wires. It is not used often today, for while it is an excellent insulator, rubber is damaged by environmental elements such as heat and chemicals. Plastics have replaced rubber in many applications because they are more durable. Most wires today use nylon, Teflon, or PVC. Sockets and connectors are usually made with these same materials.

WIRE

The primary purpose of a wire is to conduct electric current from one location to another. While aluminum or other metals are sometimes used, most wires in electronic products have copper as conductors. As mentioned earlier, copper is the preferred choice considering conductivity, cost, durability, and a few other factors.

FIGURE 5-9
SOLID AND STRANDED WIRES

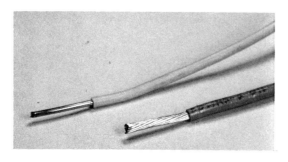

FIGURE 5-10
FIBER-OPTIC CONDUCTORS AND CONNECTORS

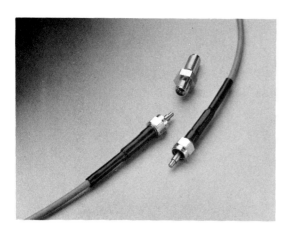

Types of Wires

The simplest wire is called **bus wire.** It is a single, solid copper conductor with a tinned coating. The silver-colored tinned coating protects the copper from deterioration and makes soldering easier.

Solid wire, shown at the top in Figure 5–9,

has a single conductor covered with insulation and can be found in a variety of places. Thin, solid insulated wire is often used inside electronic products, while a heavier type is used for wiring houses. Wire for use within the walls of a house generally consists of a solid black insulated wire, a solid white insulated wire, and a solid uninsulated wire, all within an outer insulation. An advantage of solid wire is that it will hold its position once it is put in place. A disadvantage is its inflexibility.

When flexibility is needed, stranded wire, like that at the bottom in Figure 5–9, is used. For example, stranded wire is used on lamp cords. As Figure 5–9 shows, each conductor consists of a number of small conductors twisted together to form one large conductor. As greater flexibility is required, more and finer strands are used. Electric power tools generally use a three-conductor cable — that is, three separate insulated and stranded conductors within one insulated cable. As for the lamp, two conductors provide power. The third conductor grounds the tool for safety reasons.

There are many other wire types that you should learn to identify, which you can do by looking at a wire catalog. Some you may have already seen are television twin-lead wires, coaxial cables for CB antennas, braided wire for grounds, and shielded cables for microphones.

Here are three important points to remember about wire:

1. The diameter of the conductor depends on the amount of current expected.

2. Large current requires large conductors.

3. Stranded conductors are used in applications requiring flexibility.

Copper wires are beginning to be replaced, in some applications, by glass fibers.

This segment of electronics is called fiber optics, and it utilizes conductors made with a bundle of fine glass fibers rather than the traditional solid or stranded metal conductor. Figure 5–10 shows typical fiber-optic cables and connectors. Information is transferred by using photons (light) rather than electrons (electricity), with significantly lower energy losses. The theory and applications of fiber optics are topics deferred to other courses.

Determining Wire Size

Numbers are assigned to conductors to identify them by size. A wire's size relates to its cross-sectional area — that is, how much room there is for current to flow — and is called a wire's **gauge.** A low gauge number indicates a very thick wire, and a high number indicates a very thin wire. A wire gauge chart is given in the Appendix.

The wires in the walls of most homes are probably no. 12 or no. 14, while the wires in a radio are probably no. 22 stranded. Power line wires could be no. 0, and wires inside some small electronic products may be no. 46 solid.

You may need to determine wire size for two purposes: (1) to find out the size of the wire required for a given application and (2) to find out the size of a wire you are working with. To determine the size required, you look in a chart like the one in the Appendix. It identifies the size of conductor required for a given amount of current. While you can always use a larger size of wire, do not use wire that is too small. A wire too small for its application can easily overheat.

To determine the size of a wire you are working with, you measure it with a device called a wire gauge. A wire gauge is a thin sheet of metal with numbered holes. All you need to know to use it is whether the wire is solid or stranded and which is the smallest slot

**FIGURE 5-11
DETERMINING WIRE SIZE**

that the wire will fit into, as shown in Figure 5–11.

Wire Insulators

As mentioned earlier, the purpose of insulation on a conductor is to prevent the conductor from touching another conductor or the chassis. The type of material used for wire insulation depends on the environment, and the thickness of the insulation depends on the voltage that will be expected. Rubber is used when extra flexibility is needed in high-voltage circuits, as in meter leads. Nylon, Teflon, or PVC are usually used in most other electronic applications because of their durability under many difficult conditions.

The coils of wire inside doorbell units have special insulation. Since many turns of wire must be close together and since the wire is not likely to be touched or rubbed, enameled wire is used. This enamel is a thin coat of special paint that is very durable. If the wire is to be connected or soldered, the enamel must be removed with a solvent or sanded clean with fine sandpaper.

Shielded cables also have special insulation. They begin with one or more conductors, each insulated with rubber or Teflon. They are then wrapped together with a layer of cloth. The cloth provides flexibility, and it also protects the individual insulators from contact with the next layer, a tight mesh of fine braided wire or a layer of foil, both of which are called shields. The shield is used to protect the small signals that pass through the inner conductor from interference by large signals that may be in the local area. Shields are usually covered with a cloth wrapper, which, in turn, is covered with an outer or final insulation of plastic or rubber.

The wire used for cable TV is called coaxial cable, and it is specifically designed to carry that type of signal. Coaxial cables come in a variety of impedances (a measure of resistance) and are assigned numbers such as RG8U or RG59U. They usually consist of a single-stranded conductor that is covered with milky white plastic insulation, then a shield, and, finally, a black plastic cover. Coaxial cables have specific characteristics, and the proper one must be used in order for the system to work effectively.

Colors and Stripes

The color of a wire's insulation usually indicates its use or function. A lamp cord, for instance, may be the same color on both sides, an indication that it does not matter which way the plug goes in. A car radio often has various wire colors, one for power and one for each speaker, for example. The instructions for the radio explain the difference, so you should make a practice of reading instructions that come with a product.

Transformers are often sealed units with many wires coming out of them. Rather than write numbers on the wires or attach tags that may fall off, manufacturers use different colors of insulation on the wires. Again, you must check the manufacturer's instructions to determine the correct connections.

Thus, different colors are used for wire insulation in order to distinguish one conductor from another when you cannot see both ends at the same time. When a cable has many conductors, stripes are used to avoid duplication. For example, one conductor may be red, another red with a white stripe, another red with a blue stripe, and so on. There are many possibilities.

Learning the meaning of colors on conductors is important for the proper operation of circuits and for your safety. The meaning changes from one situation to another. For example, some use black for ground, some use green, and others use white. You must know what a color means before you use a wire. Therefore, always check the manufacturer's instructions.

SUMMARY

All matter is made of atoms, which, in turn, are made of protons, neutrons, and electrons. Normal atoms have equal numbers of protons and electrons, and the number of electrons an atom has determines the type of element the atom is. It could be iron, oxygen, copper, or one of more than a hundred different elements. In electronics, we are concerned with electrons and how they can be moved from one atom to another.

Through external forces, electrons can be removed from some atoms and added to other atoms. An atom that has more electrons than it should have is said to be negatively charged, since electrons are negative. If an atom has a deficiency of electrons, it is said to be positively charged.

There is a tendency for electrons to move from a place where there is an excess to a place where there is a deficiency of electrons. This desire for balance causes an object with a deficiency to attract one with an excess — that is, unlike charges attract. In contrast, an object with an excess of electrons does not want more, so like charges repel.

In electricity, charges are moved by a force called voltage (measured in volts). There are many ways to produce this voltage or electricity. Batteries produce it with chemicals. Electricity for homes is produced by mechanical power plants, although solar cells are becoming a popular source, too. Voltage levels can range from 1.5 volts for a flashlight battery to over a million volts from lightning.

Electricity is produced by moving electrons from one place to another through conductors, which are materials that allow their electrons to be moved with relative ease. In a conductor, a negative charge moving at one end releases some excess electrons at the other end as a positive atom replaces its defi-

ciency by taking electrons from the other end. In this process, free electrons shift toward the positive end, causing a current flow. This electron current flows from negative to positive.

Conductor materials may be silver, copper, gold, aluminum, or many other materials. Copper is most often used because it is a good compromise between cost and quality. The diameter, or size, of a conductor depends on the amount of current expected to pass through it. As the cross-sectional area of a conductor increases, the ease with which electrons pass through increases.

Conductors are covered with insulators in order to protect them from contacting other conductors. Insulators are materials that strongly oppose the movement of electrons. While glass, cloth, paper, and rubber are good insulators, most conductors are covered with plastics like nylon, Teflon, or PVC. Plastic is an excellent insulator because it is flexible and can withstand a variety of environments. Insulators use different colors for two basic reasons. First, colors may identify a wire as a ground, a high-voltage wire, or a wire for some other circuit. Second, colors can be used to distinguish one wire from many others within a cable.

CHAPTER 5

REVIEW TERMS

alternating current (AC): current that changes direction back and forth

atom: smallest particle of an element

conductivity: ease with which current can flow

conductor: material that carries and directs the flow of current

current: flow of electrons from negative to positive

direct current (DC): current that travels in one direction only

dry-cell battery: battery made with solid elements, such as zinc and carbon

electromotive force (emf): voltage or force that moves electrons

electron: negatively charged particle orbiting an atom nucleus

element: basic substance such as copper, glass, and oxygen

gauge: size of wire; related to its cross-sectional area

generator: electromechanical device for producing electricity

insulator: material that inhibits the flow of current

molecule: smallest quantity of a compound of elements

negative ion: atom with excess electrons

neutron: neutral (no charge) particle in an atom nucleus

nucleus: central part of an atom

piezoelectric effect: method of producing electricity from pressure

positive ion: atom with electron deficiency

proton: positively charged particle in an atom nucleus

resistivity: extent to which a material resists electron flow

temperature coefficient: mathematical relationship between temperature and resistance; factor affecting conductivity

volt: unit of measure of voltage

voltage: force that moves electrons

wet-cell battery: battery made with liquid elements, such as an acid

REVIEW QUESTIONS

1. Define electron.

2. Define current and direct current.

3. Define voltage, and name the unit of voltage.

4. Define conductor, and name some examples of conductors.

5. Define insulator, and name some examples of insulators.

6. State the two basic laws of charges.

7. Describe how electron current flows in a conductor.

8. Describe three ways to produce voltage.

9. Describe three voltage sources and their voltage level.

10. Explain how a wire gauge is used.

11. Compare electron current with conventional current.

12. Discuss the difference between a negatively charged atom and a positively charged atom.

13. Explain what is meant by electromotive force.

14. Discuss the factors that affect the quality of a conductor. Name four conductors in order of their quality.

15. Discuss the factors that affect the quality of an insulator. Name four insulators in order of their quality.

16. What is the relationship between conductor diameter and current rating of a wire?

17. What is the relationship between insulator thickness and voltage rating of a wire?

RESISTORS AND THE COLOR CODE

Define resistance and resistor.

Identify fixed and adjustable resistors by type.

Determine resistor value by color code.

Determine resistor tolerance and range by color code.

RESISTANCE

The objective of this chapter is to introduce the wide variety of resistors used in the electronics industry. The two basic resistor types, fixed and adjustable, are described.

Electric current flows from one location to another through conductors, which allow the free movement of electrons. Insulators are used whenever free electron movement is not wanted, since insulators inhibit this motion. Insulators have **resistance,** which is defined as the property of a circuit that opposes the flow of current. Resistance is measured in **ohms** abbreviated Ω, the Greek capital letter omega). All materials have resistance. It is small and undesired in conductors, while it is large and desired in insulators.

In Chapter 5, we discussed conductors and the conductivity of various materials. As explained there, resistivity is the inverse of conductivity. One material is a better conductor than another because its conductivity is higher and its resistivity is lower. For example, silver has a lower resistivity than copper, and hence, silver is a better conductor than copper.

The net resistance of a conductor depends upon four factors: the type of material, the cross-sectional area, the length, and the temperature. The wire table in the Appendix indicates the resistance of copper conductors at normal room temperature, by diameter or wire size. According to the table, number (#) 22 copper wire has a resistance of 16.5 ohms per 1000 feet at 25 degrees centigrade ($^\circ$ C). To determine the resistance of a specific-length conductor, one simply multiplies the resistance per 1000 feet by the length. For example, the resistance of 100 feet of #22 copper wire is determined as follows:

$$\frac{\Omega}{1000 \text{ ft}} \times \text{length} = \text{resistance}$$

$$\frac{16.5 \ \Omega}{1000 \text{ ft}} \times 100 \text{ ft} = 1.65 \ \Omega$$

Similarly, a 200-foot piece of #22 wire has a resistance of 3.3 ohms, which is calculated as follows:

$$\frac{16.5 \ \Omega}{1000 \text{ ft}} \times 200 \text{ ft} = 3.3 \ \Omega$$

If the wire size is changed from #22 to #12, the resistance will change from 3.3 ohms to 0.324 ohm, as shown in the following calculation:

$$\frac{1.62 \ \Omega}{1000 \text{ ft}} \times 200 \text{ ft} = 0.324 \ \Omega$$

Thus, we see that as wire length increases, resistance increases. Also, as wire cross-sectional area increases, resistance decreases.

Suppose we look up the resistivity of glass in the Appendix. We know that it has a resistivity of 10^{10} ohm-meters (abbreviated Ωm). In contrast, the resistivity for copper is only about 10^{-8} ohm-meters. These values give you some idea of the range of resistance between a good insulator and a good conductor.

Resistors

Resistors, introduced in Chapter 4, are small components with two leads. In a DC circuit, current goes in one end and comes out the other, and it can go either way with the same amount of effort. Resistors, as their name implies, resist (or oppose) this current flow. Resistors come in values from less than 1 ohm to 10 million ohms and higher.

Since conductors and insulators have resistance, you may wonder just what is different about a resistor. The difference is that a **resistor** is a component designed to produce a specific amount of resistance. Manufactured resistors are made of carbon composition, carbon film, metal film, wire, and other materials.

Purpose of Resistance and Resistors

Resistors are used to control the amount of current in a circuit. Resistors are also used to produce specific voltages in a circuit. Resistor applications will be covered in Chapter 8. Here, resistors are mentioned simply to consider their purpose.

To explain the purpose of resistance, we consider an analogy. If you were to fall while running across a gym floor, you would probably burn your leg. The burn is caused by friction — the resistance to the free movement of your leg on the floor. Similarly, when electrons pass through conductors, some heat is caused by the resistance of the conductors. In most conductors, this heat is insignificant and unnoticed. But in others, it is significant, and sometimes, this heat is desirable. The heating elements of stoves and toasters utilize this characteristic. The heating element in a toaster is simply a conductor with enough resistance to get warm as current passes through it. The material used is nichrome wire, which can withstand this heat without melting or breaking.

Incandescent light bulbs also use the resistance principle to operate. Some wires, such as tungsten, glow as electric current passes through them. Some heat is also produced, so these thin filaments of wire are enclosed inside a glass bulb to protect them from the air. If oxygen were made available to them, they would immediately burn.

FIXED RESISTORS

Fixed resistors have one value, and they are produced in many ways. The production method depends on factors such as precision, stability, and cost. The more exact the resistor's value is, the more it will cost. Also, the more stable the resistor is, the more it will cost.

Carbon Resistors

Carbon resistors are often used in radios and televisions. They are usually a single color and have colored stripes near one end. **Carbon resistors** are made from a composition of carbon and Bakelite, which have been pressed together under heat. Carbon is a fair conductor, while Bakelite is a poor conductor. So the value of resistance is controlled by the ratio of carbon to Bakelite. In this way, 10-ohm and 10,000-ohm carbon resistors can be made with the same physical size. The difference is inside.

Carbon resistors usually come in sizes from about 0.2 to 22 million ohms. Their physical size does change with their **power rating** — that is, how much heat they can give off (or dissipate). The normal power ratings are ⅛, ¼, ½, 1, and 2 watts (abbreviated W). An assortment of carbon resistors is shown in Figure 6–1. The smallest resistor in the photograph is ¼ watt, and the largest is 2 watts.

Wire-Wound Resistors

As their name suggests, **wire-wound resistors** are made with a wire wound around a core. The largest two resistors in Figure 6–2 are wire-wound power resistors and can safely dissipate the heat of 5 watts. But they are not the only type of wire-wound resistor. Some are **precision resistors,** made from wire molded in a special epoxy. They are not designed for high power dissipation; in fact, their power dissipation is 1 watt or less. Precision wire-wound resistors are used when a *specific* value is required and when this value must be *stable* under many conditions. The tolerance

FIGURE 6-1
CARBON COMPOSITION RESISTORS

FIGURE 6-2
FIXED RESISTORS

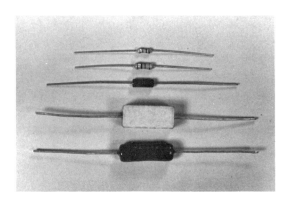

Other Fixed Resistors

The original precision resistor, the wire-wound, has been joined by others. Carbon film and metal film resistors are the most common. The two small resistors in Figure 6–2 are examples. These resistors have been introduced because of a need to overcome some technical limitations of wire-wound and carbon resistors. Pulse response and physical size are just a few of the problems that designers encounter with wire-wound resistors, for example.

New manufacturing and chemical techniques have made these new resistors possible. As with all other resistors, their names give an indication of how they are made. For example, a **carbon film resistor** is made by coating a small insulating rod with carbon. The thickness of the coating determines the resistance. Since carbon is a conductor, a thick coating produces a lower resistance than a thin coating. These resistors are also called **flameproof resistors.** Flameproof resistors will neither initiate nor support combustion.

Resistor networks such as those pictured in Figure 6–3 are frequently used on printed circuit boards. A network is a convenient way to install a number of interconnected resistors quickly and in a smaller space than an equal number of individual resistors would require. Since the sealed unit does not reveal its internal connections, a schematic is required. Figure 6–4 shows some network configurations.

ADJUSTABLE RESISTORS

There are times in the construction of an industrial electronic device when an **adjustable resistor** is needed — that is, a resistor for which the resistance can be changed. Often, this situation arises when a designer may not know the exact value needed or when the

of a carbon resistor is usually from 5% to 20%, with the lower number being the better. Precision wire-wound resistors come in values from 0.005% to 1% and have a very low temperature coefficient. The middle resistor in Figure 6–2 is a precision wire-wound resistor.

FIGURE 6-3
RESISTOR NETWORKS

FIGURE 6-3
RESISTOR NETWORKS

value may have to change as the electronic device ages.

Power Resistors

The schematic symbol of an adjustable resistor is shown in Figure 6–5. It consists of a resistor with a third connection that can be moved from one end to the other. As shown in Figure 6–6, the wire-wound adjustable power resistor resembles the schematic.

Wire-wound power resistors are available in values up to 1500 watts. You know how hot a 60- or 100-watt lamp can get, so you can imagine how much heat these resistors can tolerate. Wire-wound power resistors are made of a special wire that is wound around a ceramic core and coated with a high-temperature enamel. They are usually mounted in a way that allows for ventilation for cooling. Always keep your hands away from them; while shock is the first danger, severe burns are another possibility.

The use of the adjustable resistor is quite simple. If the resistor is 1500 ohms from one end to the other, then it is 750 ohms from one

FIGURE 6-4
SCHEMATICS FOR RESISTOR NETWORKS

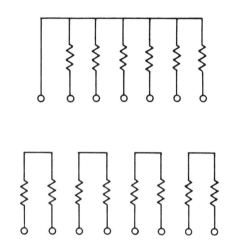

FIGURE 6-5
ADJUSTABLE-RESISTOR SYMBOL

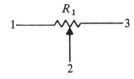

end to the middle. One-third of the way is 500 ohms, and two-thirds of the way is 1000 ohms. If you are not sure of the exact value your circuit needs but think that it is about 1000 ohms, you can install a 1500-ohm resistor and move the slider two-thirds of the way from one end.

Whenever you use an adjustable resistor, you must reconsider the resistor's power dissipation. As the amount of a resistor that is used becomes smaller, the power dissipated by that section increases. Care must be taken not to exceed a resistor's power-handling capability. We will discuss this issue again later.

FIGURE 6-6
ADJUSTABLE POWER RESISTOR

FIGURE 6-7
POTENTIOMETERS

It should also be noted that these adjustable components are usually located inside a product so that curious and untrained owners do not change settings that are best left alone.

Potentiometers

Figure 6–7 shows another very common type of adjustable resistor, the potentiometer. A **potentiometer** is a three-terminal component

FIGURE 6-8
POTENTIOMETER CONSTRUCTION

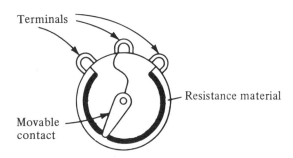

Terminals

Resistance material

Movable contact

having a fixed amount of resistance between the two end terminals and an adjustable amount between either end and the center terminal. See Figure 6–8. Adjustment is made by turning the knob of a shaft or by moving a slider. The volume controls on a radio or a television are potentiometers.

A **rheostat,** like a potentiometer, has adjustable resistance. The rheostat, though, is used in high-power applications. There are two main differences between potentiometers and rheostats. Potentiometers have three terminals, as described; rheostats usually have only a center terminal and one end terminal. In addition, potentiometers are usually made of carbon or fine wire, but rheostats are made of heavy wire, like other power resistors.

The basic potentiometer is called a one-turn potentiometer, even though its rotation is a little less than 360°. Some electronic devices use 3-, 5-, 10-, or 20-turn potentiometers.

The basic potentiometer just described is said to have a linear taper, because equal degrees of change in position cause equal changes in contact-to-end resistance. Volume controls, however, are nonlinear potentiometers, because the resistance change is not in direct proportion to the position change. That is, the relationship between the voltage inside and the sound coming from the speakers is not linear. A nonlinear (or logarithmic) potentiometer is used to provide a control with positions that are in proportion to the sound level.

Potentiometers can be made from several materials. Carbon and wire-wound potentiometers were the original and most common types. But as with fixed resistors, new film types are now available. Figure 6–9, for instance, shows miniature multiturn potentiometers designed for direct installation on printed circuit (PC) boards.

RESISTOR COLOR CODE

One of the most basic parts of the language of electronics is the resistor color code. You will use this code so much in the early part of your career that you will memorize it automatically. The **color code** uses colors in place of numbers to describe the values and other specifications of resistors. In this section, we will consider the most common resistor color code. We will see how the code works and how resistance is determined.

Value

Resistors come in a wide range of values expressed in ohms, which will be described in Chapter 8. For many years, colors have been used instead of numbers to identify resistors up to 2 watts. The colors represent numbers. Colors are used because of the possibility that part of a printed number will wear off and the value will be misinterpreted. If part of the number became chipped, for example, an 8 might look like a 3, or 1000 might look like 100. Then, the wrong-value component will be installed, which, in turn, can lead to improper operation or even cause damage. Colors also make it easier to position a resistor, since the color bands on a resistor go all the way around it. Thus, there is no need to be concerned

FIGURE 6-9
PC BOARD POTENTIOMETERS

about placing a resistor so that numbers can be seen. In addition, colors are easy to see and recognize.

There are various versions of the resistor color code. The most common has four stripes. It is interpreted as follows: The first part of a resistor's description is its value in ohms. To determine this value, place the resistor so that the colors begin at the left, as shown in Figure 6–10A. The first color band represents the first number, the second band represents the second number, and the third band represents the **multiplier** — that is, the number of zeros to add. Table 6–1 shows which number is represented by each color. (The fourth band represents the tolerance of the resistor, which we will discuss later.)

The resistor in Figure 6–10A has stripes

FIGURE 6-10
SAMPLE RESISTORS

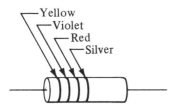

A. Resistor value of 4700 ohms

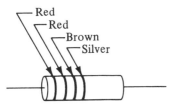

B. Resistor value of 220 ohms

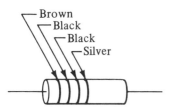

C. Resistor value of 10 ohms

that are yellow, violet, red, and silver. The first number, then, is 4, the second is 7, and the multiplier, or number of zeros, is 2. (We ignore the silver band for now.) Thus, the resistor has a value of 4700 ohms, as shown by the following calculation:

$$47 \times 100 = 4700 \ \Omega$$

The red, red, and brown resistor in Figure 6–10B has a value of 220 ohms. The first two numbers are 2, while the multiplier, or number of zeros that you add, is 1. Thus, we have the following calculation:

$$22 \times 10 = 220 \ \Omega$$

Some people get confused by resistors with colors like brown, black, and black, like the one in Figure 6–10C. This resistor has a value of 10 ohms. That is, brown stands for a 1, the first black stands for a 0, and the next black indicates that no further zeros are added. Thus, the calculation is as follows:

$$10 \times 1 = 10 \ \Omega$$

As these examples have shown, most resistor values have two digits with a number of zeros added. This procedure moves the decimal point to the right. What about moving the decimal point to the left? This procedure is represented with the third stripe by using gold or silver, as shown in Figure 6–11. Table 6–1 indicates that gold is used for a multiplier of 0.1. This coding works as follows: Green, blue, and gold stripes stand for 5 followed by 6 and multiplied by 0.1, which indicates a value of 5.6 ohms. Similarly, a yellow, violet, and silver resistor has a value of 0.47 ohm.

Tolerance

The **tolerance** of a resistor is a manufacturer's statement of how close to the exact value a resistor is when manufactured and should remain under normal use. Tolerance numbers are also indicated by color. The colors and their numbers are given in Table 6–1. For example, a brown, black, red, and silver resistor has a value of 1000 ohms and a tolerance of 10% according to Table 6–1. Thus, the resistor should be within 10% of 1000 ohms. Since 10% of 1000 ohms is 100 ohms, the resistor value can be between (1000 − 100 ohms) and (1000 + 100) ohms, or somewhere between 900 and 1100 ohms. Since the expected value can be above or below the indicated value, the tolerance is written as ±, meaning "plus or minus." This resistor, then, is described as 1000 ohms ± 10%.

An orange, white, yellow, and gold resistor is described as 390,000 ohms ± 5%. Its 5%

> **TABLE 6-1**
> **RESISTOR COLOR CODE**

Color	Number	Multiplier	Tolerance (%)
Black	0	1.	
Brown	1	10.	1
Red	2	100.	2
Orange	3	1,000.	
Yellow	4	10,000.	
Green	5	100,000.	0.5
Blue	6	1,000,000.	0.1
Violet	7	10,000,000.	0.05
Gray	8		
White	9		
Silver		0.01	10
Gold		0.1	5

tolerance indicates that it can vary as much as 19,500 ohms.

Let us determine the tolerances for the resistors shown in Figures 6–10A and 6–10B. The resistor in Figure 6–10A will be within 470 ohms, or 10%, of 4700 ohms. The resistor in Figure 6–10B will be within 22 ohms, or 10%, of 220 ohms.

After we find the value and tolerance of a resistor, we can determine its range. The red, red, brown, and silver resistor of Figure 6–10B is within 22 ohms of 220 ohms. That is, it can be as low as 198 ohms or as high as 242 ohms, as shown by the following calculations:

$$220 \ \Omega \ - \ 22 \ \Omega \ = \ 198 \ \Omega$$
$$220 \ \Omega \ + \ 22 \ \Omega \ = \ 242 \ \Omega$$

If a 220-ohm resistor has a tolerance of 5%, it has a range of 209 to 231 ohms, as shown by the following calculations:

$$220 \ \Omega \ \times \ 0.05 \ = \ 11 \ \Omega$$
$$220 \ \Omega \ - \ 11 \ \Omega \ = \ 209 \ \Omega$$
$$220 \ \Omega \ + \ 11 \ \Omega \ = \ 231 \ \Omega$$

If the tolerance is 20%, the range is from 176 to 264 ohms:

$$220 \ \Omega \ \times \ 0.20 \ = \ 44 \ \Omega$$
$$220 \ \Omega \ - \ 44 \ \Omega \ = \ 176 \ \Omega$$
$$220 \ \Omega \ + \ 44 \ \Omega \ = \ 264 \ \Omega$$

Thus, we see that the lower the tolerance percentage is, the closer a resistor is likely to be to its indicated value. While high percentages are desirable in examinations, low percentages are desirable in tolerances of components. And as the tolerance percentage goes down, the quality and cost of the component go up.

Five-Stripe Resistors

The addition of a fifth stripe allows for greater precision in the specification of resistance. In this color code, one stripe is still used for tolerance and one for the multiplier, but three are used to indicate value. Thus, the resistance value is given to three significant figures rather than the two significant figures in the

FIGURE 6-11
LOW-VALUE RESISTORS

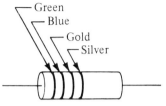

Resistor value of 5.6 ohms

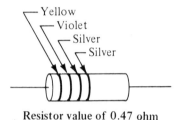

Resistor value of 0.47 ohm

FIGURE 6-12
FIVE-STRIPE RESISTORS

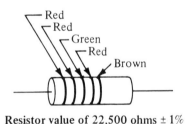

Resistor value of 22,500 ohms ± 1%

traditional four-stripe code. Consider the resistor in Figure 6–12. Its code is as follows:

1st color = red = 2
2nd color = red = 2
3rd color = green = 5
4th color = red = 2 (zeros)
5th color = brown = 1

The value of this resistor is 22,500 ohms ± 1%.

As another example, consider a gray, red, green, yellow, and green resistor. Its stripes indicate the following values:

1st color = gray = 8
2nd color = red = 2
3rd color = green = 5
4th color = yellow = 4
5th color = green = 5

The value of this resistor is 8,250,000 ohms ± 0.5%.

SUMMARY

Resistance is the property of a circuit that opposes the flow of electric current. It is present in all conductors and insulators to some degree. High resistance makes an object a good insulator, while any resistance in a conductor reduces the quality of that conductor. The resistance in an incandescent lamp filament makes it light up, while resistance in a toaster element produces heat.

A resistor is a component intentionally designed to resist current. Resistors are generally used to limit or control current or to produce desired voltage levels.

There are two basic resistor types, fixed and adjustable. Fixed resistors have a specified resistance value and are made in a variety of ways. The most common is a carbon resistor. It is manufactured by compressing a mixture of carbon and Bakelite into a solid component with two leads. The ratio of these materials, one a conductor and the other an insulator, determines the component's resistance, which is indicated by color bands on the outside.

Wire-wound resistors have a fixed value that is determined by the size and length of a wire wrapped around a core. Large wire-wound resistors are called power resistors because of their ability to tolerate and dissipate large amounts of heat. Smaller wire-wound resistors are called precision resistors because their absolute value is easy to predict and control. Other types of precision resistors are carbon film and metal film. These resistors are smaller, more precise, and less expensive than wire-wound resistors.

Adjustable resistors come in two basic forms: those that can be occasionally adjusted by loosening a band and moving it and those that can be continually adjusted by turning a knob on the end of a shaft. The second type is similar to a volume control. These continually adjustable resistors are called potentiometers

and can be wire-wound or film, can rotate or slide, and can be stacked like the volume and tone controls on a car radio.

Colors are used on resistors to represent numbers indicating the amount of resistance. The basic resistor color code in Table 6–1 is used to find the value of a resistor.

Once you determine a resistor's value, you must determine its tolerance and range. Since components are never perfect, you will need to know just how close a given resistor is expected to be to its specified value. Tolerances are indicated by a color band on a resistor; the corresponding values are given in the resistor color code. The resistor's range is derived from the tolerance and the specified value.

CHAPTER

REVIEW TERMS

adjustable resistor: resistor for which the resistance can be changed

carbon film resistor: low-power precision resistor made by coating a small insulating rod with a thin conductive carbon coating

carbon resistor: resistor with a power rating of up to 2 watts made from a composition of carbon and Bakelite

color code: system in which colors are used to indicate the value, tolerance, reliability, and other specifications of components such as resistors

fixed resistor: resistor with one value

flameproof resistor: film resistor designed to neither initiate nor support combustion

multiplier: third stripe on a four-stripe resistor used to indicate the location of the decimal point for the resistance value

ohm: unit of measure for resistance

potentiometer: three-terminal resistor with a constant value of resistance between two terminals and an adjustable amount by control to the third terminal

power rating: characteristic of resistors describing how much heat they dissipate

precision resistor: resistor with a tolerance of less than 1%

resistance: property of a circuit that opposes the flow of current

resistor: passive component designed to provide a specific amount of resistance

rheostat: two-terminal adjustable resistor designed for high-power applications

tolerance: specification that describes the probable maximum error between the actual and indicated values of a component

wire-wound resistor: fixed-value resistor with turns of wire enclosed in a heat-dissipating epoxy

REVIEW QUESTIONS

1. Define resistance, and name the unit of resistance.

2. Define resistor.

3. Discuss the resistor color code.

4. Describe tolerance.

5. Define adjustable resistor.

6. Describe how resistance affects a circuit.

7. Describe how resistors are used.

8. Describe some uses for adjustable resistors.

9. Describe how to determine a resistor value by using the color code.

10. Describe how to determine a resistor range by using the color code.

11. What does a fifth stripe on a resistor mean?

12. What materials are used to produce the resistance in fixed resistors?

13. What does the term *flameproof* mean?

REVIEW PROBLEMS

1. What is the indicated value of a resistor with brown, black, yellow, and silver stripes?

2. What is the indicated value of a resistor with yellow, violet, orange, and gold stripes?

3. What is the indicated value of a resistor with orange, white, black, and silver stripes?

4. What is the indicated value of a resistor with green, blue, red, and gold stripes?

5. What are the values of the resistors in Figures 6–13A and 6–13B?

6. What are the tolerances of the resistors in Figures 6–14A and 6–14B?

7. What are the minimum and maximum limits of a resistor with brown, black, yellow, and silver stripes?

8. What are the minimum and maximum limits of a resistor with yellow, violet, orange, and gold stripes?

FIGURE 6-13

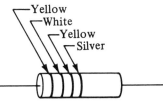

A. Resistor 1

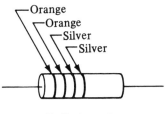

B. Resistor 2

FIGURE 6-14

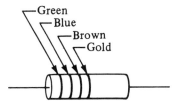

A. Resistor 1

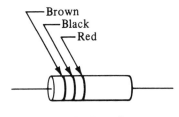

B. Resistor 2

9. What are the minimum and maximum limits of a resistor with orange, white, black, and silver stripes?

10. What are the minimum and maximum limits of a resistor with green, blue, red, and gold stripes?

11. What are the values, tolerances, and ranges of the resistors in Figures 6–15A and 6–15B?

12. What is the value of a resistor with brown, black, green, brown, and brown stripes?

13. What is the value of a resistor with yellow, violet, red, orange, and red stripes?

14. Determine the tolerance and range of the resistor described in Problem 12.

15. Determine the tolerance and range of the resistor described in Problem 13.

FIGURE 6-15

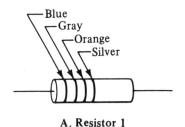

A. Resistor 1

B. Resistor 2

MULTIMETERS

State the basic functions of a multimeter.

Measure resistance with a multimeter.

Measure DC voltage and current with a multimeter.

Determine true error and true accuracy of a multimeter.

INTRODUCTION TO METERS

The objective of this chapter is to introduce the most common piece of test equipment used in electronics, the multimeter. Procedures for measuring resistance, voltage, and current with a multimeter are examined.

The most commonly measured circuit dimensions, or **parameters,** are voltage, current, and resistance. They are measured because someone needs information in order to make a decision. For example, you might measure the voltage at an outlet in your home because a radio is not working and you want to see whether any voltage is present. The radio may not be working because a circuit breaker in your home has tripped and the circuit is off. In this case, you are checking simply for the presence of any voltage; you assume that, if there is any, it is probably the right amount.

Sometimes, you measure a voltage to determine the exact amount. For instance, suppose you have a device that will work only if the line voltage is between 115 and 122 volts. In this case, you need a number for the magnitude — amplitude or amount — of voltage.

Circuit current is measured for a number of reasons. When circuit breakers keep interrupting the current, the current amount must be measured to determine whether it is too much. The resistance of resistors is also measured to determine whether the value is within the color-coded range and, if so, what these values are.

Most of the time, you will be taking measurements in order to determine how a circuit is working. For example, you may be testing a new circuit to ensure that it is working according to design standards. At other times, you may troubleshoot a defective circuit in order to locate the cause of a problem. Good skill with meters and a good understanding of the circuit are necessary if you are to locate

FIGURE 7-1
MULTIMETER

the defect. Thus, competency in using meters is essential for success in electronics.

Meter Scales and Pointers

One of the most common technician errors is to incorrectly read analog meter indications. That is, the pointer is at a specific location on the scale, but the technician interprets it incorrectly. While you may not be familiar with meter scales such as the one in Figure 7–1, you are probably familiar with automobile speedometer scales. Both scales are similar.

FIGURE 7-2
AUTOMOBILE SPEEDOMETER

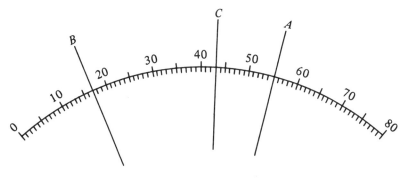

Miles per hour

The speedometer in Figure 7–2 has a full-scale value of 80 miles per hour — that is, the highest value is 80. Pointer A in Figure 7–2 is indicating 55 miles per hour. Pointer B is indicating 17 miles per hour, and pointer C is indicating 43 miles per hour. On this scale, each small division represents 1 mile per hour, and the larger divisions indicate 5 miles per hour.

Voltmeter scales are about the same. For example, the meter in Figure 7–3A has a full-scale value of 10 volts. Pointer A in Figure 7–3A is indicating 6.4 volts. Pointer B is indicating 1.3 volts, and pointer C is indicating 9.2 volts. On this meter, each small division has a value of 0.2 volt. Keep the following point in mind.

Key point: The value of each small division depends on the full-scale range of the meter.

You can change the full-scale range of most meters with switches while continuing to use the same scale and pointer. A meter that can be used to measure 1, 10, or 100 volts full scale may look like the meter in Figure 7–3B.

On such a meter, there is only one set of numbers, and you have to take care of the decimal point yourself.

For instance, if the meter in Figure 7–3B is on the 10-volt, full-scale range, pointer A is indicating 8.3 volts. However, if the meter is on the 100-volt, full-scale range, pointer A is indicating 83 volts. Pointer A is 0.83 volt on the 1-volt range. Similarly, pointer B is indicating 0.17 volt on the 1-volt range, 1.7 volts on the 10-volt range, and 17 volts on the 100-volt range.

One problem that many people have while reading meter scales is determining the value of each small division. Note that the value of each division changes with each range. So the easiest way to proceed is to begin at the right end of the scale and call that number the maximum value of the full-scale range.

Imagine that the meter in Figure 7–3C is on the 500-volt range. In this case, the 50 at the right end represents 500, the 40 represents 400, and so forth. We want to know the indicated value. The value indicated by pointer A is between 20 and 30 on the scale, or between 200 and 300 volts. Here is one approach for finding the indicated value.

1. Determine the voltage between the known values on each side of the pointer:

300 V − 200 V = 100 V

2. Count the number of divisions between those two values:

10

3. Divide the answer in step 1 by the answer in step 2 to find the voltage of each small line, or division:

$$\frac{300 \text{ V} - 200 \text{ V}}{10} = \frac{100 \text{ V}}{10}$$
$$= 10 \text{ volts per division}$$

4. Determine the voltage known to that point, that is, the number to the left of the pointer:

200 V

5. Determine the voltage above that point by multiplying the number of divisions remaining to get to the pointer by the value of each division:

3 divisions × (10 V/division) = 30 V

6. Add the answers in steps 4 and 5:

200 V + 30 V = **230 V**

Basic Measurement

The first step in measuring voltage is to select a meter that can measure voltage. This meter may be a plain voltmeter or a multipurpose meter. The next step is to switch to a voltage position, or mode of operation. You must then determine whether the voltage is AC or DC and switch the meter accordingly.

You cannot weigh a 250-pound person on a 100-pound scale. Likewise, you cannot measure more than 100 volts on the 100-volt range of a meter. You must be sure that that meter's full-scale range is greater than the voltage to be measured. If you are not sure, start with a very high range and work your way down.

Key point: The higher you can get the pointer on the scale, the more accurate your results will be.

FIGURE 7-3
VOLTMETER SCALES

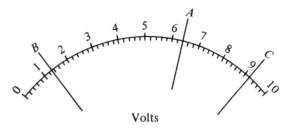

A. Full-scale value of 10 volts

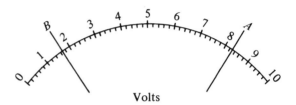

B. Full-scale value of 1, 10, 100 volts

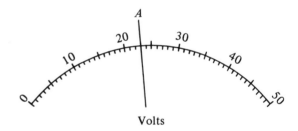

C. Full-scale value of 500 volts

FIGURE 7-4
MEASURING VOLTAGE AND CURRENT

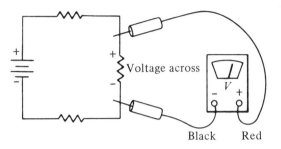

A. Voltage

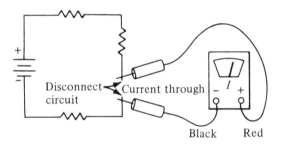

B. Current

To measure voltage, you must connect the meter across the circuit as shown in Figure 7–4A. No disconnections are necessary. If it is a DC circuit, **polarity** — that is, plus or minus poles — must be observed: Minus (−) goes to negative, and plus (+) goes to positive.

Current measurement is similar to voltage measurement in some ways. You must obey the rules about type of meter, AC or DC, about full-scale range, about polarity for DC circuits, and about safety. However, current measurement is different from voltage measurement in one respect. In current measurement, the circuit must be interrupted and the meter inserted in the path of the current, as shown in Figure 7–4B.

The subjects of voltage, current, and re-

sistance measurement will be discussed in more detail later in this chapter. However, you must remember this one point.

> **Key point:** You measure voltage across a circuit, and you measure current through a circuit.

MULTIPURPOSE METERS

Most meters used by technicians are multipurpose meters, or **multimeters,** which are designed to measure voltage, resistance, and current. Figure 7–5 shows two types of multimeters.

Types of Multimeters

There are two basic types of multimeters, analog and digital. The analog meter on the left in Figure 7–5 has long been called a **VOM** because it measures volts, ohms, and milliamperes. The meter on the right is called a **DMM** for digital multimeter. Analog meters have been around for years, are the most common, and are the type that you have been reading about so far. Digital multimeters are a more recent product and are rapidly gaining in popularity.

Analog and digital multimeters are similar in that they are portable instruments designed for measuring voltage, current, and resistance. But that is where the similarity ends. **Analog multimeters** indicate value with a moving pointer, which moves a distance that is in proportion to some measured value. You must determine which scale to read, the significance of the numbers for that range, and the indicated value. This process presents quite a few opportunities for errors. **Digital multimeters,** however, indicate a measured value with digits or numbers. Since specific numbers are displayed, there is little chance of making a reading error. That does not mean, however, that errors will not be made by connecting the leads to the wrong place or selecting the wrong range. A more detailed description of analog and digital meters is given in Chapter 22.

Functions

You must decide which of the values — voltage, current, or resistance — you want to measure, that is, which function you want the meter to perform. Then, you must inform the meter by selecting the appropriate switch position or meter jack. The meter must also be informed whether the current or voltage is AC or DC. The multimeters in Figure 7–5 can perform five basic measurements: DC voltage, AC voltage, DC current, AC current, and resistance.

Ranges

You must also inform a meter of the full-scale range you need for each measurement you intend to make. The meter at the left in Figure 7–5 has eight voltage ranges, which you can select with the rotary switch. The ranges are 0.3, 1, 3, 10, 30, 100, 300, and 1000. As mentioned earlier, you want to use the lowest range possible without causing the pointer to go beyond the high end. To prevent the pointer from going beyond the high end and causing meter damage, start at the highest range and switch down to the best range. Also, remember that the higher you can get the pointer on the scale, the more accurate your results will be.

A close look at a multimeter scale shows that different scales are often used for AC and DC measurements. In the scale in Figure 7–1, the DC graduations are on the second line from the top, and the AC graduations are on the third line. (The top line is used for resistance and will be described later.)

FIGURE 7-5
ANALOG AND DIGITAL MULTIMETERS

An analog multimeter does not give exact numbers for every range. For example, the 2.5- and 250-volt, full-scale ranges are represented by the 250 at the top (right) end, while the 50 at the top end can stand for 50 or 500 volts. You must select the correct range and function with switches and the correct jacks for the leads.

Digital multimeters eliminate most of the steps just described. Scales and pointers are replaced by a display that makes it next to impossible to incorrectly read the range and numbers. The technician must, however, select the correct function. Many digital multimeters also have autoranging, an electronic circuit that selects the most appropriate range for the meter and prevents range overloading.

Leads and Switches

Voltage, current, and resistance measurements require the connection of two leads to the meter and to the circuit. Black is used for negative (−), and red is used for positive (+). It is usual practice to connect the leads to the meter first. This procedure is a safety precaution. If you connect the leads to the circuit first, when you pick up the other ends, you will be holding the exposed ends of live wires, and you could receive a shock.

Safety tip: Connect the leads to the meter first.

If the function and range of a meter are determined by switches, the question of where to connect the test leads is an easy one to answer. If, however, there are no switches, the range and function must be selected with the jacks. Some meters may use only two jacks for most of the measurements and additional jacks for special ranges such as high voltage or high current. For example, the meter on the left in Figure 7–5 has only two jacks, and all function and range selections are made with a rotary switch. The meter on the right uses three jacks and a row of push-button switches. Function is selected with jacks and some switches, while range is selected with the other switches.

In summary, keep in mind that you must select the proper function and safest range before you connect the leads to the circuit. Also, remember that function and range are selected with jacks, switches, or a combination of both.

USING A MULTIMETER

Once you are familiar with ranges, functions, scales, and the other features of a multimeter, you are ready to make some measurements. In the following discussion, resistance measurement comes first because it allows you to become familiar with the use of a meter without risk of shock or meter damage. Then, voltage measurement is considered. Voltage is the most frequently measured circuit parameter. Finally, current measurement, the most diffi-

cult of the multimeter applications, will be examined.

Measuring Resistance

The measurement of resistance must be done when there is no power supplied to the circuit. In addition, the component being measured cannot be completely installed in a circuit while you are making a measurement. If a resistor by itself is to be measured, there is no problem. However, if the resistor is connected in a circuit, the circuit must be turned off, and one lead of the resistor must be disconnected from its adjoining circuitry.

The reason for disconnecting one lead of an installed resistor is quite simple. If you measure its resistance while it is connected, your results may be wrong. In this case, you are measuring the resistance of the item being tested and also the resistance of all other parts connected to the circuit. So, before measuring an installed resistor, disconnect one lead to ensure that the only possible complete circuit with the meter will be through the component being tested.

Note also that you will damage a meter if you attempt to measure resistance in an energized circuit.

Safety tip: Never measure resistance in an energized circuit.

Always remember to turn the circuit off first. Furthermore, do not energize a circuit you are working on unless it is absolutely necessary.

Now that the component is disconnected, it is ready for testing. Place the meter in a safe position in the middle of the workbench and away from the edge. Manufacturer instructions will state which position, upright or horizontal, should be used for obtaining accurate readings. Once the meter is correctly positioned, you can select the proper function, in this case, ohms. General instructions for the use of analog and digital multimeters are given in the subsections that follow. Also, be sure to read the directions provided by the manufacturer.

Analog Meter Use. For resistance measurement with an analog multimeter, the range switch is different from the range switch for voltage or current measurement. For resistance measurement, the positions are marked $R \times 1$, $R \times 100$, $R \times 10,000$, and so on. If you have no idea what the resistance is going to be, select the middle resistance range.

The next step in measuring resistance is to **zero the meter.** This adjustment must always be made before measuring resistance. First, connect the leads to the meter, black to negative, or common, and red to positive, or ohms. Then, touch the free ends of the leads together to show the meter what 0 ohms looks like. The pointer should move to 0 on the ohms scale. On most meters, the ohms scale is usually the upper one, and 0 is usually at the right end, as shown in Figure 7–6. If the pointer is not at 0, turn the zero control until it is. When you separate the leads, the pointer should move to the other end.

Each model of meter has its own method of zeroing. Therefore, you should determine the proper method for zeroing any meter before you use it. Here are the two important points to remember.

> **Key points:** Most meters must be zeroed for resistance measurement. The zero must be rechecked each time the range is changed.

If you do not know the procedure for the meter you are using, read the manual or ask your instructor.

Now that you have selected the ohms function, selected the middle range, and zeroed the meter, you are ready to measure. Just connect the leads across the resistor. Be sure

FIGURE 7-6
ZEROING A METER

to keep your fingers away from the ends of the leads so that you are not part of the circuit. The best range for resistance is the one that places the pointer near the middle of the scale. Switch ranges until you find the best range; then, zero the meter again.

To determine the resistance, read the number indicated by the pointer. See Figure 7–7A. Then, multiply that number by the range. In this case, the pointer is on 48, and the range is $R \times 100$. Thus, the resistance value is the number indicated by the pointer multiplied by 100, or 48×100. The resistance is 4800 ohms.

Another point to note here is that prefixes are used in electronics to simplify numbers. For example, k (**kilo-**) represents 1000, and M (**mega-**) represents 1,000,000. Figure 7–7B indicates a value of $7.2 \times 100,000$, or 720,000 ohms. In the prefix system, this value is 720 kilohms (abbreviated kΩ) or 0.72 megohm (abbreviated MΩ). Similarly, a pointer indication of 2.2 on the 10,000 range represents 22,000 ohms, or 22 kilohms. You will often hear technicians call this value 22k, with k here meaning kilohms. Prefixes will be discussed again in Chapter 8.

FIGURE 7-7
METER INDICATIONS

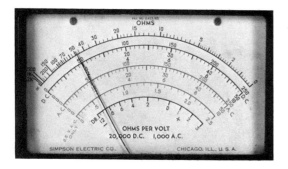

A. Indication of 480 ohms

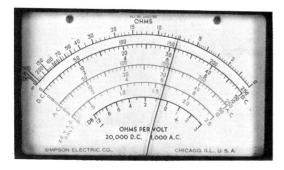

B. Indication of 720 kilohms

Digital Meter Use. Resistance measurement with a digital meter is somewhat simpler than measurement with an analog meter. Resistance ranges are not multipliers as with analog multimeters but, rather, are upper limits such as with voltmeters. That is, a 200 range on a DMM indicates that this range is appropriate for resistances up to 200 ohms. Once a range has been selected and the meter connected across the unknown resistance, the resistance value is read directly from the digital display. Prefixes are included on most displays. The best range is the one that provides the most number of digits or significant figures without

exceeding the limit of the range. This reading is often indicated by a 1 at the left position and no other digits.

When you have completed your resistance measurement, disconnect the leads and turn the meter off. If there is no off position, select the highest voltage range.

Review of the Steps. In review, the steps for resistance measurement with an analog or digital multimeter are as follows:

1. Place the meter in a safe position.
2. Select the resistance function, and select the middle range, whenever possible.
3. Zero the meter if it is analog.
4. De-energize the circuit.
5. Disconnect one component lead if the component is in a circuit.
6. Connect the meter to the component.
7. For an analog meter, select the range that puts the pointer near the middle of the scale. Then, disconnect the leads and check zero.
8. For a digital meter, select the lowest range.
9. Reconnect the leads, and read the indicated value.
10. Multiply the indicated value by the analog meter's range to determine the component's resistance. For a digital meter, read the value directly from the display.
11. Disconnect the meter, and reconnect the circuit.
12. Switch the meter off.

Measuring Voltage

Voltage measurement is probably the most common test that a technician performs. It is easy to do right — and just as easy to do wrong. Doing it right means safely using the meter and correctly determining the value of the voltage. Doing it wrong means concluding an incorrect value or injuring yourself, the meter, or the circuit under test. Therefore, it is important to learn an acceptable procedure for voltage measurement from the start.

Begin by placing the meter near the circuit to be tested in a position where it is likely to be safe, such as in the middle of a workbench. With the switches, let the meter know that it is measuring voltage (volts) and whether the voltage is AC or DC. You will be measuring DC voltage here. Place the range switch on the highest range the meter has. You are measuring voltage because you do not know what it is. Since you do not know what it is, you cannot know which ranges are safe. Therefore, select the highest range so that the meter is not overloaded.

Connecting the leads is the next step. There is a safe procedure for this step also.

Safety tip: Connect the leads to the meter first.

If you were to connect the leads to the circuit first, you would be holding live wires in your hand when you tried to connect the leads to the meter. These wires would have an unknown and possibly lethal voltage across them. Therefore, always connect the leads to the meter first. The black wire is generally used for negative, or common, and the red wire is generally used for positive. Use only one hand. Keep your other hand away from the circuit to avoid becoming part of the circuit yourself.

After the leads are connected to the meter, connect them to the circuit under test. It is a good idea to turn the circuit power off at this time. If the circuit cannot be turned off, connect the leads as follows: The black or common lead is connected first because the common side of a circuit is the safest. After this connection is made, make the last connec-

tion, red to the live side of the circuit. If a pointer moves in the wrong direction, switch the leads, because the polarity is wrong. Digital meters indicate wrong polarity with a − before the value. You can energize the circuit once the leads are connected.

The next step is to switch to lower ranges until you find the best range. The higher you can get the pointer on the analog scale, the more accurate your reading will be. The best range on a digital meter is the one with the most significant figures. So the best range is the lowest range you can use. But be careful. If you select too low a range, you will overload the meter and probably damage it.

Now that you have the best range, read the value indicated. You can read it directly from a digital display. With an analog meter, begin with the number at the top end of the scale, which represents the full-scale value of the range you are on. Then, count down that scale, using the numbers on the same line with the full-scale value. Stop at the first set of numbers below the pointer, and then count up to the pointer position.

After you have determined the voltage, de-energize the circuit being tested. If you cannot de-energize the circuit, disconnect the red lead from the circuit and then the black lead. Then, disconnect the leads from the meter. Finally, switch the meter to the highest voltage range so that it is ready for the next person.

Review of the Steps. In review, the steps for voltage measurement are as follows:

1. Turn off the power to the circuit.
2. Place the meter in a safe position.
3. Select the voltage function and AC or DC.
4. Switch to the highest voltage range.
5. Connect the leads to the meter
6. Connect the black lead to the circuit common.

FIGURE 7-8
CURRENT MEASUREMENT

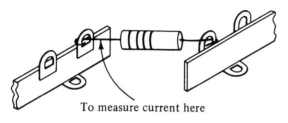

To measure current here

A. Locating the leads

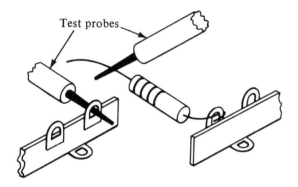

Test probes

**B. Disconnecting the component and
inserting probes**

7. Connect the red lead to the live side of the circuit.

8. Restore the power to the circuit.

9. Switch down to the best range.

10. Read the indicated value.

11. Turn off the power to the circuit.

12. Disconnect the red lead from the live side of the circuit.

13. Disconnect the black lead from the common side of the circuit.

14. Disconnect the leads from the meter.

15. Switch the meter to its highest voltage range.

Measuring Current

The procedures for current measurement are much like the procedures for voltage measurement. Place the meter in a safe position, select the current function (amperes or milliamperes), and indicate whether the current is AC or DC. You will measure DC here. Begin with the highest range. Connect the leads to the meter in the same order described for voltage measurement, and switch down to the best range. But now, the procedure differs from that for voltage measurement. For current measurement, the circuit must be opened and the meter inserted into the circuit. You measure voltage *across* a circuit, and you measure current *through* a circuit, as shown earlier in Figure 7–4.

Once you have the meter in a safe position, in the current function, and on the highest range, deactivate the circuit to be tested. Then, locate the wire or component lead through which you want to measure the current. Figure 7–8A shows where you disconnect the circuit for a current measurement. The lead can be disconnected from the circuit by cutting or desoldering. After this disconnection is made, the meter can be connected as shown in Figure 7–8B.

Note that the meter completes the broken connection by being placed into the circuit, just as water or gas meters are placed into circuits in homes. All the current must flow through the meter to get from one side of the connection to the other. Thus, current meters are never connected across a component; the circuit must be broken and the meter inserted into the break. The red lead is connected to positive and the black lead to negative. If you do not know the polarity, you will find out the polarity when the circuit is energized.

Once the meter is connected, energize the circuit. If the meter is connected properly, the pointer will move up the scale. If it moves in the wrong direction, de-energize the circuit and interchange the leads. Now, energize the circuit, select the best range, and read the indicated value (in amperes, abbreviated A, or in milliamperes, abbreviated mA). Next, de-energize the circuit. Finally, return the circuit to its original condition, and place the meter on its highest voltage range.

Review of the Steps. In review, the steps for current measurement are as follows:

1. Place the meter in a safe position.
2. Select the current function and AC or DC.
3. Switch to the highest current range.
4. De-energize the circuit.
5. Disconnect the circuit at the point where the current is to be measured.
6. Connect the leads to the meter.
7. Connect the leads to the circuit.
8. Energize the circuit.
9. Switch to the lowest safe range.
10. Read the indicated value.
11. De-energize the circuit.
12. Return the circuit to its original condition.
13. Switch the meter off.

SPECIFICATIONS

Meters are not perfect. They get damaged from misuse, and batteries become weak. Even new meters have limitations on how precise their measurements can be. While you will not be involved in precision measurements at this time, you will be involved with them in your work as a technician. Therefore, you must be able to determine the capabilities of a meter by reading the specifications. Two important specifications are accuracy and sensitivity. We will consider these features in turn.

Accuracy

The **accuracy** of a meter indicates how close to the true or actual value a meter reading will be. For example, if you get on a scale and it indicates that you weigh 165 pounds when you actually weigh 150, the scale is 15 pounds in error. If we compare the error to the value the scale should read, we find that there is a 10% error:

$$\frac{15 \text{ V (error)}}{150 \text{ V (actual)}} = 0.10 = \textbf{10\%}$$

We say that the accuracy of the scale is within 10%.

If the scale indicates 155 pounds when you weigh 150, the error is 5 pounds. The accuracy, or the percentage of error, of the scale is 3.3%:

$$\frac{5 \text{ V (error)}}{150 \text{ V (actual)}} = 0.033 = \textbf{3.3\%}$$

In other words, the closer the indicated value is to the true value, the lower the percentage of error is.

Analog meter accuracy is described as follows: A 2% meter is more accurate than a 4% meter because it has less error. You are used to a high percentage on a test being good because it indicates how well you did. But a low percentage wrong can be interpreted in the same way as a high percentage right. Thus, for an analog meter, the lower the percentage of error, the more accurate its readings will be.

If we attempt to measure a voltage on the 100-volt range of a 3% accuracy meter, our answer will be correct within 3% of 100 volts, or within 3 volts. Since the error can be too high or too low, it is written as ±3 volts.

Suppose we want to measure 60 volts on the 100-volt range of a 3% meter. The pointer can indicate any value from 57 to 63 volts. While the true value is 60, the pointer will only come within 3%, or ±3 volts, of the full-scale range. However, the idea of error becomes a bit more complicated here. Since the error is 3 volts out of an actual value of 60 volts, the true error is

$$\frac{3 \text{ V (error)}}{60 \text{ V (actual)}} = 0.05 = \mathbf{5\%}$$

One more example may be useful. Suppose we want to measure 20 volts on the 100-volt range of a 2% meter. The meter will give a reading within 2% of 100 volts, or within 2 volts:

$$100 \text{ V} \times 0.02 = 2 \text{ V}$$

But the error is 2 volts out of the actual value of 20 volts, and so the percentage of true error is

$$\frac{2 \text{ V (error)}}{20 \text{ V (actual)}} = 0.10 = \mathbf{10\%}$$

If we change to the 50-volt range, the ability of the meter improves. The accuracy is 2% of 50 volts, or 1 volt:

$$50 \text{ V} \times 0.02 = 1 \text{ V}$$

The meter reading will be within ±1 volt of the actual value. The percentage of true error, then, is

$$\frac{1 \text{ V (error)}}{20 \text{ V (actual)}} = 0.05 = \mathbf{5\%}$$

If the meter has a 25-volt range, the accuracy will be 2% of 25 volts, or 0.5 volt:

$$25 \text{ V} \times 0.02 = 0.5 \text{ V}$$

The meter reading will be within ±0.5 volt of the actual valve. The percentage of true error will be

$$\frac{0.5 \text{ V (error)}}{20 \text{ V (actual)}} = 0.025 = \mathbf{2.5\%}$$

As the calculations show, the lower the range used, the more accurate the results will be. This statement is true for both voltage and current measurements with meters that specify a full-scale accuracy value. Always check to see whether the accuracy of the instrument you are using is a percentage of the actual value or a percentage of the full-scale value. Some meters state accuracy one way, and some state it the other way. The preceding calculations have shown that there is a difference. And as mentioned earlier, the lower the meter accuracy percentage is, the more accurate the meter is.

Digital meters often describe accuracy as a percentage of the indicated value with an additional possible error in the last digit because of rounding off. For instance, a digital meter with an accuracy of ±1%, ±1 digit indicates that a 200-volt range may be off by 3 volts. That is, the 1% accuracy gives a 2-volt error, and the last-digit error amounts to another 1 volt. We always assume that errors add, not cancel, so the result is a 3-volt error out of an actual value of 200 volts, for an actual accuracy of 1.5%. To summarize, the accuracy specification tells you how much you can depend on the meter — that is, how close to the true value the meter's reading will be.

Sensitivity

Meter sensitivity is another specification you need to be familiar with. However, we will consider this specification only briefly here. After we have discussed series-parallel circuits in Chapter 16, we will take up the subject of sensitivity again and in greater detail.

You have heard people described as sensitive. What is generally meant is to what extent they respond or react to a given situation or event — a stimulus. Some people who are called sensitive notice the finer points of music or art, while other people who are called sensitive are easily upset by criticism. In both cases, the people are showing a response to a certain stimulus. Meter sensitivity also refers to a relationship between a stimulus and a response.

A voltmeter takes energy from the circuit being tested to make the pointer move up the scale. The more sensitive a meter is, the less energy it takes from the tested circuit. Meter sensitivity becomes a concern because circuits cannot always afford an energy loss. For example, if the meter is not sensitive enough for the circuit being tested, the voltage the meter is attempting to measure changes because of the meter. Thus, the circuit no longer functions as it should. Removing the meter returns everything to normal. In such a case, what is needed is a more sensitive meter.

Analog voltmeter **sensitivity** is a specification used to determine the resistance a meter will represent to a circuit under test while the meter is on a given range. This specification is described in ohms per volt and is an indication of the demand a meter will place upon a circuit. For instance, consider a meter with a specified sensitivity of 1000 ohms per volt. On the 10-volt range, this meter will have the following resistance:

$$1000 \ \Omega/V \ \times \ 10 \ V \ = \ 10,000 \ \Omega$$

In other words, this meter looks like 10,000 ohms to the circuit being tested. Similarly, on the 250-volt range, this meter looks like 250,000 ohms to the circuit being tested.

A meter with a higher sensitivity may be rated at 20,000 ohms per volt. On the 250-volt range, this meter will look like 5,000,000 ohms to the circuit it is connected across.

Digital voltmeter sensitivity is usually a constant value for all ranges, typically, 10 megohms.

In summary, the higher the ohms-per-volt

rating a meter has, the more sensitive it is and the less it will disrupt a circuit.

Other Factors

While accuracy and sensitivity are critical factors in the selection of a meter, there are other factors to consider. However, most of them involve a detailed understanding of electronics. We will consider only a few factors here.

Frequency response is one factor to consider. Frequency response describes the way a meter responds to different frequencies. It also indicates the range of frequencies over which you can expect the meter to provide accurate information.

Another factor in meter selection involves the way it performs its function inside. Does it use tubes or transistors? Does it measure AC voltage by changing it to DC? Does it have a warm-up time?

As you become more familiar with meter specifications, you will discover that meter selection involves a trade-off. For instance, if you want more accuracy or sensitivity, you may have to sacrifice frequency. If you want more frequency response, you may have to sacrifice something else. No one meter provides the best of everything. The important question you must be able to answer is: To what extent can I state that the value I have arrived at is correct?

SUMMARY

Meters are used for measuring different factors, or parameters, of electric and electronic circuits. The most common parameters measured are voltage, current, and resistance. Measurements tell a technician whether a circuit is working properly, and if it is not, why it is not. These tests usually lead to a decision about whether a product should be shipped or redesigned or possibly even scrapped. Thus, the importance of making good measurements cannot be overstressed.

The analog meter consists of a pointer that deflects somewhere along a number scale. Most errors in measurement result from an inaccurate interpretation of the indicated value. The process of meter reading is easy, but carelessness can lead to error. Digital meters eliminate this problem by displaying numbers that can be read directly. However, digital meters do not eliminate many other types of errors that can be made.

The multimeter provides technicians with three meters in one: It can measure voltage, resistance, and current. It is often called a VOM or a DMM. This meter is the most common type of test instrument you will see.

While the arrangement of the scales, switches, and jacks varies from one meter to another, most meters have a few things in common. With all meters, you can follow these steps: Place the meter in a safe position. Connect two leads to the meter. Select the function as voltage, resistance, or current. And select the best range. You must also zero the meter for resistance measurement. The three types of measurement have some differences. For example, voltage is measured across a circuit, while current is measured through a circuit. Current measurement requires a circuit disconnection. Resistance measurement requires that the circuit be de-energized and that one lead of the component tested be disconnected.

Meters vary in their ability to measure circuit parameters. Manufacturers' specifications describe this ability in terms of accuracy, sensitivity, frequency response, and a variety of other characteristics. Meter accuracy describes how close a meter indication is to the true value. As the accuracy percentage gets lower, meter quality improves. Meter sensitivity describes the likelihood of the meter having an undesirable effect on the circuit under test. This effect, called loading, occurs when the meter takes power from that circuit. Sensitivity is expressed in ohms per volt, and a high number for sensitivity is best since it indicates less loading.

Despite the care taken to select the best meter for the specific test, measurement ultimately depends on the meter user. Most of the time, it is not the meter that makes the error but the user. Good measurement requires careful thought.

CHAPTER 7

REVIEW TERMS

accuracy: specification of a meter indicating how close to the true or actual value a meter reading will be

analog multimeter: multimeter that displays measured values with the displacement of a pointer over a calibrated scale

digital multimeter (DMM): multimeter that displays measured values directly with a numerical display

kilo: prefix used to represent "times one thousand"

mega: prefix used to represent "times one million"

multimeter: test instrument capable of measuring voltage, current, and resistance

parameter: circuit dimension such as voltage or current

polarity: two points of a voltage source normally identified as positive and negative

sensitivity: specification of an analog voltmeter used to determine the resistance a meter will present to a circuit being tested

VOM: analog multimeter measuring volts, ohms, and milliamperes

zero a meter: adjustment made to certain meters prior to application in which the leads are connected together to represent 0 volts or ohms

REVIEW QUESTIONS

1. Describe a multimeter.

2. Describe the accuracy of a meter.

3. Discuss the sensitivity of a meter.

4. Describe what is meant by the function of a meter.

5. Describe the order for connecting leads when measuring voltage.

6. Describe the method for selecting the best range when measuring voltage.

7. Describe what to do with meter switches and leads before putting a meter away.

8. Explain why a meter has various ranges.

9. Describe the procedures for measuring the value of a resistor not connected to a circuit.

10. Describe the procedures for measuring the value of a resistor in a circuit.

11. Describe the procedures for measuring the value of the voltage in a circuit.

12. Describe the procedures for measuring the current in a circuit.

13. What is the best voltage range for an analog voltmeter? For a digital voltmeter?

14. Explain the conflict between accuracy and sensitivity when selecting a voltmeter range.

FIGURE 7-9

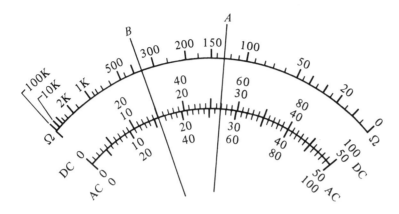

REVIEW PROBLEMS

1. Determine the value indicated by pointer *A* on the meter in Figure 7–9. The meter is on the DC function and the 1-volt range. What is indicated if the meter is on the 500-volt range?

2. Determine the value indicated by pointer *A* on the meter in Figure 7–9 if it is in the resistance function and on the *R* × 1-ohm range. What is indicated if the meter is on the *R* × 1-kilohm range?

3. Determine the value indicated by pointer *B* on the meter in Figure 7–9 if it is in the DC current function and on the 5-milliampere range. What is indicated if the meter is on the 0.1-ampere range?

4. What value is indicated by pointer *B* in Figure 7–9 if the meter is on the *R* × 10-ohm range?

5. What value is indicated by pointer *B* on the meter in Figure 7–9 if the meter is on the 5-volt AC range?

6. What value is indicated by pointer *A* on the meter in Figure 7–9 if the meter is on the 10-milliampere range?

7. On a DMM with a specified accuracy of 2%, 1 digit indicates a voltage of 0.575 volt. Between what two limits does the correct value fall?

8. On a DMM with a specified accuracy of 1%, 1 digit indicates a resistance of 1.27 kilohms. Between what two limits does the correct value fall?

9. Repeat Problem 7 for 0.5%.

10. Repeat Problem 8 for 2%.

11. What is the true error of an analog voltmeter indication of 9 volts on the 50-volt range of a 2% meter?

12. What will be the true error of the meter indication for Problem 11 if the range is changed to 10 volts, full scale?

13. What is the true error of an analog voltmeter indication of 24 volts on the 200-volt range of a 1% meter?

14. What will be the true error of the meter indication for Problem 13 if the range is changed to 50 volts, full scale?

15. What is the resistance of a 20,000-ohms-per-volt meter on the 100-volt range? On the 25-volt range?

16. What is the resistance of a 10,000-ohms-per-volt meter on the 100-volt range? On the 25-volt range?

OHM'S LAW AND SERIES CIRCUITS

State the relationship between voltage, current, and resistance in a simple circuit.

Define and identify series circuit.

State four characteristics of a series circuit.

Calculate the total resistance and current of a series circuit.

Calculate the voltage across each resistor and the total voltage in a series circuit.

Calculate the power consumed by each resistor and the total power in a series circuit.

OHM'S LAW

The objective of this chapter is to introduce basic electrical theory. In previous chapters, we focused on the products, equipment, and components used in electronics and what many of them do. Now, we will consider how many of these items work. We begin with a discussion of how each component works when connected with another in a **circuit** — that is, the path the electrons take through sources and components. Then, we will see what happens when more components are added.

An introduction to electrical theory begins with Ohm's law, the most important and most fundamental subject in the study of electronics.

Ohm's law: The current in a circuit is directly proportional to the applied emf and inversely proportional to the opposition in the circuit.

A familiar example will help explain Ohm's law. If a lamp is connected across a battery, the lamp will light because electric current flows through it. The amount of current depends on two factors, the amount of voltage in the battery and the amount of resistance in the lamp. If the battery voltage (the applied emf) is increased, the current will increase. If the lamp resistance (the opposition in the circuit) is increased, the current will decrease.

Application of Ohm's Law

The rule for Ohm's law can also be stated as a mathematical formula:

$$I = \frac{V}{R} \quad \text{or} \quad I = \frac{E}{R}$$

The amount or intensity of electrical current is

FIGURE 8-1
OHM'S LAW

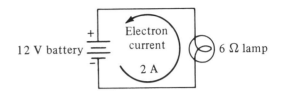

represented by I and is expressed in **amperes** (abbreviated A). The emf (electromotive force), or voltage, is represented by V and is expressed in **volts** (abbreviated V). Historically, electromotive force was represented by the letter E. Today, the trend is to use the letter V for voltage at the source as well as voltage appearing across circuit components. The opposition or resistance of the circuit is represented by R and is expressed in ohms (abbreviated Ω).

We can use the formula to determine the amount of current in a circuit. For example, the lamp in Figure 8–1 has a resistance of 6 ohms and is connected across a 12-volt battery. Then, the circuit has a current flow of 2 amperes. That is:

$$I = \frac{V}{R} = \frac{12 \text{ V}}{6 \text{ } \Omega} = \textbf{2A}$$

We can also use the formula to determine voltage or resistance in a circuit. We simply transpose the factors in the formula. For example, the basic formula is

$$I = \frac{V}{R}$$

By transposing factors, we obtain the following two formulas:

$$R = \frac{V}{I} \quad \text{and} \quad V = I \times R$$

Now, let's use these two formulas. Suppose a car radio draws 1.2 amperes from a 12-volt battery. The circuit must have a total resistance of 10 ohms, as shown by the following equation:

$$R = \frac{V}{I} = \frac{12 \text{ V}}{1.2 \text{ A}} = \mathbf{10 \; \Omega}$$

A 1000-ohm resistor that has 0.01 ampere flowing through it must have 10 volts across it, as shown by the following calculation:

$$V = I \times R = 0.01 \text{ A} \times 1000 \; \Omega = \mathbf{10 \; V}$$

When applying Ohm's law, we must always use **absolute units** — that is, units without prefixes. Suppose the resistance is 1 kilohm. If we use the value of 1 for resistance, we will get the wrong answer. Since kilohm means 1000 ohms, we must use the absolute-units value of 1000 ohms.

Here are some conversion factors that will be helpful in applying Ohm's law:

1 microampere (μA) = 0.000001 A

1 milliampere (mA) = 0.001 A

1 kilohm (kΩ) = 1000 Ω

1 megohm (MΩ) = 1,000,000 Ω

The prefix **micro-** is represented by the Greek letter μ (mu); the prefix **milli-** is represented by the letter m. The preceding conversion factors show their numerical values.

Here are some examples that show how the conversion factors can be used:

25 kΩ = 25,000 Ω

1.5 MΩ = 1,500,000 Ω

8 mA = 0.008 A

20 mA = 0.02 A

10 μA = 0.00001A

In other words, always convert the given units

to absolute units of amperes, ohms, and volts before applying Ohm's law.

Let's do some examples of conversions. Consider a circuit with 48 volts in series with 1.2 megohms. The current is

$$I = \frac{V}{R} = \frac{48 \text{ V}}{1.2 \text{ M}\Omega} = \frac{48 \text{ V}}{1.2 \times 10^6 \; \Omega}$$
$$= 40 \times 10^{-6} \text{ A} = \mathbf{40 \; \mu A}$$

A circuit with 0.5 milliampere through a 100-kilohm resistor has the following voltage:

$$V = I \times R$$
$$= (0.50 \times 10^{-3} \text{ A}) \times (100 \times 10^3 \; \Omega)$$
$$= \mathbf{50 \; V}$$

If 44 volts is connected across a resistor and 220 milliamperes of current flows, the resistance value can be computed as follows:

$$R = \frac{V}{I} = \frac{44 \text{ V}}{220 \times 10^{-3} \text{ A}}$$
$$= 0.2 \times 10^3 \; \Omega = \mathbf{200 \; \Omega}$$

Electric Power

Common items in most homes are 60-watt and 100-watt lamps. The **watt** (abbreviated W) is the unit used to express the power consumed by an electric device. **Power** consumed is the time rate at which work is done. It is based on the voltage across and the current through a device and is calculated from the following power equation:

$$P = V \times I$$

Let's use the power equation and Ohm's law in some calculations. Suppose a 240-ohm lamp is connected across a 120-volt circuit. The lamp will cause a current flow of 0.5 ampere, calculated as follows:

$$I = \frac{V}{R} = \frac{120 \text{ V}}{240 \; \Omega} = \mathbf{0.5 \; A}$$

This circuit will consume a power of 60 watts:

$$P = V \times I = 120 \text{ V} \times 0.5 \text{ A} = \textbf{60 W}$$

As another example, if the voltage is 48 volts and the current is 40 microamperes, the power is

$$P = V \times I = 48 \text{ V} \times 40 \text{ μA}$$
$$= 1920 \times 10^{-6} \text{ W}$$
$$= 1920 \text{ μW} = \textbf{1.92 mW}$$

If the current is 10 milliamperes and the voltage is 9 volts, the power is

$$P = V \times I = 9 \text{ V} \times 10 \text{ mA}$$
$$= 9 \text{ V} \times 10 \times 10^{-3} \text{ A}$$
$$= 0.09 \text{ W} = \textbf{90 mW}$$

The power equation, like Ohm's law, has two other variations. These formulas are

$$I = \frac{P}{V} \quad \text{and} \quad V = \frac{P}{I}$$

Let's do another example. A 1000-watt toaster is plugged into a kitchen outlet. The voltage available at the outlet is 120 volts. How large must the fuse or circuit breaker be? The current through the circuit is

$$I = \frac{P}{V} = \frac{1000 \text{ W}}{120 \text{ V}} = \textbf{8.3 A}$$

Thus, the circuit will need a fuse with a rating larger than 8.3 amperes just for the toaster to operate.

There are other formulas for determining power. If the value for the voltage is unknown, the step of calculating the voltage first is eliminated by using this formula:

$$P = I^2R$$

Let's use this formula for the earlier example with 1.2 megohms of resistance and 40 mi-

croamperes of current. The power is

$$P = I^2 \times R = (40 \times 10^{-6} \text{ A})^2$$
$$\times (1.2 \times 10^6 \text{ Ω})$$
$$= 1600 \times 10^{-12} \text{ A} \times 1.2 \times 10^6 \text{ Ω}$$
$$= 1920 \times 10^{-6} \text{ W} = \textbf{1.92 mW}$$

If the current is 100 milliamperes and the resistance is 1.5 kilohms, the power is

$$P = I^2 \times R = (100 \times 10^{-3} \text{ A})^2 \times 1.5$$
$$\times 10^3 \text{ Ω}$$
$$= (1 \times 10^2 \times 10^{-3} \text{ A})^2 \times 1.5$$
$$\times 10^3 \text{ Ω}$$
$$= (1 \times 10^{-1} \text{ A})^2 \times 1.5 \times 10^3 \text{ Ω}$$
$$= \textbf{15 W}$$

The two other forms of this equation are

$$R = \frac{P}{I^2} \quad \text{and} \quad I = \sqrt{\frac{P}{R}}$$

Another and more complex formula for determining power will be introduced later when we discuss AC circuits. For now, we confine our study to voltage, resistance, current, and power in a variety of DC circuits.

SERIES CIRCUITS

The word *series* means "one after another." In major league baseball, the World Series consists of a number of games played one after another. The second game cannot begin before the first is completed, and two cannot be played at the same time. That is, a certain order must be followed.

Electronic components can be connected to each other in series. Series connection has nothing to do with *what* the components are but rather with *how* they are connected. In a **series circuit,** components are connected one

FIGURE 8-2
SERIES CIRCUIT (ONLY ONE PATH)

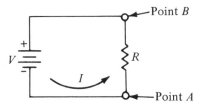

FIGURE 8-3
SERIES CIRCUITS

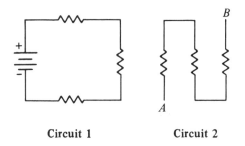

Circuit 1   Circuit 2

FIGURE 8-4
CIRCUITS THAT ARE NOT SERIES

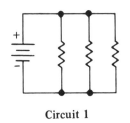

Circuit 1

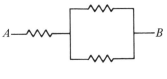

Circuit 2

after another. Series connection is much like a one-lane road where there is only one path. If one car stops on the road, all others must stop. Similarly, if the current is interrupted at one place in a series circuit, all current stops. As shown in Figure 8–2, current must pass point A to reach point B.

Recognizing Series Connections

We must follow the complete path of a circuit to decide whether it is a series circuit. The circuit must have only one path for current flow if it is to be a series circuit. If we find any point where there is more than one way for current to flow, the circuit is not a series circuit. It is some other type of circuit.

Before we begin to study the circuit, we must choose a starting point. If we can find an end, we start there. Also, the negative terminal of a battery is a good place to begin.

Both of the circuits in Figure 8–3 are series circuits. Each circuit has only one path for current. For example, we can follow the current path from negative to positive through three resistors in circuit 1. There are no alternative paths. Circuit 2 is also series since there is only one path from point A to point B. However, the circuits in Figure 8–4 are not series circuits. Each circuit has sections that result in more than one path for the current to follow.

So, the way to identify a series circuit is to trace the current path from one end to the other. If there is only one path with no alternatives along the way, it is a series circuit.

Resistors in Series

Imagine a 1000-ohm resistor connected in series with a 2000-ohm resistor. An electron going through this circuit must work its way through 1000 ohms and then through 2000 ohms, for a total of 3000 ohms. This example leads to an important rule.

Rule: Resistances in a series circuit are added.

They are added because the current must pass through all of them.

A formula is used to determine the total resistance in a series circuit. In words, it states the following rule.

Rule: The total resistance in a series circuit is equal to the sum of the separate resistances in the circuit.

The formula is as follows, where R_T represents total resistance:

$$R_T = R_1 + R_2$$

For the example just described,

$$R_T = 1000 \ \Omega + 2000 \ \Omega = \mathbf{3000 \ \Omega}$$

If there are more than two resistors, this formula can be extended as follows:

$$R_T = R_1 + R_2 + R_3 + \cdots$$

The three dots following R_3 in the formula mean that the addition of resistances continues for all the resistors in the circuit.

The resistance of the circuit in Figure 8–5 can be determined with this formula. We have the following calculation:

$$R_T = R_1 + R_2 + R_3 + R_4$$
$$= 100 \ \Omega + 200 \ \Omega + 150 \ \Omega + 50 \ \Omega$$
$$= \mathbf{500 \ \Omega}$$

Here is an important point to keep in mind.

Key point: The total resistance in a series circuit is always greater than the resistance of the largest resistor.

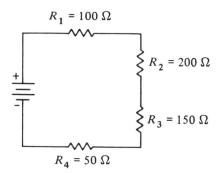

FIGURE 8-5
RESISTANCES IN SERIES

$R_1 = 100 \ \Omega$

$R_2 = 200 \ \Omega$

$R_3 = 150 \ \Omega$

$R_4 = 50 \ \Omega$

RULES AND CALCULATIONS

A good understanding of series circuit operation is essential in building and troubleshooting circuits. There are two parts to this understanding: the rules that show how a series circuit operates and the formulas used to determine all circuit values. We will use the formulas to evaluate circuits in this section. We will also discuss the rules that determine how these circuits work.

Resistance

The first step in evaluating a series circuit is to determine its total resistance. It is a good idea to identify the resistors as R_1, R_2, and so forth, if this labeling has not already been done, since it will help us keep track of the calculations.

Let's begin with the pictorial of the circuit shown in Figure 8–6A. The usual first step in beginning a study of the circuit is to draw a schematic of the circuit by using the appropriate schematic symbols. A battery symbol is commonly used to represent any DC voltage source. The schematic will look like Figure 8–6B.

**FIGURE 8-6
SAMPLE CIRCUIT**

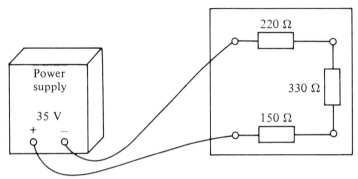

A. Pictorial of circuit

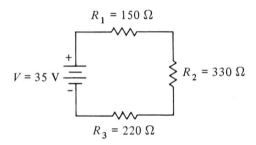

B. Schematic of circuit

The next step is to write the formula, first as the basic formula and then with the appropriate values (and their units of measure) replacing the letters. The formula and the substitutions are

$$R_T = R_1 + R_2 + R_3$$
$$= 150 \ \Omega + 330 \ \Omega + \mathbf{220 \ \Omega}$$

Then, we determine the answer, including the correct units of measure, and distinguish the answer in some way from the formula and calculations. That is, we write the answer on a separate line, we box the answer, or we use some other method of emphasis. Notice, in this book, that we use boldface type to distinguish the answer from the calculations. The reason we emphasize the answer is to set it apart from all the other numbers in a calculation. As you progress in your work in electronics, you will have pages filled with calculations. If you do not separate answers from intermediate calculations, you could easily become confused and make unnecessary errors.

So, the complete solution to the problem is

$$R_T = R_1 + R_2 + R_3$$
$$= 150 \ \Omega + 330 \ \Omega + 220 \ \Omega = \mathbf{700 \ \Omega}$$

Current

The next step in evaluating a series circuit is to determine the amount of current that is flowing. Before we do so, we must remember that there is only one value for current in a series circuit. There is only one value because of the following rule.

Rule: The current is the same in all parts of a series circuit.

Let's consider an analogy. If we count 500 cars passing a point on a highway in 1 hour, we expect a friend down the road to count the same number. That is, the rate of flow is the same along all parts of the road.

Determining current flow is the same as counting cars, except that we are counting electrons in current flow. One ampere of current means that approximately 6.24×10^{18} electrons are passing a given point in the circuit in 1 second. (The number 6.24×10^{18} is expressed in scientific notation. The 10^{18} means 10 multiplied by itself 18 times. That is, if the number 6.24×10^{18} were to be written in the usual fashion, the decimal point in the number would be moved 18 places to the right.) This number of electrons (6.24×10^{18}) is called a coulomb. So, 1 ampere equals 1 coulomb per second.

The amount of current in the series circuit in Figure 8–6B is calculated by using Ohm's law. However, notice that total voltage and total resistance values must be used when determining total current. The calculation is

$$I = \frac{V}{R} = \frac{35V}{700 \ \Omega} = \textbf{0.05 A}$$

Voltage

While there is only one value (35 volts) of *applied* voltage in this circuit, it is *dropped* across each resistor in the circuit as the current flows along. That is, voltage across a re-

sistor can cause a current flow, and, similarly, current through a resistor can cause a voltage drop. These voltage drops can be calculated with Ohm's law. The form we use is easy to select since V is the unknown. It is common practice to use subscripts for each value of voltage we are calculating to keep from mixing them up. Thus, voltage across R_1 can be called V_1. Since there is only one value for I, it can be called I_T, for total current. The voltage across R_1 in Figure 8–6B is calculated as follows:

$$V_1 = I_T R_1 = 0.05 \text{ A} \times 150 \ \Omega = \textbf{7.5 V}$$

The same formula is used for the other voltage drops in Figure 8–6B:

$$V_2 = I_T R_2 = 0.05 \text{ A} \times 330 \ \Omega = \textbf{16.5 V}$$
$$V_3 = I_T R_3 = 0.05 \text{ A} \times 220 \ \Omega = \textbf{11 V}$$

Here is another important rule in series circuits.

Rule: The total voltage in a series circuit is equal to the sum of the separate voltage drops.

Thus, we can check our values by adding them:

$$V_T = V_1 + V_2 + V_3$$
$$= 7.5 \text{ V} + 16.5 \text{ V} + 11 \text{ V} = \textbf{35 V}$$

Notice, in the preceding calculations, that the voltage drop across a resistor is in proportion to its size. Equal-size resistances have equal voltage drops. Large resistances have large voltage drops, and small resistances have small voltage drops.

Consider again the circuit in Figure 8–6B with the values changed as follows:

$$V_T = 76 \text{ V} \qquad R_2 = 47 \text{ k}\Omega$$
$$R_1 = 33 \text{ k}\Omega \qquad R_3 = 15 \text{ k}\Omega$$

The unknown values are calculated as follows:

$$R_T = R_1 + R_2 + R_3$$
$$= 33 \times 10^3 \; \Omega + 47 \times 10^3 \; \Omega + 15$$
$$\times 10^3 \; \Omega$$
$$= \mathbf{95 \; k\Omega}$$

$$I_T = \frac{V_T}{R_T} = \frac{76 \; V}{95 \times 10^3 \; \Omega} = \mathbf{0.8 \times 10^{-3} \; A}$$

$$V_1 = I_T \times R_1$$
$$= 0.8 \times 10^{-3} \; A \times 33 \times 10^3 \; \Omega$$
$$= \mathbf{26.4 \; V}$$

$$V_2 = I_T \times R_2$$
$$= 0.8 \times 10^{-3} \; A \times 47 \times 10^3 \; \Omega$$
$$= \mathbf{37.6 \; V}$$

$$V_3 = I_T \times R_3$$
$$= 0.8 \times 10^{-3} \; A \times 15 \times 10^3 \; \Omega$$
$$= \mathbf{12 \; V}$$

Check the answer:

$$V_T = V_1 + V_2 + V_3$$
$$= 26.4 \; V + 37.6 \; V + 12 \; V = \mathbf{76 \; V}$$

Power

A circuit designer selects the correct resistance so that a circuit will operate as desired. But resistance is not the only consideration. The resistor must also be able to withstand the power level that it will be subjected to. This level can be determined quite simply by using the power formula. Once this level is determined, a resistor with twice this power rating should be selected in order to ensure a good margin of safety.

The safety margin is needed because of the **square-law effect,** which can be described as follows: Power in a circuit is equal to voltage times current. Also, the voltage in a circuit causes the current. Therefore, if the applied voltage in a circuit doubles, the current will also double. The result will be a power increase of four times the original value. In other words, the power will increase by the square of the voltage increase.

To calculate the power of the first resistor in Figure 8–6B, we use the power formula. Notice that we use the voltage across and the current through that resistor for our calculations, which are as follows:

$$P_1 = V_1 I_T = 7.5 \; V \times 0.05 \; A = \mathbf{0.375 \; W}$$

Since resistors come in ⅛-, ¼-, ½-, 1-, 2-, 5-, and 10-watt ratings, a 1-watt resistor is a good choice for this situation. While a resistor with a higher power rating could be used, it would cost more, be larger, and occupy more space.

The second power level in Figure 8–6B is

$$P_2 = V_2 I_T = 16.5 \; V \times 0.05 \; A$$
$$= \mathbf{0.825 \; W}$$

A 2-watt resistor is a good choice here.

The third power level is

$$P_3 = V_3 I_T = 11 \; V \times 0.05 \; A = \mathbf{0.55 \; W}$$

A 2-watt resistor will provide a good margin of safety here.

There are two ways to determine total power in a series resistive circuit. The first method uses the following rule.

Rule: The total power (P_T) in a DC circuit is equal to the total voltage (V_T) times the total current.

Therefore, we have the following calculation:

$$P_T = V_T I_T = 35 \; V \times 0.05 \; A = \mathbf{1.75 \; W}$$

The second method uses another rule.

Rule: The total power in a series circuit is equal to the sum of the separate powers.

Therefore, we have the following calculation:

$$P_T = P_1 + P_2 + P_3$$
$$= 0.375 \text{ W} + 0.825 \text{ W} + 0.55 \text{ W}$$
$$= 1.75 \text{ W}$$

These two approaches also provide an opportunity to check calculations. The calculations are correct if these two values for P_T agree.

The preceding examples have shown just about all there is to do in calculating all the values for a series circuit. The calculations can be more time-consuming if there are more resistors, and they can be a little more difficult if you begin with color codes only. Also, large-value resistors can sometimes be a problem if you forget a few zeros. However, these situations do not require more theory, only more care.

Solution of a Series Circuit

Suppose a circuit has four resistors in series with a 44-volt power supply. The resistor colors are as follows:

—R_1, blue, gray, brown, and silver;

—R_2, gray, blue, brown, and silver;

—R_3, green, blue, brown, and gold;

—R_4, brown, black, brown, and silver.

Resistor values are determined by using the color code in the Appendix. The pictorial of the circuit is shown in Figure 8–7A. The schematic for the circuit is shown in Figure 8–7B.

Our task is to determine all values of voltage, current, resistance, and power. The steps are given next.

1. Compute total circuit resistance:

$$R_T = R_1 + R_2 + R_3 + R_4$$
$$= 680 \text{ }\Omega + 860 \text{ }\Omega + 560 \text{ }\Omega + 100 \text{ }\Omega$$
$$= \mathbf{2200 \text{ }\Omega}$$

2. Compute the total current:

$$I_T = \frac{V_T}{R_T} = \frac{44 \text{ V}}{2200 \text{ }\Omega} = \mathbf{0.02 \text{ A}}$$

3. Compute the circuit voltage drops:

$$V_1 = I_T R_1 = 0.02 \text{ A} \times 680 \text{ }\Omega = \mathbf{13.6 \text{ V}}$$
$$V_2 = I_T R_2 = 0.02 \text{ A} \times 860 \text{ }\Omega = \mathbf{17.2 \text{ V}}$$
$$V_3 = I_T R_3 = 0.02 \text{ A} \times 560 \text{ }\Omega = \mathbf{11.2 \text{ V}}$$
$$V_4 = I_T R_4 = 0.02 \text{ A} \times 100 \text{ }\Omega = \mathbf{2 \text{ V}}$$

$$V_T = V_1 + V_2 + V_3 + V_4$$
$$= 13.6 \text{ V} + 17.2 \text{ V} + 11.2 \text{ V} + 2 \text{ V}$$
$$= \mathbf{44 \text{ V}}$$

4. Compute the power consumed by each resistor:

$$P_1 = V_1 I_T = 13.6 \text{ V} \times 0.02 \text{ A} = \mathbf{0.272 \text{ W}}$$
$$P_2 = V_2 I_T = 17.2 \text{ V} \times 0.02 \text{ A} = \mathbf{0.344 \text{ W}}$$
$$P_3 = V_3 I_T = 11.2 \text{ V} \times 0.02 \text{ A} = \mathbf{0.224 \text{ W}}$$
$$P_4 = V_4 I_T = 2 \text{ V} \times 0.02 \text{ A} = \mathbf{0.04 \text{ W}}$$

5. Calculate the total circuit power:

$$P_T = P_1 + P_2 + P_3 + P_4$$
$$= 0.272 \text{ W} + 0.344 \text{ W} + 0.224 \text{ W}$$
$$+ 0.04 \text{ W}$$
$$= \mathbf{0.88 \text{ W}}$$

$$P_T = V_T I_T = 44 \text{ V} \times 0.02 \text{ A} = \mathbf{0.88 \text{ W}}$$

Notice that our values for V_T and P_T check.

FIGURE 8-7
SAMPLE CIRCUIT

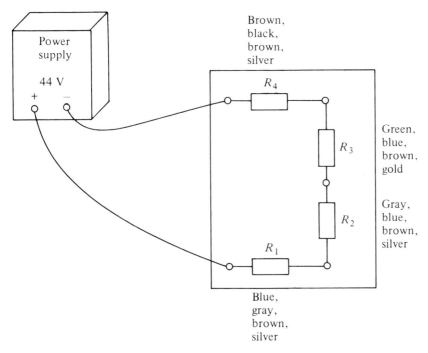

A. Pictorial of circuit

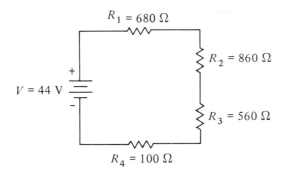

B. Schematic of circuit

Here is another example. A three-resistor series circuit has the following values:

$R_1 = 47\ \Omega$ $R_3 = 60\ \Omega$
$R_2 = 33\ \Omega$ $V_T = 35\ V$

Calculate the resistance and current:

$$R_T = R_1 + R_2 + R_3$$
$$= 47\ \Omega + 33\ \Omega + 60\ \Omega = \mathbf{140\ \Omega}$$
$$I_T = \frac{V_T}{R_T} = \frac{35\ V}{140\ \Omega} = \mathbf{0.25\ A}$$

Calculate the voltages:

$$V_1 = I_T \times R_1 = 0.25\ A \times 47\ \Omega$$
$$= \mathbf{11.75\ V}$$
$$V_2 = I_T \times R_2 = 0.25\ A \times 33\ \Omega$$
$$= \mathbf{8.25\ V}$$
$$V_3 = I_T \times R_3 = 0.25\ A \times 60\ \Omega = \mathbf{15\ V}$$

Check the answer:

$$V_T = V_1 + V_2 + V_3$$
$$= 11.75\ V + 8.25\ V + 15\ V = \mathbf{35\ V}$$

Determine the power:

$$P_1 = V_1 \times I_T = 11.75\ V \times 0.25\ A$$
$$= \mathbf{2.94\ W}$$
$$P_2 = V_2 \times I_T = 8.25\ V \times 0.25\ A$$
$$= \mathbf{2.06\ W}$$
$$P_3 = V_3 \times I_T = 15\ V \times 0.25\ A$$
$$= \mathbf{3.75\ W}$$
$$P_T = V_T \times I_T = 35\ V \times 0.25\ A$$
$$= \mathbf{8.75\ W}$$

Check the answer:

$$P_T = P_1 + P_2 + P_3$$
$$= 2.94\ W + 2.06\ W + 3.75\ W$$
$$= \mathbf{8.75\ W}$$

USES OF SERIES CIRCUITS

One good example of a series circuit is an arrangement of resistors called a **voltage divider.** It is used when many different voltages are needed. They are produced by connecting a group of resistors in series with a large DC source; the resistors provide the needed voltage drops, which are added to each other. The circuits connected to the divider increase the complexity of the design. We will consider these complexities in a later chapter.

Another good example of a series circuit is a string of Christmas tree lights. The so-called one-wire set shown in Figure 8–8 is a simple series circuit. Each lamp operates on 15 volts, and there are eight lamps. The string is designed to be connected to a 120-volt circuit. Since it is a series circuit, all lamps go off if one burns out or is disconnected.

Suppose one lamp in a one-wire set burned out, and someone decided to remove that socket and have only seven lamps. Then, the 120 volts would divide seven ways instead of eight. So, the lamp voltage would increase from the desired 15 volts each to 17 volts each. This increase in voltage would also increase the power. Consequently, another lamp would burn out.

The two-wire sets of Christmas tree lights use 120-volt bulbs that do not go out when one is removed. They are not connected in series but are connected in parallel. Parallel circuits will be discussed in Chapter 14.

FIGURE 8-8
TYPICAL SERIES CIRCUIT

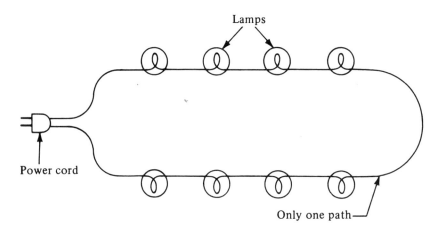

Lamps

Power cord

Only one path

SUMMARY

Series circuits provide a basic introduction to electrical theory. This theory is based on Ohm's law, which describes the relationship between voltage, current, and resistance in a circuit. Ohm's law states that an increase in the voltage in a circuit increases the current. Also, an increase in the resistance in a circuit decreases the current. The formula for Ohm's law is $I = V/R$. You can determine any one of these factors by using Ohm's law as long as you know the other two factors.

The power consumed in a circuit is calculated by using the power equation. Circuit power is equal to voltage times current. An important point to keep in mind is that doubling the voltage in a circuit, without changing the resistance, also doubles the current. In that case, the power will increase by four times its original value, or by the voltage increase squared. This effect is called square-law effect.

A series circuit is simply a group of components connected (electrically) in a row, one after another. The best way to think of series circuits is by the rules that describe them, which are as follows:

1. There is only one path for current in a series circuit.
2. The current is the same in all parts of a series circuit.
3. The total resistance in a series circuit is equal to the sum of the separate resistances.
4. The total voltage in a series circuit is equal to the sum of the separate voltage drops.
5. The total power in a series circuit is equal to the sum of the separate powers.

The solution of a series circuit consists of determining all individual and total values of resistance, voltage, current, and power. Some of these values must be identified before you begin. The others are determined by calculations, using formulas described in this chapter. Your ability to use these formulas is essential for your success in electronics. We note here that these values can also be determined by measurement, the topic of the next chapter.

Series circuits are everywhere. For instance, some Christmas tree lights are wired this way. The series circuit provides a simple way to operate these low-voltage lamps with line voltage. Voltage divider circuits are also series circuits; they produce a variety of lower voltages from a large-voltage source. Television power supplies use such voltage dividers. Most of the time, however, you will be concerned only with simple series circuits, such as a resistor in series with a transistor.

You will find that advanced electronics theory is explained mainly with series circuits. For example, the operation of a transistor is best explained by series circuits. They also provide the easiest way to understand many other circuits. So, series circuit theory is not only an introduction to more advanced work — it is also a foundation.

CHAPTER

REVIEW TERMS

absolute units: units without prefixes, such as volts, ohms, and amperes

ampere: unit of measure for current

circuit: path electrons take through sources and components

micro: prefix used to represent "times one millionth"

milli: prefix used to represent "times one thousandth"

Ohm's law: relationship between voltage, resistance, and current — the current in a circuit is directly proportional to the applied voltage and inversely proportional to the opposition in the circuit

power: time rate at which work is done; the product of voltage and current in a circuit, expressed in watts

series circuit: arrangement of a circuit in which current must flow through every component

square-law effect: relationship between source voltage and power — doubling the voltage doubles the current, resulting in a fourfold power increase; the power increases by the square of the voltage increase

volt: unit of measure for voltage or emf

voltage divider: series arrangement of resistors connected to a large DC source and used to produce proportionally lower amounts of an available voltage

voltage drop: voltage resulting from current flow through a resistor

watt: unit of measure for power

REVIEW QUESTIONS

1. State Ohm's law.

2. Define coulomb.

3. Define series circuit.

4. State four characteristics of a series circuit.

5. Describe how current flows in a series circuit.

6. Describe electric power, and name the unit of power.

7. Describe the effects of increasing resistance in a series circuit.

8. Describe the effects of increasing total voltage in a series circuit.

9. Explain the difference between electromotive force and voltage drop.

10. Describe the relationship between voltage drop and resistor size in a series circuit.

11. What factors affect power in a series circuit?

12. What is the effect on circuit power if the circuit resistance is reduced by 50%?

13. What is the effect on circuit power if the applied emf is doubled?

14. Indicate how each of the following prefixes can be represented in scientific notation: *milli-, micro-, mega-, kilo-,* and *pico-.*

15. Review the schematic diagram of a product, and identify sections of the circuit that are connected in series.

REVIEW PROBLEMS

1. What is the current in the circuit shown in Figure 8–9 if the total voltage is 9 volts and the resistance is 1.5 kilohms? What is the power?

2. How will the power change if the voltage in the circuit for Problem 1 is increased to 18 volts?

FIGURE 8-9

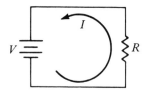

3. What is the current in the circuit of Figure 8–9 if the voltage is 24 volts and the resistance is 1.2 megohms? What is the power?

4. How will the circuit for Problem 3 change if the voltage is reduced to 12 volts?

5. What is the voltage in the circuit of Figure 8–9 if the current is 16 milliamperes and the resistance is 470 ohms?

6. What is the voltage in the circuit of Figure 8–9 if the current is 250 microamperes and the resistance is 1.5 megohms?

7. What is the circuit power for the circuit of Problem 6?

8. What is the resistance in the circuit of Figure 8–9 if the voltage is 15 volts and the current is 3 microamperes?

9. What is the resistance in the circuit of Figure 8–9 if the voltage is 175 volts and the current is 14 milliamperes?

10. Using Figure 8–10, determine all values of resistance, current, voltage, and power. The values are as follows: V_T = 40 volts, R_1 = 2200 ohms, R_2 = 3300 ohms, R_3 = 1000 ohms, and R_4 = 1500 ohms.

11. Calculate all values for the circuit described in Problem 10 for an increase in voltage to 60 volts.

12. Calculate the changes in values for the circuit described in Problem 10 with R_3 shorted.

13. Determine all values of resistance, current, voltage, and power for the circuit in Figure 8–10 if the values are as follows: V_T = 240 volts, R_1 = 330 ohms, R_2 = 250 ohms, R_3 = 470 ohms, and R_4 = 150 ohms.

14. Calculate all values for the circuit described in Problem 13 for a decrease in voltage to 180 volts.

FIGURE 8-10

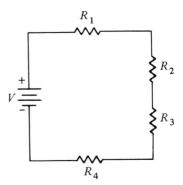

15. Calculate the changes in values for the circuit described in Problem 13 with R_4 shorted.

16. What is the effect on individual lamp power if one of the eight 15-watt lights in a 120-volt series string is shorted?

SERIES CIRCUIT TESTING

OBJECTIVES

Measure the voltages in a series circuit.

Measure the current in a series circuit.

Measure the resistance in a series circuit.

Determine if a series circuit is operating properly.

Locate defects in a series circuit.

TEST AND MEASUREMENT

The objective of this chapter is to introduce some basic methods of circuit testing and evaluation. A major portion of a technician's time is spent performing tests. **Testing** can be defined as the thoughtful use of instrumentation to evaluate the operation of components and products. Testing includes **measurement,** which is the determination of magnitudes that define a parameter.

Let's consider an example. When you are ill, your doctor may perform some tests, which may include the measurement of your temperature, your blood pressure, and your weight. These measurements are made to determine their values so that they can be compared with what is normal or typical for a person of your age.

Suppose your doctor determines that your temperature is 102°. This number is an indication that something is wrong, but it does not tell the doctor *what* is wrong. The temperature test confirms that you feel ill, but it does not explain *why* you are ill. More tests would be necessary. X rays, blood tests, or other tests would be performed to determine the *cause* of your illness.

Testing performs a similar function in electronics. For example, a friend may ask you to fix his television. You ask, "What is wrong?" He answers simply, "It doesn't work" or "It is broken." But you may be able to get a description of **symptoms** — that is, behavior characteristics of a circuit or component resulting from a malfunction. Some symptoms of a malfunctioning TV set may be no picture or distorted sound. Symptoms along with an understanding of how the circuit works will then help you select and perform tests. From there, you will try to find the cause of the problems.

So, we look for symptoms, which lead us to tests. Our knowledge about the use of test equipment ensures our making accurate measurements. The results of the measurements give us numbers describing what is, which we can compare with numbers that describe what should be. More tests and an analysis of the circuit then help us locate causes, which must be corrected.

Test Procedures

The medical evaluation of a person can range from a check of temperature to a few dozen tests conducted over several days in a hospital. Likewise, the evaluation of a stereo can range from determining whether it plays to analyzing its frequency response, power output, distortion, and many other standard parameters. Whether tests are simple or complex, there are acceptable and unacceptable procedures you must know about.

Two basic concerns in testing are safety and quality. Safety comes first: your personal safety, the safety of what you are using, and the safety of what you are working on. The procedures outlined in this text always stress safety. As mentioned before, most accidents can be avoided. The key to safety is to think before you act.

Quality is also important. The quality of your measurements is based on your ability to select, use, and interpret the readings of test instruments. And the measurements that you determine must be correct. Someone is going to make decisions on the basis of your measurements. If the measurements are wrong, the decisions will probably be wrong. So, safety and quality should be the focus of all test procedures.

Before beginning any test, you must know what parameter you wish to measure: voltage level, resistance, current, or some other feature. In addition, you must know how accurate your results have to be. Generally speaking, 3% accuracy is close enough for voltage, current, and resistance measurements. Most multimeters give results that are good. However, if greater accuracy is required, better instruments must be used. Since these instruments are more delicate and more expensive, they are only used when greater accuracy is required.

Quality measurement requires careful thought. Just because you read a certain value does not mean that it is right. And the fact that you used a new or expensive meter does not guarantee perfect results, either. The meter could be defective, connected in the wrong place, or misread. Thus, you should always ask yourself: Do the results make sense? Many errors that technicians make would have been detected if they had asked themselves whether the results seemed reasonable. Therefore, you must learn to detect, and correct, your own test procedure errors.

Keeping Records

Most technicians keep records of test results. These records will be helpful to someone who, at a later date, may want to know the results of the tests or how they were conducted. Therefore, many companies require their technicians to keep notebooks of test procedures and results.

Notebooks should provide an accurate record of how tests were conducted and what the results were. Some problems take a long time to solve. Therefore, it makes sense to record the solution so that the next time a similar problem comes along, the answer is available. Notebooks are also a necessary part in the documentation of patent applications. Companies rely on technicians' and engineers' notebooks to show the development of a product.

Only some basic features of notebooks will be described here; the subject of lab reports will be discussed later in Chapter 23.

Bound notebooks are best since the pages cannot get lost or out of order. Your name should be on the cover, and each page should be dated. While the type of entry varies with the project, some information is common to all projects. Always record what you were trying to do, the equipment or materials you used, what you actually did, the data you obtained, and your results or conclusions.

The notebook is not filled with finished reports. It is primarily used to record data while working through a project. Thus, it becomes a source of information for final reports. Examples of tables from a notebook that might be used for a final report are included in the next section of this chapter.

EVALUATING
A SERIES CIRCUIT

The evaluation of a circuit consists of a series of tests to determine just what the circuit is doing. Basically, circuit evaluation means measuring all the values of voltage, current, and resistance. This evaluation leads to the real issue, which is: Is the circuit doing what it is supposed to be doing? All circuit troubleshooting begins with evaluation.

Circuit Specifications

Before you can test a circuit to see whether it is working properly, you must know what it is supposed to do. There are two ways to determine these factors: by calculating all values based on a knowledge of circuit theory and by using manufacturers' specifications. The calculation method is reasonable for simple circuits, and it is the method used for most of the circuits in this text. In addition, this method provides a good understanding of how a circuit works because it includes a theoretical analysis — calculating values — and a practical analysis — building and testing the circuit.

In the real world of electronics testing, though, technicians usually use specifications. A television repair technician, for example, uses schematics, which have typical voltages and resistances noted at a number of places on the circuit. Most electronics manufacturers ship manuals with their products to provide technical information for service technicians. These manuals describe how to install, test, align, and repair these products, and they provide a complete parts list.

Once you have the specifications for the circuit parameters, you must take measurements to determine whether these values are what they should be. Next, you must decide how close to the specified voltage the measured voltage has to be for it to be acceptable. You must understand the use of the circuit before you can answer this question. Finally, if the measured voltage is far beyond its specified value, you must determine the cause of the problem. We will discuss this procedure later in the chapter. For now, we will focus on specified values. Note that specified values may also be called calculated values since, in many cases, the specifications are arrived at by calculation. Measured values, however, are always arrived at by measurement.

Some technicians work with engineers in the development of new products. In the typical procedure, the engineer designs a circuit and gives the technician a sketch. The technician then builds the circuit, tests it, and gives the results to the engineer. Finally, the engineer decides whether to modify the design or send it on to the production people. We will follow that procedure here. The circuit we are to evaluate is shown in Figure 9–1.

Since we are not given any specifications, we must calculate them. The calculation procedure is the same as the one we discussed in the preceding chapter. The calculation steps are given next.

FIGURE 9-1
CIRCUIT TO BE EVALUATED

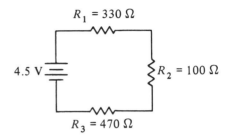

1. Compute the total circuit resistance:

$$R_T = R_1 + R_2 + R_3$$
$$= 330 \ \Omega + 100 \ \Omega + 470 \ \Omega = \mathbf{900 \ \Omega}$$

2. Compute the total current:

$$I_T = \frac{V_T}{R_T} = \frac{4.5 \ V}{900 \ \Omega} = \mathbf{0.005 \ A}$$

3. Compute the circuit voltage drop:

$$V_1 = I_T R_1 = 0.005 \ A \times 330 \ \Omega = \mathbf{1.65 \ V}$$
$$V_2 = I_T R_2 = 0.005 \ A \times 100 \ \Omega = \mathbf{0.5 \ V}$$
$$V_3 = I_T R_3 = 0.005 \ A \times 470 \ \Omega = \mathbf{2.35 \ V}$$
$$V_T = V_1 + V_2 + V_3$$
$$= 1.65 \ V + 0.5 \ V + 2.35 \ V$$
$$= \mathbf{4.5 \ V} \qquad \text{(answer checks)}$$

4. Compute the power consumed by each resistor:

$$P_1 = V_1 I_T = 1.65 \ V \times 0.005 \ A$$
$$= \mathbf{0.00825 \ W}$$
$$P_2 = V_2 I_T = 0.5 \ V \times 0.005 \ A$$
$$= \mathbf{0.0025 \ W}$$
$$P_3 = V_3 I_T = 2.35 \ V \times 0.005 \ A$$
$$= \mathbf{0.01175 \ W}$$

5. Calculate the total circuit power:

$$P_T = P_1 + P_2 + P_3$$
$$= 0.00825 \text{ W} + 0.0025 \text{ W}$$
$$+ 0.01175 \text{ W}$$
$$= \mathbf{0.0225 \text{ W}}$$
$$P_T = V_T I_T = 4.5 \text{ V} \times 0.005 \text{ A}$$
$$= \mathbf{0.0225 \text{ W}} \quad \text{(answer checks)}$$

Although we will not be measuring power in this experiment, these power calculations tell us that a ⅛-watt rating, or larger, will be safe for all resistors in this experiment. (Recall that resistor ratings should be at least twice the value of the expected power.) The calculation portion of the evaluation is now complete.

Measuring Voltage

Before we can test this circuit, we must assemble it. There are many ways assembly can be done. Since this is an experimental circuit, it probably should be assembled in such a way that it can be easily disassembled. This procedure is called **breadboarding,** and it usually means that leads are not cut or wrapped together but rather are held together with clip leads or a little solder.

There are two sources we can use for voltage, a power supply or batteries. Batteries will be used here. Our breadboarded circuit should resemble the one shown in Figure 9–2. Notice that this figure also shows how to measure V_2, the voltage across resistor R_2.

The procedure for measuring voltage was described in Chapter 7. To summarize:

1. Place the meter in a safe position.

2. Select DC voltage.

3. Select a high range.

4. Connect the leads to the meter, then to the circuit.

FIGURE 9-2
CONNECTIONS FOR MEASURING V_2

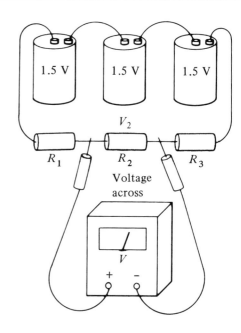

5. Select the best range.

6. Carefully read the indicated value.

7. Disconnect the leads in a particular order.

8. Switch to a high range, and prepare to measure the next voltage. Note that the power was not turned off here because this circuit is a low-voltage circuit.

9. When you are finished with all voltage measurements, switch to the highest voltage range before putting the meter away.

You should keep a record of your calculated values and your measurements in a table in your notebook for future reference. You will need this information later for your final report. For example, a table recording calculated values of voltage might look like Table 9–1.

TABLE 9-1
SAMPLE TABLE OF SPECIFIED
(OR CALCULATED) VOLTAGES

Parameter	Value (in Volts)
Battery 1	1.55
Battery 2	1.45
Battery 3	1.50
V_T	4.50
V_1	1.65
V_2	0.5
V_3	2.35

Measuring Current

The procedure for measuring current was described in Chapter 7. The key difference between measuring current and measuring voltage is that the circuit must be broken, as shown in Figure 9–3. Since this circuit is a series circuit, the current is the same in all parts. Therefore, the circuit can be broken anywhere. However, be careful not to leave out or bypass any components when installing the meter.

To summarize the important points in current measurement:

1. Place the meter in a safe position.
2. Select DC current.
3. Select the highest current range.
4. Disconnect a battery lead to de-energize the circuit.
5. Separate any two leads in the circuit.
6. Install the meter.

Before you reconnect the battery to energize the circuit, have your instructor check to ensure that the meter is properly connected.

Now, you can energize the circuit and check the meter for proper polarity. When you have correct polarity, select the best range,

FIGURE 9-3
CONNECTIONS FOR MEASURING I_T

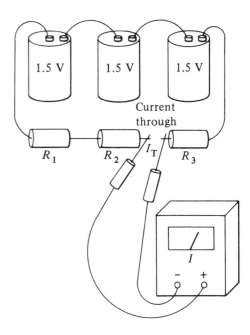

read the value, and enter it in your notebook table. Your table should look something like the following one:

Parameter	Value
I_T	5 mA

You may want to measure the current in other parts of the circuit to be sure that it is the same in all places. It should be. When you are finished, de-energize the circuit, disconnect the meter, place it on the highest voltage range, and set it aside.

Measuring Resistance

Resistance measurement was also described in Chapter 7. The important point to remember here is to make sure that one lead of the

FIGURE 9-4
CONNECTIONS FOR MEASURING R_1

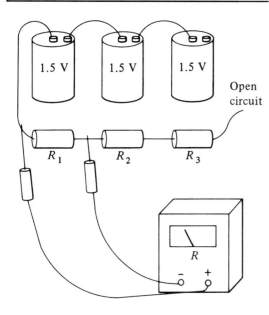

resistor being measured is disconnected from the rest of the circuit, as shown in Figure 9–4. This procedure prevents other parts from being included in your measurements. You must also make sure that one side of the power source is disconnected so that the circuit is de-energized.

Once the circuit is ready, follow these steps:

1. Place the meter in a safe position.
2. Select the ohms function.
3. Select the middle range.
4. Zero the meter.
5. Connect the meter across the resistor.

Make sure your fingers are away from the connections so that you do not become part of the circuit. The procedure for measuring R_1 is shown in Figure 9–4.

Remember that the best range for resis-

tance on an analog meter is the one that places the pointer closest to the center of the scale. After you measure R_1, enter the value in your notebook table. Then, measure and record R_2 and R_3. Your table should look like the following one:

Parameter	Value (in Ohms)
R_1	330
R_2	100
R_3	470

The measurement portion of circuit evaluation is now complete. We can disassemble the circuit and put the parts away. However, the evaluation itself is not complete. Now, we put all our results together in order to arrive at some conclusions.

Drawing Conclusions

After we have measured the parameter values for the circuit, we must decide whether the circuit is working properly. The first step in making this decision is to put all our information together so that we can make a comparison. That is, we want to compare what should be (the specified or calculated values) with what is (the measured values).

If everything in the circuit is working properly — if everything is exactly as it should be — our results will look something like the ones in Table 9–2.

Results are usually not that close, though. It is not unusual for measured battery voltage to be a little higher than calculated voltage. If each battery provided 1.6 volts, for example, the total voltage would measure 4.8 volts. The other voltages would increase in proportion, and so would the current. This result would be considered a reasonable variation for this circuit, however, and we could conclude that the circuit was working properly.

TABLE 9-2
SAMPLE TABLE OF CALCULATED AND
MEASURED VALUES

Parameter	Calculated Value	Measured Value
V_T	4.50 V	4.5 V
V_1	1.65 V	1.65 V
V_2	0.5 V	0.5 V
V_3	2.35 V	2.35 V
I_T	5 mA	5 mA
R_1	330 Ω	330 Ω
R_2	100 Ω	100 Ω
R_3	470 Ω	470 Ω

Other variations are also possible. A 100-ohm, 10% resistor may measure any value between 90 and 110 ohms and still be within tolerance. If we need a value closer to 100 ohms, then we probably would use a 5% resistor. It should measure between 95 and 105 ohms. So variations as much as 10% between calculated and measured values are to be expected in a circuit like this one. Furthermore, when we consider the real accuracy of the meter, we can get even more variation.

Here are some examples of obvious errors that you should notice. A voltage that is twice or half what it should be is an indication of a problem. Zero voltage or current also indicates a problem. Detecting obvious problems will not be too difficult. Deciding whether or not a small variation is a problem is much more difficult.

You must know the details of the circuit and its use in order to decide if a given variation is acceptable. There are times when 20% variation is acceptable, and yet there are times when 2% variation is not. However, the components and equipment discussed here will have a reasonable variation of 10% to 20%. When measured results exceed these variations, troubleshooting must be done.

TROUBLESHOOTING

Troubleshooting is a process undertaken to find the cause of a problem. Troubleshooting is not done in a random manner; it is done in a step-by-step procedure after giving some thought about the symptoms. A good troubleshooter has a fairly good idea of the cause of a problem before making any measurements. A television technician, for example, can often pinpoint a problem to a specific area before removing the back of the set.

Getting a good description of the symptoms helps you determine a variety of information. For example, you can locate the circuits that are not working properly and those that are. You can identify the circuits where the cause of a problem might be and eliminate the other circuits. A good technician looks for a cause of a problem by finding and eliminating those places where the problem could not be and evaluating those places where it could be.

Troubleshooting a complex circuit can be a difficult and time-consuming process. However, most problems can be narrowed down to one or two components in a circuit. The problem will usually be a short circuit, an open circuit, or a component significantly changed in value. The procedures that follow are the basic ones in troubleshooting.

Locating Open Circuits

An open circuit is probably the most common defect that you will find. An **open circuit** is a break in the normal conducting path that interrupts the flow of current. When you turn a switch off, you are intentionally opening the circuit to stop the flow of current. Circuit breakers and fuses open circuits to stop the excessive current caused by a defect. A burned-out light bulb is simply one in which the filament wire has deteriorated from long

FIGURE 9-5
OPERATION OF A LAMP

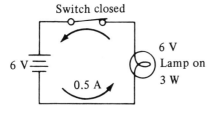

A. Closed switch

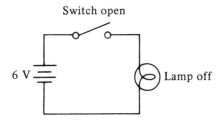

B. Opened switch

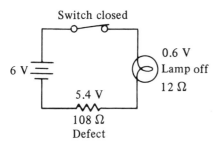

C. Switch with resistance in series

use and has broken, interrupting the flow of current.

A circuit does not have to have an absolute break to be considered open. What matters is the **relative value** of the resistance of this open circuit — that is, the value of the component in relation to the values of those around it. This point is an extremely important one, and the best way to consider it is through an example.

Suppose a 6-volt, 3-watt lamp is operated with a battery and a switch. We can calculate the values of current and resistance from the voltage and power given. Since $P = VI$, we can determine the current as follows:

$$I = \frac{P}{V} = \frac{3 \text{ W}}{6 \text{ V}} = \textbf{0.5 A}$$

The resistance is

$$R = \frac{V}{I} = \frac{6 \text{ V}}{0.5 \text{ A}} = \textbf{12 } \Omega$$

In Figure 9–5A, the closed switch allows 0.5 ampere to flow, which in turn causes the light to go on. The 0.5 ampere through the lamp and the 6 volts across it cause a dissipation of 3 watts, and the lamp turns on. Opening the switch will reduce the current through the lamp and the voltage across it to zero, as shown in Figure 9–5B. The lamp will turn off. This type of open circuit is easy to see.

Suppose, however, that the switch is not opened. Rather, a resistance is placed in series with the lamp. That resistance need not be a resistor; it may simply be something that has resistance, something that opposes the flow of current. That something may be a loose connection, corrosion, or a frayed conductor. Let's assume that this defective part of the circuit has a resistance of 108 ohms, as shown in Figure 9–5C. This new actual circuit can be analyzed in the following way:

$$R_T = R_1 + R_2 = 108 \text{ } \Omega + 12 \text{ } \Omega$$
$$= \textbf{120 } \Omega$$
$$I_T = \frac{V_T}{R_T} = \frac{6 \text{ V}}{120 \text{ } \Omega} = \textbf{0.05 A}$$
$$V_1 = I_T R_1$$
$$= 0.05 \text{ A} \times 108 \text{ } \Omega = \textbf{5.4 V}$$
$$V_2 = I_T R_2 = 0.05 \text{ A} \times 12 \text{ } \Omega$$
$$= \textbf{0.6 V}$$
$$P_2 = V_2 I_T = 0.6 \text{ V} \times 0.05 \text{ A} = \textbf{0.03 W}$$

As we see, this defect causes the current to be drastically reduced, and most of the voltage appears across the defect. Consequently, the lamp power will be so small that the lamp will probably not even glow. Essentially, the lamp will be off.

Here are several important points to remember about open circuits:

1. A perfect open circuit has infinite opposition, or resistance, causing no current to flow.

2. The absolute resistance value of a series defect is not as important as its size in relation to the remainder of the circuit. A 100-ohm defect will not be noticeable in series with 1,000,000 ohms, but it will be totally disruptive in series with 1 ohm.

3. A circuit is essentially open when only a small fraction of the intended current is flowing.

4. Since the voltage drops in a series circuit are in proportion to each resistance, essentially all of the applied voltage is dropped across the open part of the circuit.

An open in a circuit can be found with a voltmeter. Just measure the voltage across all the components in the circuit, except the source. The component or connection with all the voltage across it is open. This result can be demonstrated by reassembling the series circuit we used earlier in the chapter. If we break the circuit anywhere to measure the current and then measure the voltage across the break, it will be 4.5 volts.

For example, suppose R_1 in Figure 9–1 is open. Then, the 4.5 volts from the power supply will appear across the two ends of R_1. Connecting a voltmeter across R_1 will be the same as connecting it across the power supply. Therefore,

$$V_1 = V_T = \textbf{4.5 V}$$

The voltages across R_2 and R_3 will both be zero:

$$V_2 = I_T \times R_2 = 0 \text{ A} \times 100 \ \Omega = \textbf{0 V}$$
$$V_3 = I_T \times R_3 = 0 \text{ A} \times 470 \ \Omega = \textbf{0 V}$$

Check:

$$V_T = V_1 + V_2 + V_3$$
$$= 4.5 \text{ V} + 0 \text{ V} + 0 \text{ V} = \textbf{4.5 V}$$

If R_3 is open, the values are

$$V_1 = 0 \text{ V} \qquad V_2 = 0 \text{ V} \qquad V_3 = 4.5 \text{ V}$$

Locating Short Circuits

An open circuit was described as a relatively high resistance unintentionally connected in series with components in a circuit. An open will just about stop the flow of current. A short circuit is described in the opposite way. A **short circuit** is an alternate path of relatively low resistance that increases and bypasses the flow of current. A short circuit often increases current in a place where it is not wanted.

As shown in Figure 9–6, if two wires from a lamp cord touch each other, the current will bypass the lamp by taking the shorter and eas-

FIGURE 9-6
SHORT CIRCUIT

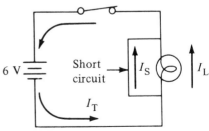

I_S = short-circuit current
I_L = lamp current

FIGURE 9-7
LOW RESISTANCE CAUSING SHORT

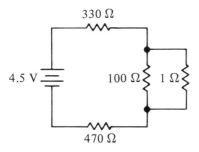

ier path and go directly back to the outlet. The voltage at the wall outlet will not see the resistance of the lamp. It will see the much lower resistance of the two touching wires. As a result of the lower resistance, the current will be greater than normal, and the fuse will interrupt the circuit.

If we connect a wire across the 100-ohm resistor in the circuit of Figure 9–1, the resistor is essentially eliminated. Why should electrons struggle through a 100-ohm resistor when they can easily find their way through an alternate path — especially when the alternate is a conductor of close to 0 ohms resistance? Thus, the wire creates a short circuit, an alternate path of relatively low resistance that increases and bypasses the normal flow of current.

Consider again the series circuit in Figure 9–1. Figure 9–7 shows what the circuit looks like if a 1-ohm wire is placed across the 100-ohm resistor. The current will take the path of least resistance — the 1-ohm wire — and ignore the 100-ohm resistor. The *before* calculations have already been done. The following *after* calculations show quite a change:

$$R_T = R_1 + R_2 + R_3$$
$$= 330 \ \Omega + 1 \ \Omega + 470 \ \Omega = \textbf{801} \ \Omega$$

$$I_T = \frac{V_T}{R_T} = \frac{4.5 \ V}{801 \ \Omega} = \textbf{0.005618 A}$$

$$V_1 = I_T R_1 = 0.005618 \ A \times 330 \ \Omega$$
$$= \textbf{1.854 V}$$

$$V_2 = I_T R_2 = 0.005618 \ A \times 1 \ \Omega$$
$$= \textbf{0.005 V}$$

$$V_3 = I_T R_3 = 0.005618 \ A \times 470 \ \Omega$$
$$= \textbf{2.641 V}$$

The calculations show that the short has caused an increase in current and a redistribution of the voltage drops. Notice that the voltage V_2 across the shorted component is essentially zero.

As another example, if R_1 in Figure 9–1 is shorted and the other resistors are normal, the calculations give the following results:

$$R_T = R_1 + R_2 + R_3$$
$$= 0 \ \Omega + 100 \ \Omega + 470 \ \Omega = \textbf{570} \ \Omega$$

$$I_T = \frac{V_T}{R_T} = \frac{4.5 \ V}{570 \ \Omega} = \textbf{0.00789 A}$$

$$V_1 = I_T R_1 = 0.00789 \ A \times 0 \ \Omega$$
$$= \textbf{0 V}$$

$$V_2 = I_T R_2 = 0.00789 \ A \times 100 \ \Omega$$
$$= \textbf{0.789 V}$$

$$V_3 = I_T R_3 = 0.00789 \ A \times 470 \ \Omega$$
$$= \textbf{3.708 V}$$

Here are several important points to remember about short circuits:

1. A perfect short circuit has zero resistance and no voltage drop.

2. The absolute resistance value of a shorting defect is not as important as its size in relation to that of the component it is across. Placing 1,000,000 ohms across 100 ohms will cause no noticeable effect. Placing 100 ohms across 1,000,000 ohms will

reduce the net value to about 100 ohms and will totally change the circuit.

3. A circuit is essentially shorted when most of the current takes an unintended alternate route.

4. Since the voltage drops in a series circuit are in proportion to their resistance, essentially zero voltage is dropped across a short.

Because of this last point, shorts should be easy to find. If current is flowing in a circuit, any component with a zero voltage drop is probably shorted. Some components are supposed to be shorts. For example, closed switches and good connections should appear as shorts. Similarly, open switches and poor connections should look like opens. So, shorts and opens can be found with voltage tests.

There is one more complication that must be mentioned. Short circuits tend to cause current increases, which, in turn, cause fuses to open or components to heat up and open. Thus, shorts can cause opens. If you fix the open without fixing the short, the open will reoccur. Therefore, before you replace a defective component, determine and repair the cause of its defect.

Locating Other Problems

Shorts and opens are the easiest defects to locate because their symptoms are so obvious. A component in a series circuit should have a voltage in proportion to its resistance. If the voltage across it is zero and the other components share the drops, the component is probably shorted. Likewise, if all the applied voltage is dropped across one component and none is dropped across the others, the component is probably open. These guesses can and should be verified by resistance measurements.

There is, however, another type of prob-lem. A resistor can have a disproportionate voltage drop — more or less than it should have. Two equal resistors in series should share the applied voltage, so each resistor should drop half of the total. But if one has 70% and the other has 30%, something is wrong. Either the one with 70% has more resistance than specified, or the one with 30% has less resistance than specified. In this situation, voltage measurement has made you aware of the problem. Resistance measurement will locate the problem for you.

This discussion leads to two points about drawing conclusions. The first point is that components are not simply short, open, or good; they can also change value and be out of tolerance. The second point is that defects in one component produce symptoms in others. The two resistors in series just described present a good example. If you conclude that a resistor with too little voltage across it is defective, you could be wrong. A resistor in series with it may have increased in value, and you may be blaming the wrong component.

SUMMARY

Circuits are tested to determine whether they are working according to specification or to determine why they are not working as they should. Testing consists of using test instruments to measure circuit parameters. An analysis of the data obtained helps in determining the cause of problems. Fixing defects is not the difficult part of a job; finding the cause is the difficult part.

Test procedures are based on two concerns. Safety is always the first concern — personal safety, product safety, and test equipment safety. The second concern is the quality of your data or conclusions. You should be able to guarantee the accuracy of your results.

Most technicians keep records. Records of test results are kept as the tests are conducted. When the tests are completed, reports are prepared from these records.

The actual measurement of circuit parameters was described in Chapter 7. Voltage is measured across a component, and current is measured through a component. Resistance is measured by de-energizing a circuit and disconnecting one lead of the component under test.

Test and measurement results are compared with the circuit specifications in order to determine whether the circuit is working properly. (If circuit specifications are not available, you must calculate them yourself.) When everything in a circuit is working properly, specified values and measured values will be approximately the same. Results far out of the range of the specifications show obvious problems. Results that are only slightly off are usu-

ally acceptable. However, you need a good understanding of a circuit before you can decide the range of acceptable variation.

Troubleshooting is the process of circuit analysis in which symptoms are used to find causes. Shorts and opens are the most frequent causes of circuit problems. An open interrupts the flow of current, and an open component has the total applied voltage across it. A short increases the flow of current and may cause an open. Shorted components have zero voltage across them.

Another frequent cause of circuit problems is a component that has changed value. Defects like this are more difficult to find because they are not as obvious as shorts or opens. The symptoms of these defects appear in other parts of a circuit and can lead the technician astray. Therefore, be sure to learn the cause of the problem before changing components.

CHAPTER

REVIEW TERMS

breadboarding: assembling experimental circuits so that they can be easily modified

measurement: determination of magnitudes that define a parameter

open circuit: break in a circuit or an unintended high series resistance that interrupts the flow of current

relative value: value of a circuit component in relation to the values of those around it

short circuit: unintended alternate path of relatively low resistance that increases and bypasses the flow of current

symptoms: behavior characteristics of a circuit resulting from a malfunction

testing: use of instrumentation to evaluate components; the process of observing symptoms

troubleshooting: process of measurement and analysis to determine a malfunction's cause

REVIEW QUESTIONS

1. Define testing.

2. Define measurement.

3. Define troubleshooting.

4. Define short circuit.

5. Define open circuit.

6. Describe how to measure voltage in a series circuit.

7. Describe how to measure current in a series circuit.

8. Describe how to measure resistance in a series circuit.

9. Describe the effects of a short in a series circuit.

10. Describe the effects of an open in a series circuit.

11. Explain why records must be kept, and describe the type of records that are kept.

12. Name and describe two reasons for conducting tests.

13. What is meant by the terms *symptoms* and *cause?*

14. Describe the procedures used to locate possible shorts in a series circuit.

15. Describe the procedures used to locate possible opens in a series circuit.

REVIEW PROBLEMS

1. Is the circuit shown in Figure 9–8 working properly? If not, what is wrong with it?

2. Is the circuit in Figure 9–8 working properly if V_1 = 0 volts, V_2 = 12 volts, and I_T = 12 milliamperes? If not, what is wrong?

3. Is the circuit in Figure 9–8 working properly if V_1 = 0 volts, V_2 = 12 volts, and I_T = 0 milliamperes? If not, what is wrong?

4. What is wrong with the circuit in Figure 8–7B if V_1 = 12.1 volts, V_2 = 15.2 volts, V_3 = 9.9 volts, and V_4 = 1.8 volts?

FIGURE 9-8

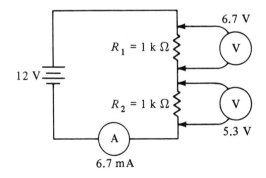

5. What is wrong with the circuit in Figure 8–7B if $V_1 = 0$ volts, $V_2 = 0$ volts, $V_3 = 44$ volts, and $V_4 = 0$ volts?

6. What is wrong with the circuit in Figure 8–7B if $V_1 = 0$ volts, $V_2 = 24.9$ volts, $V_3 = 16.2$ volts, and $V_4 = 2.9$ volts?

7. What causes the lamp in Figure 9–5A to be off when the voltage across it is 6 volts?

8. What causes the lamp in Figure 9–5A to be off when the battery measures 6 volts and the lamp measures 12 ohms?

9. What minimum and maximum current do you consider acceptable for the circuit of Figure 8–7B if the resistors all have 10% tolerance? 5%?

10. What minimum and maximum current do you consider acceptable for the circuit of Figure 8–7B if the power supply has an output variation of no more than 2%? 1%?

11. What is the maximum current you will accept under the combined effects described in Problems 9 and 10?

12. What is the minimum current you will accept under the combined effects described in Problems 9 and 10?

TOOLS AND HARDWARE

Identify common hand tools used in electronics.

Describe the proper use of common hand tools.

Identify common hardware and describe its use.

Identify common lamps, connectors, and similar items used in electronic devices.

TOOLS

The objective of this chapter is to introduce many of the hand tools and assembly hardware you will be working with as a technician. Your quality as a technician will depend partially on your knowledge of hand tools. You should know their names and use the correct tool for a given task. You also must take proper care of your tools. We begin this section, therefore, with a look at some of the common hand tools used in electronics.

Common Hand Tools

Screwdrivers come in many different sizes and shapes, as shown in Figure 10–1. The choice provides you with a best match of screw slot to blade width as well as shaft length to available space. The most common blade tip is the flat blade, shown in the center in Figure 10–1. The screwdriver at the top of the figure is a Phillips, recognized by its crossed-shaped head. A holding type of screwdriver is shown at the bottom of the figure. It holds screws with a split blade that expands when the lower part of the plastic handle is moved toward the screw. Screwdriver handles are usually plastic, to provide insulation and durability.

Pliers are used for holding small parts and cutting wires. They are made from metal and often have insulating handles, which provide a comfortable grip. This feature is especially important for individuals who work all day with them in an assembly job. Many different jaw sizes and shapes are available to suit the conditions of the task. For example, tools like the upper one in Figure 10–2 are used for cutting small wires and component leads and are called diagonal pliers. Tools like the lower one in the figure are used for holding small parts in close spaces and are called long-nose pliers.

Needle-nose pliers have even thinner

FIGURE 10-1
ASSORTED SCREWDRIVERS

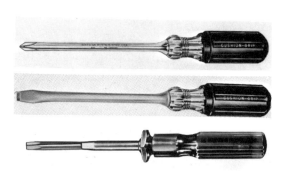

FIGURE 10-2
TWO TYPES OF PLIERS

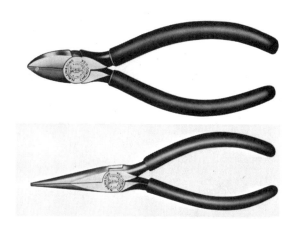

blades. They are intended for use in holding items in places where your fingers will not fit or where safety dictates that your fingers should not go. Needle-nose pliers are not intended for twisting or bending anything bigger than a wire. Use heavier pliers for heavier tasks.

Nut drivers are popular for installing and

removing hex-head screws and nuts. They are available in a variety of sizes from less than ⅛ inch to greater than ¾ inch and are used like screwdrivers. Figure 10–3 shows an assortment of nut drivers.

Wire strippers, shown in Figure 10–4, help in the removal of insulation, although this task is usually done by machine in high-production areas. Soldering aids are used for wrapping and unwrapping leads. **Heat sinks** are clamp devices used to prevent the heat caused by soldering from traveling up the lead and damaging the component being soldered.

Soldering can be done with either irons or guns. Soldering guns heat up quickly and are convenient for service technicians on the road when they need to make solder connections. Soldering irons are generally used by individuals who solder continually during the day. Different power ratings, tip sizes, and tip shapes are available for a wide range of soldering situations. The iron in Figure 10–5 has a small tip, making it convenient for work on PC boards. Using an iron much larger than required can cause the radiation of excessive heat and damage to adjacent components.

The use and care of soldering tools is discussed in the next chapter. However, here we can mention one tool used in soldering and assembly that requires the utmost care — your eyes.

Safety tip: Wear safety glasses whenever you are performing hazardous tasks or are near someone performing hazardous tasks.

Cutting, drilling, and soldering are all hazardous tasks.

Tool Kits

Many manufacturers sell tools in groups, or kits. The tool kit provides the field service

FIGURE 10-3
NUT DRIVERS

FIGURE 10-4
WIRE STRIPPERS

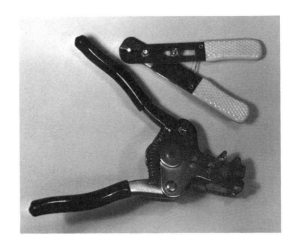

FIGURE 10-5
MICROTIP SOLDERING IRON

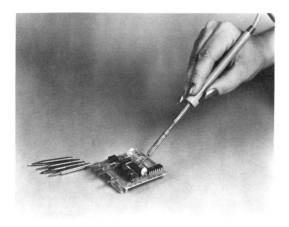

FIGURE 10-6
ALIGNMENT KIT

sions. They are generally made of nonmagnetic materials and have shafts at both ends of the handle. Alignment tools are available in a wide range of ends, which match a wide range of brands and models of equipment. Many different alignment tool kits are offered by manufacturers.

Use and Care of Tools

A review of a tool catalog will provide you with more than a knowledge of tool names. It will also make you aware of their cost. Good tools are expensive. Inexpensive tools are not worth buying because they do not work well or for very long. So, you should buy good tools and take care of them.

Screwdrivers are used for turning screws and not for punching holes or opening cans. Always use the correct type and size of screwdriver for a given task. The wrong type of screwdriver can damage the screw slot, making the screw difficult or impossible to remove with any tool. The wrong-size screwdriver may make the task more difficult, and screwdrivers that are too small may break. The insulated handle on a screwdriver will give you some protection if you must work on live circuits — a practice to be avoided if at all possible.

Wire cutters are used for cutting wires of a specific size. If you cut wires larger than the intended size, or if you cut or pull nails with your cutters, you will end up with a pair of useless cutters. Also, you will damage the cutting edges by rough use. The insulated handles on cutters provide comfort and some safety. However, never cut wires in a live circuit.

technician with a convenient way to carry tools. The kit also organizes and protects tools. One popular tool kit includes a variety of nut drivers, as shown in Figure 10–3. Sets of Allen wrenches are also available.

Alignment tools, shown in Figure 10–6, are used in the servicing of radios and televi-

HARDWARE

Thousands of little parts are used to hold an electronic product together. For instance, there are screws, which come in many lengths, diameters, materials, and head shapes. There are nuts and washers, which also have many shapes and sizes. And there are many other fasteners, such as wire clamps, which come in various sizes and shapes. This section will provide no more than an introduction to hardware because the category is so large. You will discover new and additional hardware items throughout your career.

Screws, Nuts, and Washers

Many items are held together with screws, and there are many types of screws you should recognize. Plated steel screws are the type you will work with most often, although there will be times when you will need nylon screws. You may also work with wood screws on occasion. Most of the time, though, you will be using sheet metal or machine screws. **Sheet metal screws** are designed to thread their way through metal and hold the parts without the use of a nut. A **machine screw** requires a pre-threaded hole or a nut.

Screws are also identified by their length, thickness, and number of threads per inch. For example, a ½-inch, 4–40 screw is ½ inch long, has a #4 diameter, and has 40 threads per inch. A ¾-inch, 12–20 screw is longer and thicker, and it has only 20 threads per inch. The most common screw sizes in electronic products are 4–40, 4–36, 6–32, 8–32, and 10–32.

Another way to describe a screw is by the shape of the opening on the head. The most common screws are single-slotted. Phillips screws, the next most common type, have a cross slot. Tamperproof screws are used when removal is undesired. Figure 10–7 shows various screws, nuts, and washers used in electronics assembly.

Screws may also be described by the shape of the screw head. Round-head and binder-head screws are the most common types. Flat-head and oval-head screws require countersunk holes because of their shape.

FIGURE 10-7
ASSORTED ELECTRONICS HARDWARE

Most nuts are made of plated steel and are hexagonal or square. Acorn nuts are used as an attractive way to cover the exposed end of a screw. Knurled and wing nuts are used when easy hand removal is needed.

Washers are used for a variety of purposes. Flat washers expand the contact surface area of screws and nuts and reduce scratching. Lock washers reduce the likelihood of a screw or nut turning and becoming loose. Most washers are made of steel, and their size is specified by the screw's diameter. Washers can also have internal or external teeth.

Extruded washers are made of an insulating material and are used when a screw is not supposed to touch the surface it is attached to. Care must be taken when working with extruded washers, for they are easy to forget. Reassembling a circuit without them will most likely lead to a short circuit.

Other Fasteners

Many other fasteners are used in electronic products. For example, plastic clamps are used to safely secure wires and cables on panels. Insulating standoffs are used to support and isolate high-voltage circuitry. Rubber grommets are used to protect wires as they pass through metal panels. Strain relief clamps prevent wires from being pulled from a chassis.

One good way to appreciate the extent of hardware variety is to look inside an oscilloscope. Have your instructor open one for you. Notice how the wires go from one place to another. Look at how the panels are mounted, and see what holds the circuit boards. Notice how the cathode ray tube is supported, and determine the kinds of screws that are used.

An inquisitive look inside an oscilloscope will introduce you to the use of electronics hardware, as will a look in a hardware catalog. When you are working as a technician, you will also be introduced to the specific hardware your employer uses.

CONNECTIONS

Screws and fasteners make mechanical connections. Soldering makes items secure. But it is often desirable to have connections that can be easily separated, as in television power cords and antennas, for example. Thus, electric connections use sockets, disconnects, terminal strips, and the like. Sockets are found in lamps and vacuum tubes. Disconnects are used with computer cables. Terminal strips are used in wired circuits. The quality required, the number of conductors, and the environment are some of the factors that determine the connector used in a given situation.

Sockets

Electric power is made accessible throughout a house with the use of wall sockets. Radios can be moved from one location to another quickly and safely by unplugging the power cord from the wall socket. Sockets have been used for tubes in radios and televisions be-

cause they allow easy replacement of the tubes. Desoldering is not necessary; the old tube can simply be pulled out and its replacement plugged in. While transistors do not have the failure rate that tubes have, sockets are sometimes used for them, also. Transistors can be easily damaged by the heat of a soldering iron, so a socket can be soldered in first. When the socket cools, the transistor can be plugged into it.

Many other components such as fuses and relays also use sockets. Sockets are used for two reasons. The first reason is ease of component replacement, since these components do not last very long. The second reason is to avoid possible damage during assembly. A socket can take the heat and other abuse of assembly, and the more sensitive part can be plugged in as a last step.

Integrated circuits also use sockets for assembly convenience and safety. These **DIP sockets** (for dual in-line pin) are convenient for mass production situations where circuit boards are assembled by machine and soldered by automatic systems. The integrated circuit is hand-installed by an assembler near the end of the production process. Integrated circuit sockets come in a wide and ever-increasing variety of sizes and forms. Figure 10–8 shows an example of a DIP socket.

Disconnects

The radio plug and wall socket combination in your home is an excellent example of a **disconnect pair.** Unlike the tube and its socket, they are designed to be connected and disconnected often as part of normal use. The cable between a computer printer and the computer connector is another example of a disconnect pair. There are also many examples of disconnects under the instrument panel of a car. The radio antenna, tape player speakers, and instrument lights all use some form of disconnect. Figure 10–9 shows two types of disconnects on the back of an electronic cabinet. The connectors on the left are for single-conductor shielded cables. The one on the right is a multiconductor connector, as indicated by all its pins.

Large electronic systems generally use disconnects in two situations. First, connections between cabinets are made so that they can be plugged or unplugged at the site of installation. All other wiring is inside and is done at the factory. Second, subsystems within a cabinet use disconnects so that the subsystem can be removed and another put in its place. This large-scale replacement process allows a temporary subsystem to be installed while the original is returned to the factory for repair.

Terminal Strips

The use of integrated circuits is slowly reducing the number of terminal strips used. However, they probably will not disappear altogether. **Terminal strips** are series of connection points insulated from each other. They are used to support discrete (single) components, such as resistors and capacitors, and to provide a convenient place for several connections to be made. Wired circuits often use terminal strips as a convenient place for grounds.

Screw-type terminal strips provide a simpler and less expensive disconnect pair than a plug and socket. They are generally used in situations where the characteristics of the connection are not critical. For example, stereo speakers are often connected with terminal strips. The inside leads are soldered and the outside leads are screwed on. Antenna leads for television receivers are also connected in this way. In contrast, transmitters, like CB radios, have critical connection requirements, and thus, they use plugs and sockets designed for the purpose.

FIGURE 10-8
DIP SOCKET

FIGURE 10-9
ASSORTED CONNECTORS

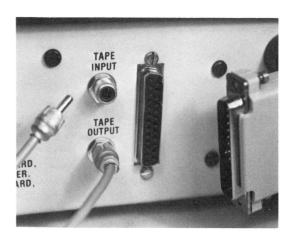

LAMPS

Electronic products use lamps for a variety of reasons. The first and most obvious reason is to allow us to see what we are doing. Dial indicator lamps are an example of this application. Lamps are also used as status indicators. For example, tape players occasionally use lamps to indicate which channel is on. Stereo radios use lamps to indicate that the program is in stereo. Some push-button switches have built-in lamps that indicate the status (on or off) of the switch.

A third and more critical use for lamps is as alarms. Many systems have internal circuitry that detects malfunctions and signals the malfunctions with lamps. Failure of these lamps may allow improper operation to go undetected. Because of this possibility, bells and buzzers are also used as backup alarms.

Most lamps used in electronic products are low-voltage incandescent lamps. Many voltages are available, although 6, 12, and 24 volts are the most common. These lamps may operate with AC or DC voltage. However, AC operation is more common because it is more often available. Some systems use neon lamps to indicate that power is on since these lamps can be operated directly from line voltage with the addition of a current-limiting resistor.

SUMMARY

Your quality as a technician will depend partially on your knowledge of tools — on how you use them and how you care for them. Good tools, though expensive, are necessary for good work. Furthermore, you should always use the right tool for the job.

Diagonal cutters and long-nose pliers are two of a technician's most important tools. Flat-blade, Phillips, and holding screwdrivers are also important tools. Soldering irons and guns are necessary for assembly and disassembly work. Nut drivers and wire strippers make assembly work easier. Good-quality tools with insulated handles are the best choice.

Tool kits are also available. Service kits usually include cutters, pliers, screwdrivers, and nut drivers. Alignment kits have the specialized tools necessary for the different brands and models of radios, televisions, and transceivers.

Hardware is used to hold products together. Sections that need not come apart may be welded, riveted, or molded together. Sections that may need to come apart are held together with screws, nuts, and washers. Screws come in a variety of diameters, threads, lengths, and head shapes, selected according to the specific application. Nuts are chosen to match the screws. Washers may be locking, flat, or insulated. Most fastening hardware is made of plated steel, although nylon fasteners are becoming popular.

While screws and fasteners facilitate the mechanical connections, plugs and sockets facilitate the electric connections. Plugs and sockets are also called connectors or disconnects. Their purpose is to provide a positive electric contact between sets of conductors that will not come apart unintentionally but will come apart when desired. Terminal strips provide a convenient way to support and connect components and wires. Without them, these connections would be electrically sound but not mechanically sound.

Lamps are used in electronic products for several reasons. For example, they provide light to show what is happening, as in dial indicator lamps. They indicate status, as in left- and right-channel displays on tape players. Finally, they function as alarms, as in oil or battery lights in cars.

CHAPTER 10

REVIEW TERMS

DIP socket: dual in-line pin socket used for convenient mounting and connection of integrated circuits

disconnect pair: plug-and-socket combination used for the convenient electrical connection and disconnection of system sections

heat sink: clamp device used to dissipate undesired heat from the leads of heat-sensitive components during soldering

machine screw: uniformly threaded screw used in the assembly of electronic products

needle-nose pliers: pliers with long, thin jaws used for holding parts in inconvenient locations

nut driver: device having various sized hexagonal openings for nuts and a screwdriver type of handle

sheet metal screw: coarsely threaded, tapered screw designed to thread the parts it secures and to hold them without the use of a nut

terminal strip: series of connection points insulated from each other and providing places where components and wires can be joined

wire stripper: tool used for the removal of insulation from wires

REVIEW QUESTIONS

1. Name three types of screwdrivers.

2. What is a nut driver? What is it used for?

3. Why are plastic handles used for hand tools?

4. Name two types of pliers.

5. What are pliers used for?

6. Compare a soldering iron and a soldering gun.

7. What type of wire cutter is most commonly used in electronics?

8. What do 4–40 and 6–32 mean in the description of screws?

9. Name four types of screws, and describe what screws are used for.

10. What are extruded washers, and why are they used?

11. Why are sockets used for components such as integrated circuits?

12. What are disconnects, and what are their advantages?

13. Why are terminal strips used?

CIRCUIT FABRICATION

Identify basic circuit types as wired, printed, integrated, or hybrid.

Describe wire and component preparation.

Describe how to install components in wired and printed circuits.

Recognize acceptable soldered connections.

METHODS OF PRODUCTION

The objective of this chapter is to introduce the common practices used today in the fabrication of electronic products. Basic circuit types are identified, and procedures for installing components in wired and printed circuits are examined.

Types of Circuits

If you look inside an old radio or television, you will see many wires and many components. This assembly is called **hard wiring,** and the circuit is called a **wired circuit.** During manufacture, assemblers of hard wiring sat at a bench and mounted each component and installed each wire, one by one — a time-consuming assembly process. Technology slowly changed the assembly procedure. For example, semiconductors required less space than tubes and led to smaller circuits, which, in turn, made assembly procedures more delicate.

The next step was the introduction of **printed circuits,** copper-clad fiberglass boards on which the circuit image has been transferred and the extra copper etched away. Printed circuits completely changed the assembly process by reducing assembly time, saving space, and leading to a much lower product cost. Consequently, a printed-circuit transistor radio sold for a fraction of the cost of hard-wired tube radios. These changes were just the beginning, though, because each transistor and resistor was still a separate, or discrete, component that had to be installed separately.

Integrated circuits created another revolution in the electronics industry. In **integrated circuits,** sophisticated processes allow layers of certain solid materials to be chemically modified into thousands of microscopic areas that behave in a variety of ways. For instance, they act as diodes, transistors, resistors, capacitors, or conductors. With these processes, complete circuits are fabricated on a chip.

The initial development cost of integrated circuits is very high, but they create great savings in assembly time when a company produces a hundred thousand or so. Integrated circuits also produce great space savings because they are so small. In addition, failure is reduced because the circuit is a solid in one piece. The increase in desk calculator capabilities, along with the decrease in size and cost, is a direct result of the development of integrated circuits.

Technology has not yet reached the state where all circuits can be integrated. High development costs also limit the use of integrated circuits to high-volume products. Therefore, you may find products that have integrated circuits in one area and discrete components in another. Circuits like these are called **hybrid circuits** since they are not all one kind or another. Most televisions, for example, use hybrid circuits. They have tubes, transistors, and integrated circuits. They also have both wired and printed circuits.

Manufacturing Systems

Electronics assemblers of years gone by often did the complete assembly of a product alone. The assembler began by collecting parts on a parts list. A layout diagram indicated where the assembler should drill and punch holes in the chassis. When the metal work was completed, large components and terminal strips were installed. Then, the wiring and other connections were made according to a schematic drawing, pictorial diagram, or run sheet. This process could take hours or even days. Because it is so inefficient, this process is only used when it cannot be avoided.

FIGURE 11-1
SURFACE-MOUNTED COMPONENT
PLACEMENT SYSTEM

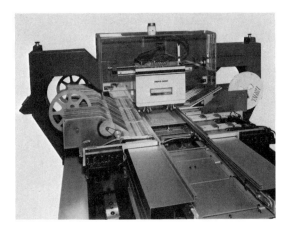

Eventually, specialists were brought into the assembly procedure. One person prepared the chassis, one installed hardware, one installed wiring, and so on. Then, slowly, machines took over. Some machines made the chassis, and other machines produced wiring harnesses. The assembler just wrapped and soldered connections.

The next change came when printed circuits were introduced. Printed circuit boards are produced by printing an image of the circuit on a sheet of copper that has been bonded to an insulator such as fiberglass. The image is printed on an acid-resistant material. When the board is placed in a chemical such as ferric chloride, the copper is etched from the board in all places except those where the image was printed. The remaining copper serves as a conductor, and holes must be drilled on the board where components are to be mounted. Once the design is complete, printed circuits can be rapidly mass-produced at great savings in space, wire, time, and, of course, dollars.

In the next phase of development, machines were used to insert components. This procedure is the one in use today. Computer-operated programmable machines insert components into circuit boards as the boards move along a production line; see Figure 11–1. Operators do not install parts onto boards. Instead, they load rolls of components into the machine, and the machine, in turn, inserts the components into the boards. With an insertion rate of 6000 components per hour, the machine far outperforms any worker.

Soldering follows assembly, and it also is done by machine. Wave-soldering machines can clean, flux, solder, and reclean hundreds of connections on one side of a board in a single brief pass. Again, the operator simply keeps the machine going.

The automated processes just described have eliminated many assembler and solder positions in electronics. However, all companies are not completely automated. Some still have people who do hard wiring, while others still have people who assemble circuit boards. Even those companies that are highly automated need people to finish the assembly by installing integrated circuits into boards and boards into systems. And, of course, technicians are needed to remove and reinstall components, for these tasks are part of the testing, troubleshooting, and repairing process. Therefore, if you intend to become a qualified technician, you will need all the assembly skills.

Specifications

A technician working for a design engineer may get a schematic drawn with pencil on a scrap of paper as instructions for an assembly project. These instructions are usually accompanied by the statement, "Let me know when you have it working." This method is adequate when you work on a one-to-one basis with the person who wants the product and when the finished product does not have to meet critical standards. But in most industrial situations, technicians need more information than a rough sketch of a circuit provides.

Most companies have detailed instructions — called specifications — for all phases of product manufacture. Drawings show how circuit boards are to be etched. Parts lists describe each component, while layout drawings and schematics show where and how each component is to be installed. In other words, there are specified procedures for each phase of assembly. If all steps are followed as outlined, the final product should work. Furthermore, quality control personnel check incoming components to ensure that they meet specifications. They also check subassemblies along the way in order to prevent defects from being passed on to larger assemblies.

Once a product is assembled, final test departments inspect the quality of assembly. This department also performs extensive tests on each product to ensure that it does what is expected. The range of tests is unlimited. Voltage, frequency, power, and many other parameters are likely to be measured and compared with specifications. Then, adjustments and alignments are also made, if necessary. Next, the product is cleaned and packaged for shipment. Finally, the technician signs a report stating that this technician has inspected and tested the product and it meets company standards.

INSTALLATION OF PARTS

Before you begin any installation, make sure that you have plenty of room, that your work space is clean, and that your work surface will prevent damage to the parts you are working with. Padded bench tops are good surfaces to work on. Also, be sure to wear safety glasses.

Further precautions may be necessary to avoid component damage from electrostatic

FIGURE 11-2
METHODS OF STATIC REDUCTION

discharge (ESD), as mentioned in Chapter 3. Figure 11–2 shows some methods taken to eliminate ESD. The worker is grounded through conductive clothes, shoes, chair, floor covering, and wrist strap. The work is performed at a grounded bench, which, in turn, has a grounded conductive cover. Remember that many components can be destroyed by static voltages too small for you to even feel.

Component Markings

Ideal assembly instructions provide pictures that show the exact location and position of each component. However, most instructions are not ideal. Moreover, the technician must

FIGURE 11-3
SEMICONDUCTOR DIODE

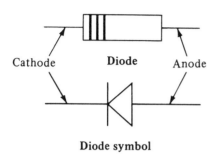

Diode symbol

FIGURE 11-4
TRANSISTOR

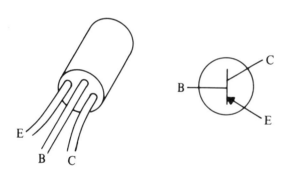

Transistor   Transistor symbol

be able to select each component on a schematic from a box containing dozens of parts. And the part must be properly connected in the circuit.

The first part of the selection process is recognizing the parts as resistors, capacitors, and so forth. The second part of the selection process is choosing the correct resistor, capacitor, or other part. Finding the correct resistor poses little problem. If, for example, R_1 is described as a 1000-ohm, 10%, 1-watt resistor on the parts list or schematic, the correct resistor can be found by consulting the color

code. For other components, though, finding the correct one can be more difficult.

Fixed capacitors come in many shapes and sizes, making their recognition difficult. Some are recognized by the units or the manufacturer's name printed on the side. Others can be identified only through experience. The value and ratings of a capacitor can be indicated in a number of ways. In some cases, the capacitance and voltage rating are printed on the side. In other cases, a color code is used. Some of the common capacitor-marking systems used today are presented in the Appendix. Electrolytic capacitors have a polarity marking also, because they can explode if installed in the wrong direction.

Some small-signal diodes, like the semiconductor diode shown in Figure 11–3, also use color codes for identification. The colors are the same as those used with resistors, and they indicate the identification number. Diodes have a very high resistance in one direction and a very low resistance in the other. Therefore, like capacitors, they must be installed in the proper direction. The end where the color stripes begin is the cathode, as Figure 11–3 shows. On some diodes, the cathode end is indicated by a black stripe.

Transistors are usually identified with a common or a manufacturer's number printed on them. The identification of their leads is critical, and there are many possibilities. In the transistor shown in Figure 11–4, the three leads — emitter (E), base (B), and collector (C) — are located by counting clockwise, beginning at the large space. Transistors will be discussed further in Chapter 29.

Integrated circuit pins are numbered counterclockwise, as seen from the top. As Figure 11–5 shows, the #1 pin is near the notch. Sometimes, a dot over #1 is used instead of the notch. Integrated circuits are discussed again in Chapter 29.

There are, of course, many other components that might be installed in a circuit, but we will not discuss them here. Some, like lamps and fuses, can be installed in either direction without affecting circuit operation. Others, like diodes and capacitors, must be installed in one specific direction. And in still other components, like transformers, the leads have different colors for particular reasons, and the leads cannot be interchanged. So, always make sure that you have the correct component. Then, determine where each lead is to be connected.

Large Components

The first step in the installation of large components is to check to see that the assembly surface is ready. A metal chassis should be machined and all burrs removed. Printed circuit boards should be cleaned and inspected for cracks.

The next step is to install the larger components, such as transformers. Screws, lock washers, and nuts are used to secure them. Rubber grommets should be installed in a chassis in any places where wires pass through.

The next items to be installed are the lamp sockets, other sockets, switches, connectors, and terminal strips. Lights and switches should be tightened from inside the chassis to avoid scratching the front. Terminal strips should be securely fastened since they are generally used for grounds. Some parts such as capacitors or power transistors may need to be insulated from the surface. Mica washers and ceramic standoffs can be used for this purpose. Once again, you must know the specific instructions for each component. When the large parts are installed, they should be neatly aligned, undamaged, and secure.

Small Components

Resistors and similar-shaped capacitors can be prepared for installation in circuit boards by bending the leads, as shown in Figure 11-6, with round-nose pliers. Even bends in the

FIGURE 11-5
INTEGRATED CIRCUIT

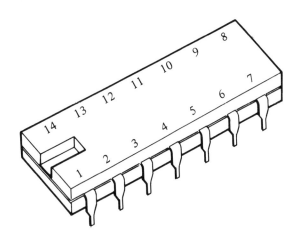

FIGURE 11-6
RESISTOR PREPARATION

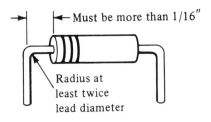

← Must be more than 1/16"

Radius at least twice lead diameter

Prepared

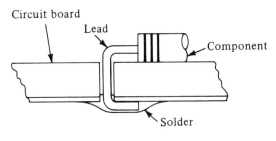

Circuit board

Lead

Component

Solder

Installed

FIGURE 11-7
WIRE-WRAP TOOL

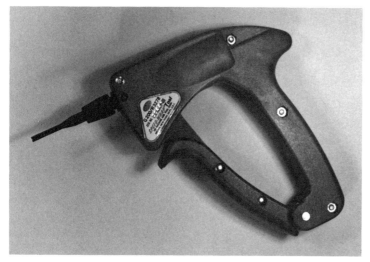
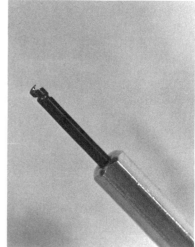

leads reduce strain on the component while providing a uniform appearance to the product. The prepared component can then be pressed into the board, its leads slightly bent, and the excess lead cut off. See Figure 11–6. Components should be secure without solder; do not rely on solder alone to hold them in place.

Semiconductors are a bit more complicated to install than resistors or capacitors. The leads must be left long enough so that there is room for a heat sink to be attached. As noted earlier, some transistors and integrated circuits can be damaged if their leads are touched with the fingers or pliers while being installed. This danger can be avoided by keeping the factory-installed grounding ring on the leads until the component is installed and soldered. Special grounded soldering irons may also be used.

There are many ways to attach a component lead to a terminal. Military products must meet very specific and strict assembly codes; industrial standards may be less specific. There is no one correct way to attach a component lead to a terminal strip. However, it is a good practice to wrap all the connections in the same way to achieve uniformity. Wire-wrap tools like those shown in Figure 11–7 improve this process. And, again, the connection must be secure before it is soldered. If it is not, the connection might physically move during the soldering procedure, resulting in a poorly soldered joint.

Figure 11–8 shows how a wire is connected to a terminal. Three points should be noted:

1. The wire is wrapped securely on the terminal.

2. The insulation approaches but does not touch the terminal.

3. A service loop is provided.

A **service loop** consists of extra wire between the cable and connection, which provides stress relief and some available wire in case disconnection and reconnection are necessary.

WIRE PREPARATION

The first step in preparing wire for connection consists of removing the insulation and ensuring that the conductor is clean. You will find that it takes some practice to develop skill in wire preparation, especially with shielded and multiconductor cables.

Stripping and Tinning Leads

Stripping is the removal of wire insulation. **Tinning** is the application of a thin coat of solder to the wire itself. Tinning keeps stranded wire from unraveling as it is wrapped and makes final soldering a little easier.

Stripping is best done with wire strippers, although some workers do it with a knife. The danger with using a knife is that the wire will be nicked where the insulation is cut. This nick will weaken the wire just at a point where it may be further weakened when wrapped around a terminal. A nick in the wire is not likely to happen if you use wire strippers. But if it does, cut the wire and strip it again.

The amount of insulation you must strip depends on the amount of bare wire you need for the connection. And the wire you need varies from one terminal to another, although $5/16$ inch is a typical amount, as shown in Figure 11–9. Once stripped, the individual strands of a twisted wire should be returned to their original shape. One twist of this wire with the fingers will make it ready for tinning.

Wires can be tinned by dipping the end in a solder pot. Even without a solder pot, tinning wires is quite easy. A soldering iron is

**FIGURE 11-8
WRAPPED CONNECTIONS**

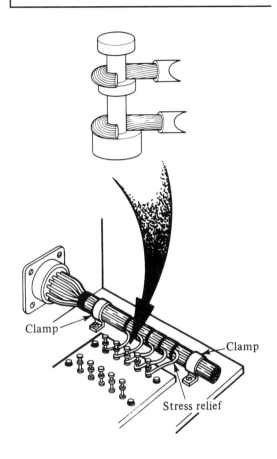

**FIGURE 11-9
PROPERLY STRIPPED WIRE**

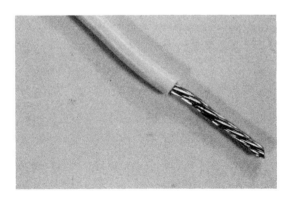

FIGURE 11-10
ATTACHING WIRES TO A CONNECTOR

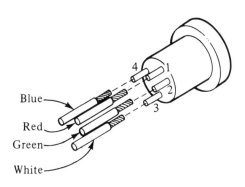

Blue
Red
Green
White

Connections

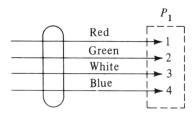

Schematic symbol

placed under the wire, and solder is placed on top of the wire. The iron should not be close to the insulation because the heat may damage it. The iron should heat the wire enough so that the wire melts the solder. Once it does, the solder will flow quickly, and the excess solder and iron can then be removed.

Multiconductor Cables

As more conductors are added, the wiring task becomes more difficult. The first step is to determine how the cable fans out — that is, where each color conductor must be connected. If the wires are to be attached to a round connector, they will probably all be the same length. Flat connectors and terminal strips require wires of different lengths. Most

schematics identify wires in a cable by showing a loop around them. The jack, plug, or terminal strip and specific contact numbers are also noted on the schematic.

Figure 11–10 gives pictorial and schematic diagrams showing how a cable is to be attached to a connector and shows how the connection is actually made. Different colors are used to identify the wires. There may be additional significance to the colors, such as voltage level, type of signal, and so forth. Pin numbers are almost always printed on the connector, as indicated in the figure.

The next step is to cut each conductor to the length necessary for reaching the correct point of attachment. Many companies prefer that technicians leave a service loop, similar to those shown in Figure 11–8, when connectors and switches are attached, in case the item must be replaced in the future. If it must be replaced, then the wires can be cut off right at the terminal, leaving enough wire to strip and tin for a new connection. One attempt at desoldering and reconnecting many conductors from a connector will convince you of the value of service loops.

Some terminal strips use screw-type connections for easy attachment and removal. It is common practice to use soldered or solderless lugs for this connection because they will not break, as a wire might, from constant attachment and removal. A simple hook connection and soldering are all that is necessary for a soldered lug. Solderless lugs, shown in Figure 11–11, are made for many sizes of conductors and are attached with a crimping tool.

Shielded and Coaxial Cables

Shielded cables have one or more conductors wrapped with cloth, copper braid or foil cloth, and an outer insulating cover. The copper braid or foil serves as an electromagnetic shield to protect sensitive circuits. Preparation of shielded cables can be easy or difficult depending on the method of attachment. If the connection is to be made to a terminal strip, you begin by removing about 2 inches of the outer insulation. To do so, you bend the cable and cut into the insulation with the end of the diagonal cutters. If you partially cut into the insulation and work your way around, you can split the outer insulation open. Avoid cutting beyond the outer insulation. This task can be difficult at first, but it will become easier with practice.

Once the necessary amount of outer insulation has been removed, unwind and cut the cloth layer under it. This step brings you to the shield, and it can be handled in two ways. If there is only one conductor inside the shield, open the shield with a soldering aid, pull the conductor through the opening, and flatten the shield. If there is more than one conductor inside, unbraid the shield with a soldering aid and twist it into a thick conductor. Then, terminate it by tinning it, covering it with insulated tubing, and attaching a lug. Once the shield is secure, the conductor or conductors inside can be connected. Like other conductors, they must be stripped, tinned, wrapped, and soldered. These steps complete the connection of the cable.

Attaching connectors presents a challenge because there is less room for variation. All the lengths must be just right. No loose strands should touch adjacent pins. Check to be sure that you have not removed too much insulation. To eliminate problems like these, place small pieces of insulated tubing up each wire, which can then slide down over the pin after the solder has cooled.

Coaxial cable, like single-conductor shielded cable, has a copper shield. But unlike shielded cable, it usually has a large conductor and a large inner insulator. Coaxial cable is generally between ¼ and ½ inch in diameter. The cable is difficult to work with because you can easily nick the center and damage the inner insulation with too much heat. It is also

FIGURE 11-11
SOLDERLESS LUGS

difficult to work with because it is usually attached to connectors and the dimensions are very critical. Coaxial cable connections are more critical than most others because the signals carried are usually quite small. Poor connections make small signals even smaller.

The best way to strip coaxial cable is with coaxial cable strippers. The method of attaching coaxial cable to a connector depends on the connector. In all cases, though, coaxial cables require high-quality soldering without damage from excess heat. Fortunately, there are special tools for cutting and stripping coaxial cable and crimping the connectors. They are recommended when many connectors are to be installed.

SOLDERING

The connections in most electronic circuits are reinforced through a process called **soldering.** The process is usually done with solder, which is made from 60% tin and 40% lead. Other ratios are used for special applications, but 60–

FIGURE 11-12
SOLDERING IRON WITH TIP ASSORTMENT

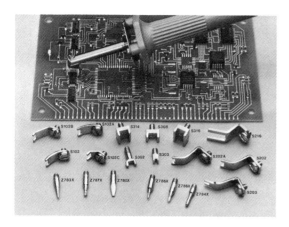

FIGURE 11-13
TEMPERATURE-CONTROLLED SOLDERING STATION

40 is the most common ratio. Solder is melted at a temperature of about 370°F (188°C) with a soldering iron. A paste called rosin **flux** is required to clean the oxide from the surfaces to be soldered — a necessary step if the solder is to adhere. Although rosin flux can be applied with a brush, this step is usually unnecessary since most solder comes as a hollowed-out wire with rosin in it. So most electronics connections are made with #20 or #22 rosin-core solder.

The soldering iron used depends on what is being soldered. A soldering gun is useful if only a few connections have to be made and excess heat is not a problem. Soldering guns reach full temperature in seconds; but because of their power and size, they can possibly damage surrounding components. Since they can also burn you very quickly, you should treat them with respect.

A 25- to 35-watt pencil iron is the most popular soldering iron for work on circuit boards. It provides sufficient heat for most connections, but it is not so large that it can touch adjacent components or damage them with too much heat. For example, a full-size, 35-watt tip, because of its surface area, can often solder what a 240-watt gun cannot. The tip used on the iron depends on the shape of the item to be heated. Most tips are made of copper and are replaceable. Figure 11–12 shows an assortment of soldering tips. The tip on the iron shown in the figure is capable of reaching all the pins of an integrated circuit at one time.

For soldering tasks, two accessories are helpful. Soldering iron holders provide a safe way to keep a hot iron from burning you or items on your workbench. Holders like the one shown in Figure 11–13 have thermostats to keep the iron at a selected temperature. A damp sponge or cloth is the second accessory; it is used to keep the tip clean.

Tinning the iron, a very important practice, means coating the tip with a thin coat of solder. Tinning is done by heating the iron, applying solder, wiping it off with a damp sponge, and adding more solder. The flux cleans the oxide from the tip, and the solder keeps oxide from building up. An unclean tip is almost impossible to solder with because the oxide reduces the transfer of heat. Note, also, that new tips must be tinned before using.

Making Good Connections

Good connections are easy to make. Be sure, though, that the connection is as secure as it can be before you begin to solder it. Do not expect the solder to hold components in place.

Apply the heat to the components being soldered and let them melt the solder. If both items being soldered are not hot enough to melt the solder, they will not hold the solder. To ensure that the components are hot enough, place the solder on one side of the components and place the iron on the other, as shown in Figure 11–14.

When the items being soldered are hot enough, the solder will flow. It will run quickly and spread over the area. Just enough solder to fill the area of contact is all that is necessary. When that amount is obtained, remove the extra solder and the iron. But do not move the connection until it cools. If the connection gets moved as it is curing (cooling), it will become dull, porous, and weak and will have a high resistance. This connection is called a cold solder joint and must be resoldered. A good solder connection will be shiny and smooth, and the solder will appear to have melted the component lead and terminal into one. Figure 11–15 shows good solder connections to turret and socket terminals. Notice the amount of solder used and the distance between insulation and connection.

In summary, making good connections requires the following steps:

FIGURE 11-14
SOLDERING A CONNECTION

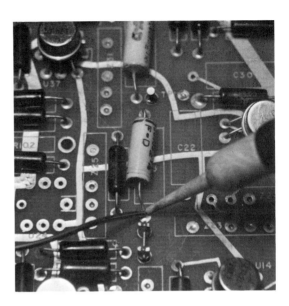

1. Make a good mechanical connection.
2. Attach a heat sink if required. (The use of heat sinks is discussed in Chapter 28.)
3. Use a tinned iron of the proper size.
4. Use rosin-core solder; never use acid-core solder.
5. Apply the iron to one side of the connection and the solder to the other. Make the connection melt the solder.
6. Remove the iron after the solder flows.
7. Keep the connection stationary until it cools.
8. Inspect the connection to ensure that you did not make a cold solder joint.

Desoldering

Soldered components must be desoldered before they can be removed. Any attempt to heat

FIGURE 11-15
GOOD CONNECTIONS

FIGURE 11-16
DESOLDERING TOOL

them and pull them apart will cause solder to splatter on you and the circuit. This practice is harmful and unnecessary, since a part being desoldered may be a good part that can be reinstalled. **Desoldering tools,** like the one pictured in Figure 11–16, are used with a soldering iron or gun for the removal of solder. The iron or gun melts the solder first. Then, the desoldering tool is used to vacuum and remove the melted solder.

Solder can also be removed with thin braid that has been coated with rosin. Its use is quite simple. Place the braid over the location to be desoldered, and place the iron on the braid. The heat will melt the solder and draw it into the braid. Then, cut off the solder-filled braid and repeat the process as required. This technique is called **wicking.**

SUMMARY

Circuits have changed over the years from a hard-wired chassis with tubes to printed circuits with transistors and, finally, to completely sealed integrated circuits. The methods of assembly have also changed. At one time, an assembler may have spent one or two days building a product alone. Then, workers became specialized. Some people installed large components, some installed small ones, and others did the soldering. Today, automatic machinery does much of the work.

Large-scale assembly systems require a organization. Thus, many companies use highly detailed instructions and specifications so that all work fits together.

Before you can install parts, you must recognize them. You must be able to separate the diodes from the capacitors. You must also be able to distinguish, say, capacitor C_1 from capacitor C_3. Specific identification is done by comparing a component's markings with the parts list. Many markings, such as resistor color codes, are standardized. Others, such as transformer wire colors, vary from one manufacturer to another. With experience, you will learn to read these markings and manufacturers' specifications.

Large components must be installed first since they are usually attached with screws and nuts. They generally are part of the structure of the product. Small components are often more delicate and more easily damaged than large components, so their installation is postponed until late in the process.

Wire preparation involves stripping and tinning — that is, uncovering the conductor and coating it with solder so that it can be easily attached to a component. The primary concern is not to cut or nick the stripped conductor, which would make it weak and more likely to break later. Multiconductor cables are difficult to handle because you can easily cut one of the individual insulators as you try to remove the outer insulator. If any of the insulators or conductors are damaged, they must be prepared again. Thus, all the conductors must be stripped again since all must be the same length.

Shielded and coaxial cables are difficult to prepare because of the shield. It must be moved aside by unbraiding it or by pulling the conductor through an opening in the side. Good connections are especially important for circuits that use coaxial cable. These circuits carry small signals, and poor connections make small signals even smaller.

Good solder connections are as easy to make as bad ones. Most technicians use 25- to 35-watt pencil irons and 60–40 rosin-core solder. A good connection requires a firm mechanical connection and enough heat so that the iron heats the connection and the connection melts the solder. After the solder begins to flow through the connection, the iron is removed. The component is left alone until it cools. Excess or unwanted solder can be removed by placing rosin-coated braid on the connection and an iron on the braid. The solder will melt and be drawn into the braid. This process is called desoldering or wicking.

CHAPTER 11

REVIEW TERMS

coaxial cable: single-conductor, shielded cable with a large inner insulator; used in radio frequency (RF) applications such as television antennas and cable TV

desoldering tool: vacuum-operated device used to lift molten solder from a terminal as part of the disconnection process

flux: pastelike mixture in the core of most solder that cleans the connections as the solder flows

hard wiring: circuit connections made with individual pieces of insulated wire

hybrid circuit: circuit made with a combination of hard wiring and printed and/or integrated circuits

integrated circuit: microscopic complex of electronic components and their connections produced on a chip

printed circuit: circuit made on a copper-clad, insulated board by transferring an image onto the board and removing the extra copper by etching; also called printed wiring

service loop: extra wire left at a terminal to facilitate future disconnection and reconnection by cutting and stripping

shielded cable: cable with one or more conductors wrapped with cloth and outer insulation

soldering: process of applying hot solder, which is a mixture of tin and lead, to connections in electronic circuits

stripping: process of removing insulation from a wire

tinning: addition of solder to a stripped wire before connection to ease the soldering process

wicking: process of removing solder by pressing braid against the connection with a hot iron

wired circuit: circuit connected with hard wiring

REVIEW QUESTIONS

1. Describe a printed circuit.

2. Describe an integrated circuit.

3. What is meant by *hard wiring?*

4. What is a service loop?

5. Describe a hybrid circuit.

6. Describe how a printed circuit board is prepared.

7. Describe the fabrication of a printed circuit.

8. What advantages does a printed circuit have over a hard-wired circuit?

9. What advantages does an integrated circuit have over a printed circuit?

10. Compare the installation of diodes with the installation of resistors.

11. What methods are used for identifying diode leads?

12. How are transistor leads identified?

13. Describe the process of preparing a wire for connection.

14. Describe how shielded and coaxial wires are prepared.

15. What must be avoided when you strip insulation from a conductor?

16. What soldering iron is most popular in electronics work? Why?

17. Describe the characteristics of good and bad solder connections.

18. Describe how to properly solder a conductor to a terminal.

19. Describe the desoldering process.

MAGNETISM AND INDUCTANCE

OBJECTIVES

Define magnetism, and describe electromagnetism.

State the characteristics of magnetic lines and magnetic fields.

Describe the left-hand and right-hand rules.

Define inductance, and describe the characteristics of inductors.

Define and calculate the time constant of an inductive circuit.

MAGNETISM

The objective of this chapter is to introduce the basic concepts of magnetism and inductance. The principles of magnetism and electromagnetism will be discussed first.

Magnets are probably familiar to you. **Magnets** are objects that possess magnetism. **Magnetism** is the ability to attract iron and steel. Magnetism is not a recent discovery. Centuries ago, people learned that magnetite, when suspended in space, aligned itself with the poles of the earth. While magnetite is a mineral that can be mined, magnets are generally manufactured from iron, steel, or metallic alloys.

Magnetism has many uses in electronics products. For example, it is the principle on which videotape, speakers, and many other products operate. Human beings are not the only species that use magnetism, though. Migrating birds use magnetic sensors to determine their direction and latitude relative to the earth. It is highly likely that extensive use of magnetism by other species will soon be discovered.

Magnetic Lines

What is fascinating about magnets is that they can sit on a bench looking as if nothing were going on. Yet an invisible field of energy surrounds them. Although it cannot be seen, it can be felt. If you approach a magnet with another magnet, there is a response. An interaction occurs between the magnetic field that surrounds each magnet. This field is created by magnetic lines that come out of one end of a magnet and enter the other.

While the magnetic lines are invisible, their effect can be seen. If you place a bar magnet under a paper and sprinkle some iron filings on the paper, a pattern will appear. Figure 12–1 shows the effect that the invisible magnetic lines have on the iron filings. Iron is easily magnetized, and the small filings respond to the strong attraction by aligning themselves with the magnetic lines. **Magnetic lines** form loops like current loops by leaving the north pole of a magnet and entering the south pole, as shown in Figure 12–2. Magnetic lines bend but never cross, and they follow the path of least opposition.

If a plain iron bar were placed at the end of the original magnet, we would notice a change. The added bar would become magnetized and would serve as an extension of the original magnet. In effect, there would simply be a larger magnet, with still only one north and one south pole. The lines would continue on through the bar because iron presents an easier path for magnetic lines than air.

Another study of magnetic lines can be made by moving, and securing, the bar slightly away from and at the side of the original magnet. Two responses occur. Magnetic lines in the area redirect themselves to follow the path of least opposition, which is through the iron bar. This response, in turn, magnetizes the iron bar with the same polarity as that of the original magnet. The bar becomes polarized south at one end because the lines are entering there, and it becomes north at the other end because the lines are leaving there. Notice that, in the original magnet, lines come out of one end because that is the north end. But in the adjacent iron bar, one end becomes north because magnetic lines come out of it. We also note that, if the iron bar is unsecured, it will quickly move toward the magnet.

In summary, here are the important points to remember about magnetic lines:

1. They form loops like current loops.
2. They leave the north pole of a magnet and enter the south.

FIGURE 12-1
PATTERN OF FILINGS AROUND A MAGNET

FIGURE 12-2
FIELD AROUND A MAGNET

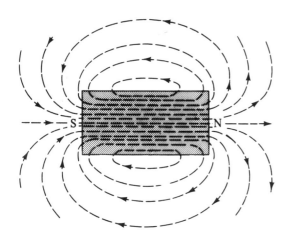

3. They follow the path of least opposition.

4. They bend but never cross.

Terms and Definitions

The original magnet in the preceding experiment is a permanent magnet; the soft iron bar is a temporary one. A **permanent magnet** is one that retains its magnetism after the magnetizing force has been removed. A **temporary**

magnet is one that does not retain its magnetism. The magnetism that remains is called **residual magnetism,** and the ability to retain magnetism is called **retentivity.** Low retentivity and little residual magnetism are characteristics of soft iron, which is used for temporary magnets. Hard steel and alloys containing nickel are often used for permanent magnets.

While magnetic lines pass through all materials, they do so with varying degrees of difficulty. **Permeability** is the ease with which magnetic lines pass through a material. **Reluctance** is the opposition to this passage. Nonmagnetic materials have a low permeability, but the permeability of magnetic materials is high. Table 12–1 compares the permeability of some common materials. The magnetic materials are easy to identify.

High-permeability materials like steel are useful as shields around electromagnetic devices such as transformers. The magnetic field produced by the transformer is absorbed by the steel casing rather than allowed to radiate into the surrounding space.

Undesired magnetic fields can affect the operation of many electronic circuits. Thus, low-permeability materials like copper are used as shields to protect inner conductors from any interference of external magnetic fields. In this case, the magnetic field is not absorbed since the shield is made of nonmagnetic copper. Instead, the magnetic field produces an electric current in the shield, which, in turn, produces an opposing magnetic field. This effect is explained in more detail later in this chapter.

A discussion of how magnets work begins by considering the composition of materials. All materials are made of millions of molecules, each one an independent magnet. Figure 12–3A shows that these molecular magnets are generally arranged in a random manner with their poles pointing in all directions. There is no net magnetic effect in any particular direction.

TABLE 12-1
PERMEABILITY OF COMMON MATERIALS

Material	Permeability
Copper	0.99999
Water	0.9999992
Air	1.00000029
Nickel	1000
Steel	1500
Iron	7000

If a magnet is brought near this material, though, magnetic lines will penetrate it. In low-permeability materials, there is no significant reaction. But in high-permeability materials, all the molecular magnets easily rearrange themselves into rows, all facing in the same direction, as illustrated in Figure 12–3B. This response occurs quite easily in iron but with some difficulty in steel. The result is a new magnet.

When the magnetizing force is removed, steel molecules will remain aligned, while the iron molecules will drift back into random positions like the molecules in a nonmagnetic material. There is another interesting feature to note. Figure 12–3C shows that, if the magnetized steel bar is cut into several pieces, several magnets will be produced.

Fields of Force

As explained earlier, magnetic lines leave the north end of a magnet and enter the south end. These lines can bend but not cross. They establish a **magnetic field of force,** or **magnetic flux,** in the general area surrounding the outside of a magnet. This field was visible in Figure 12–1. Here is an important rule about the magnetic field.

> **Rule:** The strength of a magnetic field depends on the field density — that is, on the

FIGURE 12-3
MOLECULAR MAGNETS

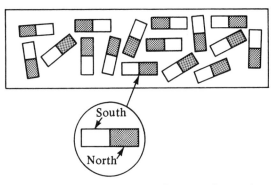

A. Unmagnetized bar: Molecules in random order

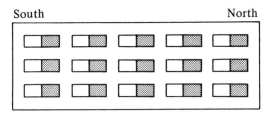

B. Magnetized bar: Molecules aligned

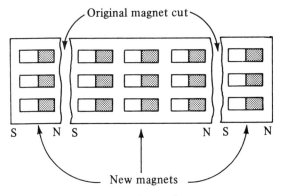

C. Small magnets from a large magnet

number of magnetic lines in a given cross-sectional area.

The number of magnetic lines is expressed in units of **webers** (abbreviated Wb). The den-

FIGURE 12-4
INTERACTION OF MAGNETIC POLES

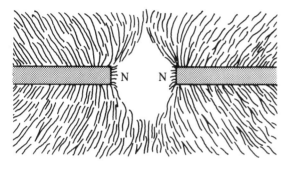

A. Like poles

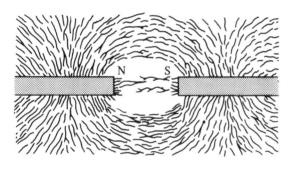

B. Unlike poles

sity of a magnetic field is expressed in units of **teslas** (abbreviated T). Thus, a **flux density** of 1 tesla is equivalent to 1 weber per square meter (abbreviated Wb/m^2).

Recall the law of charges discussed in Chapter 5. The law of charges states that like charges repel and unlike charges attract. Experiments with magnets show a similar situation.

> **Law of magnets:** Like poles repel, and unlike poles attract.

Figure 12–4A shows what happens when two north poles are placed near each other. The lines leaving one north pole do not want

to enter the other north pole; they want to enter a south pole. They also cannot cross, but they do bend. The reaction is similar to what happens when two balloons are pressed against each other. There is a repulsion and a bending of the surfaces but no crossing.

Placing two south poles near each other produces the same effect. However, if a north pole is placed near a south pole, an attraction occurs, as shown in Figure 12–4B. The lines leaving the north pole near the center of the magnet choose to enter the closest south pole, which is the south pole of the other magnet. The closest south pole, in turn, attracts those lines in an attempt to get closer to their source. The strength of this attraction is described by the following rule.

> **Rule:** The strength of magnetic attraction and repulsion is directly proportional to the flux density and inversely proportional to the square of the distance between the surfaces.

ELECTROMAGNETISM

Many electronic components, such as transformers, speakers, and antennas, operate on the principle of electromagnetism. Electromagnets are temporary magnets that get their magnetism from electric current and essentially lose it when the current is removed.

Field around a Conductor

Here is the important point to remember about current and magnetic field.

> **Key point:** Current passing through a conductor produces a magnetic field around that conductor.

This phenomenon, for example, is the basis for the workings of speakers, relays, induc-

FIGURE 12-5
MAGNETIC FIELD AROUND A
CONDUCTOR

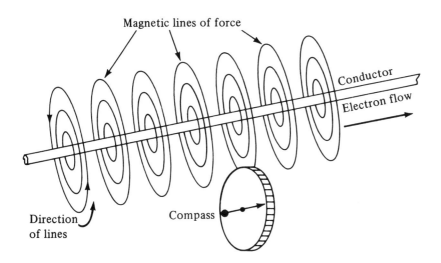

FIGURE 12-5
MAGNETIC FIELD AROUND A
CONDUCTOR

tors, and transformers. If we place a compass beside a conductor carrying a current, as shown in Figure 12–5, we can see some electromagnetic effects. The field around the conductor has a strength and direction determined by the strength and direction of the current. This field causes the compass magnet to align itself so that the magnetic lines enter the south pole and leave the north pole. The strength of the field is described by the following rule.

Rule: The field strength is strongest near the conductor and weaker as we move away from it.

Placing two current-carrying conductors adjacent to one another, as illustrated in Figure 12–6, can produce two reactions. If the currents are in opposite directions, the fields will oppose each other, forcing the conductors apart. When the currents are in the same direction, their fields will add, drawing the conductors closer together.

A coil can be made by taking one long

FIGURE 12-6
FIELDS AROUND ADJACENT CONDUCTORS

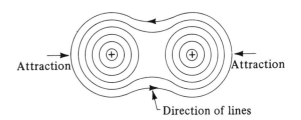

Currents in same direction

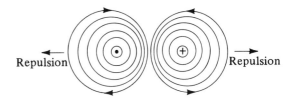

Currents in opposite direction

⊙ Current toward observer
⊕ Current away from observer

FIGURE 12-7
FIELD AROUND A COIL

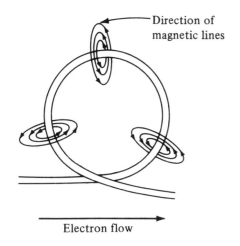

Direction of magnetic lines

Electron flow

A. Single turn

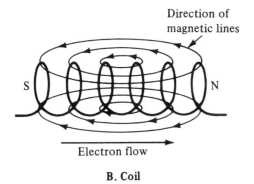

Direction of magnetic lines

S   N

Electron flow

B. Coil

conductor and wrapping it around a cylinder. Since the current in all adjacent conductors is going in the same direction, the fields will add. Figure 12–7A shows the magnetic field established around each individual turn. Figure 12–7B shows how the individual fields of adjacent turns add to produce one large overall field. The direction of these fields can be determined with two left-hand rules, as we will see next.

Left-Hand Rules

There are simple methods, called left-hand rules, to determine and remember the direction of magnetic fields in conductors and coils. Let's discuss the direction of the field around a single conductor first. We consider current as the movement of electrons from negative to positive. To determine the direction of the field, we use our left hand, as shown in Figure 12–8A. Here is the rule.

> **Left-hand rule for conductors:** Hold a conductor with your left hand so that your thumb points in the direction of current flow. Your fingers will point in the direction in which the magnetic lines are moving.

This approach also works for coiled conductors, as illustrated in Figure 12–8B. Hold any part of this coil with your left hand so that your thumb points in the direction of current flow, from bottom to top along the closer side of the coil. Your fingers point in the direction in which the magnetic field is moving, from left to right inside the coil.

The left-hand rule for coils is illustrated in Figure 12–8C. Here is the rule.

> **Left-hand rule for coils:** Hold the coil so that your fingers are pointing in the direction of current flow. That is, hold the coil so that your fingers are wrapped around the coil and pointing up in front and down in back of the coil. Your thumb then points toward the right, showing the direction in which the lines are leaving the coil.

Electromagnets

The coil just described has magnetic lines entering the left side and leaving on the right as current passes through its conductor. Since this coil has a magnetic field around it caused by an electric current, it is called an electromagnet. An **electromagnet** is a magnet whose field is caused by an electric current. **Electromagnetism** is the name given to the temporary magnetic field caused by the current.

Because of the current direction, the left side of the electromagnet in Figure 12–8C is the south pole and the right side is the north. If the battery connections were reversed, the current direction would reverse, causing the magnetic polarity to change also. Furthermore, an increase in current will cause an increase in the field strength. Increasing the number of turns in the coil will also cause an increase in the field strength. Thus, we have the following key point about the strength of the field.

> **Key point:** Field strength increases as the current is increased and as the number of turns is increased.

However, increasing the number of turns eventually reaches a point of diminishing returns. As the number of turns increases, so does the resistance, which, in turn, reduces the current.

The relationship between current and the resulting magnetic force can be determined in two steps. The first step is to calculate the magnetomotive force (F_M) produced. This force is based upon the number of turns (n) in the coil and the amount of current (I). Consider a coil with a current of 5 milliamperes (abbreviated mA) passing through 2000 turns. The force is

$$F_M = nI = 2000 \text{ turns} \times 0.005 \text{ A}$$

$$= \textbf{10 A-turns}$$

The second step is to calculate the magnetic field strength (H). This strength is directly proportional to the magnetomotive

FIGURE 12-8
LEFT-HAND RULES

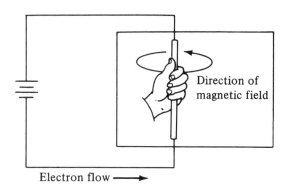

Direction of magnetic field

Electron flow ⟶

A. Left-hand rule for conductors

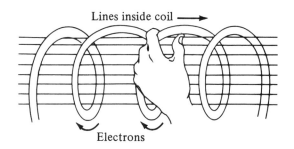

Lines inside coil ⟶

Electrons

B. Left-hand rule for coiled conductors

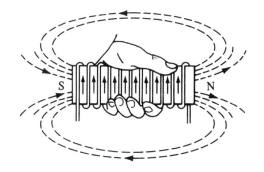

S   N

C. Left-hand rule for coils

force and inversely proportional to the length of the coil (l). Consider, again, the 2000-turn coil; its length is 20 millimeters (mm). The magnetic field strength is

$$H = \frac{F_M}{l} = \frac{10 \text{ A-turns}}{0.02 \text{ m}} = \textbf{500 A-turns/m}$$

Increasing the number of turns from 2000 to 4000 produces the following change:

$$F_M = nI = 4000 \text{ turns} \times 0.005 \text{ A}$$

$$= \textbf{20 A-turns}$$

$$H = \frac{F_M}{l} = \frac{20 \text{ A-turns}}{0.02 \text{ m}}$$

$$= \textbf{1000 A-turns/m}$$

Increasing the length of the original 2000-turn coil from 0.02 to 0.04 meter has the following effect:

$$H = \frac{F_M}{l} = \frac{10 \text{ A-turns}}{0.40 \text{ m}} = \textbf{250 A-turns/m}$$

The preceding calculations show that the strength of an electromagnet is proportional to the current, the number of turns, and the density of the coil.

Another factor affecting the field strength of an electromagnet is the quality of the core. Since flux density is a factor of permeability, and since iron has a much higher permeability than air, iron-core coils have a much stronger field than air-core coils. Iron also has an advantage over metals such as steel: low retentivity. Although we want strong magnetism present while the current is flowing, we also want the magnetism to leave when the current stops. While iron has some residual magnetism, it is far less than that of steel. For zero residual magnetism, we use an air core.

Electromagnets are sometimes used on cranes for moving scrap iron and steel. Magnetic attraction eliminates the need for jaws that would grab and hold the metal. However, there are some limitations in this application. Nonmagnetic materials such as copper and aluminum separate since they are not attracted, and hence, they must be moved separately. Later chapters will describe how meter movements and stereo speakers utilize the interaction between permanent magnets and electromagnets as a basis for operation.

Buzzers and Relays

Door buzzers use an electromagnet and switch circuit for their operation. The usual circuit has a voltage source, a normally open, push-button switch, and a buzzer. Most buzzers consist of an electromagnet in series with a normally closed switch. The total circuit is shown in Figure 12–9.

The buzzer works in this way: When the button is pressed, current flows through the coil, causing it to become an electromagnet. This electromagnet attracts the steel arma-

FIGURE 12-9
ELECTROMAGNETIC BUZZER

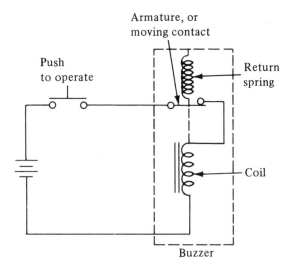

ture, which, in turn, opens the attached switch. With the switch open, the current stops, the field collapses, and the armature returns to its original position. Once the armature returns, the switch closes and the current flows again, causing the cycle to be repeated. This back-and-forth motion occurs rapidly, resulting in a buzzing sound from the armature.

A relay, as shown in Figure 12–10, is a switch that is operated electromagnetically rather than manually. It allows a device to be operated from a distance. For example, suppose we have a high-voltage device that we want to operate from another part of a building. We can run two high-voltage wires from the device to the desired location of the switch. But in this case, there is a problem about the safety of these wires. As an alternative, we can place a relay at the location of the device and run wires for operating the relay from the device to the desired location of the switch. In this situation, the wires around the building are not high-voltage circuit wires but, rather, low-voltage, relay-operating wires.

The relay acts like a buzzer in that coil current produces a field that moves an armature toward the coil. However, in the relay, the circuit does not break, and the armature stays at the coil as long as the energizing voltage is applied. In addition, the relay switch contacts are mechanically attached to the armature but are electrically insulated from it and from each other. So, while the contacts can switch high voltages and high currents, the command wires that travel to a remote-control area carry only the small voltage and current necessary to energize the relay cell.

The relay just described is a single-pole–double-throw (SPDT) model. But many variations are possible, with several contact arrangements and capacities available. For example, the starter solenoid in a car is a relay switch. Under the hood, a large battery cable connects the solenoid to the battery. The starter switch carries the small current necessary to operate the solenoid. When the solenoid operates, it sends the large current to the starter. Without a solenoid, the large-diameter cable would have to be connected to the starter switch, which, in turn, would have to be able to switch the large starter current. Instead, a relay does it. The process by which a small current controls a larger one is called **amplification.** In Chapter 29, we will consider how transistors do this task.

FIGURE 12-10
ELECTROMAGNETIC RELAY

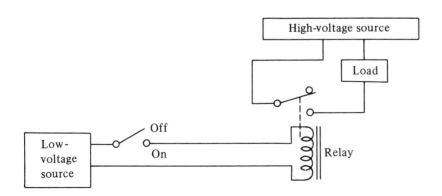

FIGURE 12-11
CURRENT INDUCED BY A MOVING
MAGNETIC FIELD

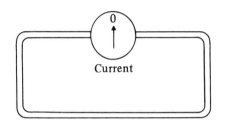

A. No current

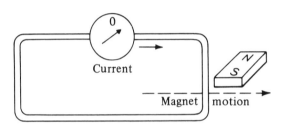

B. Induced current

INDUCTANCE

Photos and symbols of inductors and transformers were presented in Chapter 4. An introduction to their theory of operation is given here. Both components operate on a principle called inductance, which, in turn, is based on the interaction between current flow and magnetism. Thus, the explanation here begins with induced current.

Induced Current

We have seen how a current moving through a conductor creates a magnetic field around that conductor. We have also seen that the magnitude and the direction of the field are determined by the magnitude and the direction of the current. When the current begins, this field expands to a distance that is proportional to the amount of current. When the current stops, the field collapses in toward the conductor and eventually disappears. Reversing the direction of the current reverses the direction of the field.

Like so many other electrical principles, this one has an opposite. That is, current produces a magnetic field, and a moving magnetic field produces a current. If we connected the circuit shown in Figure 12–11A, we would expect the current indication to be zero since there is no voltage source. But if we moved a magnet past the conductor so that the magnetic lines crossed the conductor, the meter would show a deflection, as illustrated in Figure 12–11B. In other words, magnetic lines cutting across a conductor cause a current to flow. This current is called **induced current** or current by induction.

The direction of the induced current depends on the direction in which the magnetic lines cross. The current direction can be determined by the **right-hand rule** for induced currents. It works this way. Hold your right hand so that your thumb, first finger, and second finger point in three directions and at right angles to each other, as shown in Figure 12–12A. When your first finger points in the direction of the magnetic lines and your thumb points in the direction in which the magnet is moving (which is upward in Figure 12–12A), your second finger points in the direction of current flow. Figure 12–12A, then, shows the right-hand rule with a magnet that is moving up. The first finger is pointing in the direction of the magnetic field, and the thumb is pointing in the direction of the magnet motion. The induced current will flow in the direction in which the second finger is pointing, toward the left.

When the magnet is moving down, current direction can be determined in the following manner: Since the field is in the same direction as before, your first finger must still point away. But now your arm must rotate so that your thumb points down, since that is the direction of the magnet motion. The resulting current direction is indicated by your second finger. That is, induced current now travels toward the right, as shown in Figure 12–12B.

In summary, the direction of the induced current flow is determined by the direction in which the magnetic lines cut across the conductor. The magnitude of the current is determined by the rate at which the lines cut across the conductor.

Inductors

At this point, we combine two concepts. First, current through a conductor produces a magnetic field around that conductor. Second, passing a magnetic field across a conductor induces a current in that conductor. Our next step is to wrap that long conductor into a coil so that the field caused by one turn of wire crosses the next turn. In other words, the magnetic field around one part of the coil causes an induced current in all other parts of the coil. Since induced current forms the basis of operation of this component, the coil is now called an **inductor,** and its characteristic is called inductance.

Inductance is the property of a circuit that opposes change in the current in a circuit. It is *not* like resistance, which opposes current. Inductance does not oppose current. It opposes only a *change* in the amount of current — that is, increases and decreases. The unit of inductance is the **henry** (abbreviated H). An inductor has an inductance of 1 henry when it produces a back emf of 1 volt when the current is changing at the rate of 1 ampere per

FIGURE 12-12
RIGHT-HAND RULE FOR INDUCED CURRENT

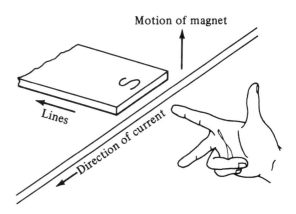

A. Magnet moving up

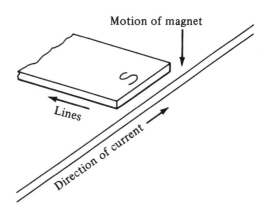

B. Magnet moving down

second. A **back emf** is a force produced within an inductor, and it is opposed to the intended current flow. Just as resistors have resistance and power ratings, inductors have inductance and current ratings. Inductors are measured in henrys, millihenrys (mH), and microhenrys (μH).

An explanation of how inductors work in-

FIGURE 12-13
SIMPLE INDUCTOR

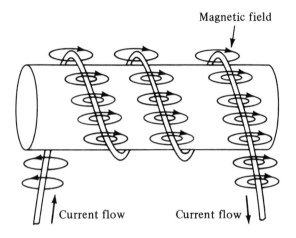

FIGURE 12-14
CURRENT CHANGE IN A RESISTIVE CIRCUIT

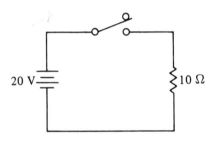

A. Simple resistive circuit

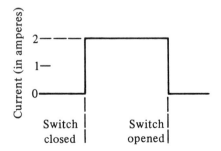

B. Graph for resistive circuit

volves a few concepts examined earlier in this chapter. For instance, we know that if a coil has no current passing through it, then there will be no field around it. But suppose we apply a voltage, which causes a current to flow. As discussed earlier, this current will cause a magnetic field to expand from each conductor, as shown in Figure 12–13. Furthermore, the field expanding from each conductor will cut across each adjacent conductor.

Now, the second concept comes in. We know that, as these magnetic lines cross the adjacent conductors, they, in turn, try to induce a current to flow. If we apply the right-hand rule to the coil in Figure 12–13, we see that the induced force is in direct opposition to the original force. The resulting current is the difference between these two.

The magnitude of this induced opposing force depends upon the magnitude of inductance (L) and the rate of change of the current as a function of time, stated as ΔI (Δ is the Greek capital letter delta) divided by Δt, or

$$\frac{\Delta I}{\Delta t}$$

Suppose a 0.5-henry inductor has a current change of 20 milliamperes in a time of 10 milliseconds. The induced counter emf is

$$V = L\,\frac{\Delta I}{\Delta t} = \frac{0.5\text{ H} \times 0.02\text{ A}}{0.01\text{ s}} = 1\text{ V}$$

As we have seen, if there is no current flowing in a coil, there will be no field around it. But when there is current flowing, there is a field. Both of these situations, without current and with current, are described as **steady states** — that is, they are conditions that do not change with time. A steady state occurs after the current has stopped or after it has been flowing for a while. Inductance occurs when the steady state changes — that is, when the current starts, stops, or changes level.

Figure 12–14A shows a simple resistive circuit with 20 volts across a 10-ohm resistor. As shown in Figure 12–14B, the current will rise to 2 amperes the instant the switch closes and will drop to 0 amperes the instant the switch opens. The current reaches steady state with no delay. Adding an inductor to this circuit will not change the steady-state values of current; they will still be 2 and 0 amperes. The inductor will simply add a delay.

When inductors are connected in series, their inductances add, which is similar to the way resistances combine. For example, suppose a circuit contained three inductors, $L_1 = 2$ henrys, $L_2 = 3$ henrys, and $L_3 = 1$ henry. Then, the total inductance, L_T, is calculated as follows:

$$L_T = L_1 + L_2 + L_3$$
$$= 2\text{ H} + 3\text{ H} + 1\text{ H} = \textbf{6 H}$$

Since the field of an inductor extends outside an inductor, this field may interact with the field of an adjacent inductor. This interaction is called **mutual inductance** (its symbol is M), and it affects the total inductance of two inductors in series. Therefore, the formula for two inductors in series is

$$L_T = L_1 + L_2 \pm 2M$$

Thus, the magnitude of mutual inductance depends on the field strength and the closeness of the two inductors. It also depends on whether the two fields aid or oppose each other. For example, the total inductance of two 100-millihenry inductors having 20 millihenrys of aiding mutual inductance is

$$L_T = L_1 + L_2 + 2M$$
$$= 100\text{ mH} + 100\text{ mH} + 2 \times 20\text{ mH}$$
$$= \textbf{240 mH}$$

If one of these inductors is rotated 180°, the mutual inductance is opposing rather than aiding. The total inductance in this case is

$$L_T = L_1 + L_2 - 2M$$
$$= 100\text{ mH} + 100\text{ mH} - 2 \times 20\text{ mH}$$
$$= \textbf{160 mH}$$

RL Time Constant

When the switch in the circuit of Figure 12–15A is open, there is no current. Closing the switch applies a voltage to the circuit that is large enough to normally cause 2 amperes to flow. However, the initial flow of current through the inductor L causes an expanding magnetic field in the inductor, which is in opposition to the intended flow. As a result of the magnetic field, a large back emf is produced across the inductor, and there is little voltage across or current through the resistor R. As the field approaches the normal position of expansion around the coil, though, the number of lines crossing the conductors decreases, and the back emf decreases. Hence, the current slowly rises to its normal level of 2 amperes, and the opposition of the inductor disappears. Figure 12–15B shows these relationships among the applied voltage, opposing induced voltage, current, and resistor voltage drop for the circuit in Figure 12–15A.

As we noted earlier, when the switch is opened in a resistive circuit, the current will immediately go to zero. But in the circuit of Figure 12–15A, the current will not immediately go to zero because of the inductor. As the magnetic field collapses in the inductor, the induced force tries to maintain the flow of current. That is, the resulting voltages are such that the voltage induced in the inductor has the same polarity as the original battery. Therefore, the inductor keeps the current going for a while. However, the induced voltage diminishes as the collapsing field diminishes, and the current eventually reaches zero after a short time delay.

This discussion leads us to an important idea — that of time constant. The **time con-**

FIGURE 12-15
RL CIRCUIT

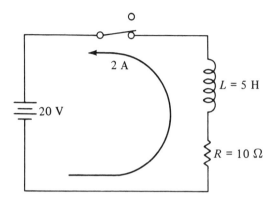

A. Current in *RL* circuit

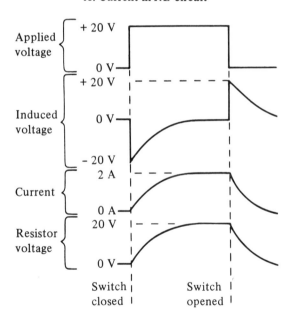

B. Relationships of components in *RL* circuit

stant of an *RL* circuit is the time it takes the current to rise to 63% or fall to 37% of its maximum value. It is directly proportional to the inductance and inversely proportional to the resistance in a circuit — that is, $t = L/R$. Figure 12–16A shows the rising current with spe-

cific percentages at 1, 2, 3, 4, and 5 time constants. Figure 12–16B shows the falling current with percentages at the same five points.

Calculating the time constant for a circuit is quite simple. For the circuit shown in Figure 12–15A, the time constant t is

$$ t = \frac{L}{R} = \frac{5\text{ H}}{10\ \Omega} = \textbf{0.5 s} $$

Thus, 0.5 second after the switch is closed, the current will have risen to 63% of 2 amperes, or to 1.26 amperes. It will take about 5 time constants, or 2.5 seconds, to reach the full 2 amperes. When the switch is opened, the current will not instantly drop to zero. It will reach 37% of 2 amperes, or 0.74 ampere, after 0.5 second, and it will reach zero after 2.5 seconds. The approximate times for rise and fall as well as current levels are given in Table 12–2.

Consider the effect of changing the circuit in Figure 12–15A to these values:

$$ V_T = 20\text{ V} \qquad L = 250\ \mu\text{H} \qquad R = 5\text{ k}\Omega $$

The circuit will now perform in this way:

$$ t = \frac{L}{R} = \frac{250 \times 10^{-6}\text{ H}}{5 \times 10^{3}\ \Omega} = \textbf{50 ms} $$

With a 50-millisecond time constant, this circuit changes more quickly than the previous one.

The next step is to determine the maximum current:

$$ I = \frac{V}{R} = \frac{20\text{ V}}{5 \times 10^{3}\ \Omega} = \textbf{4 mA} $$

After one time constant of 50 milliseconds, the circuit current rises to

$$ 0.63 \times 4\text{ mA} = \textbf{2.52 mA} $$

The remaining current values are given in Table 12–3.

FIGURE 12-16
CURRENT CHANGE IN AN *RL* CIRCUIT

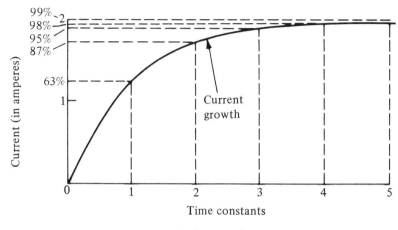

A. Current rise

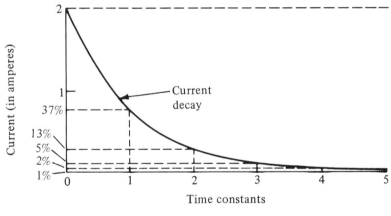

B. Current fall

Applications of Inductors

Inductors are used primarily because they oppose changes in current. In some applications, they are called chokes or filter chokes because of their ability to filter or smooth current flow. More will be said about this application in Chapter 31. Inductors are somewhat like the flywheels used on motors. The purpose of a flywheel is to help the motor maintain a smooth, constant speed. Similarly, the purpose of the inductor is to help maintain a smooth, constant current flow.

Amount of inductance required, frequency, voltage, and current are some of the factors that determine the construction of an inductor. Power supplies often use inductors for filtering current that has been changed from

TABLE 12-2
RL CIRCUIT CURRENT WITH 5 HENRYS AND 10 OHMS

Time Constant	Seconds	Switch Closed		Switch Opened	
		% Maximum	I	% Maximum	I
1	0.5	63	1.26 A	37	0.74 A
2	1.0	87	1.74 A	13	0.26 A
3	1.5	95	1.90 A	5	0.10 A
4	2.0	98	1.96 A	2	0.04 A
5	2.5	99	1.98 A	1	0.02 A

TABLE 12-3
RL CIRCUIT CURRENT WITH 250 MICROHENRYS AND 5 KILOHMS

Time Constant	Seconds	Switch Closed		Switch Opened	
		% Maximum	I	% Maximum	I
1	0.050	63	2.52 mA	37	1.48 mA
2	0.100	87	3.48 mA	13	0.52 mA
3	0.150	95	3.80 mA	5	0.20 mA
4	0.200	98	3.92 mA	2	0.08 mA
5	0.250	99	3.96 mA	1	0.04 mA

AC to DC. Since such inductors require 10 to 15 henrys for effective filtering and must carry large current, iron-core inductors are used.

Two inductors are shown in Figure 12–17. An air-core inductor is on the left, and an iron-core inductor is on the right. Air-core inductors are often used in radio circuits. Because of the frequency level, few turns are required to produce a significant amount of inductance. The reason few turns are required will be considered in Chapter 24. As Figure 4–8 showed, the symbols for inductors indicate whether they are adjustable. Inductor symbols also indicate if the core is air or iron.

Many television tuners have inductors with adjustable metallic cores. These cores (or slugs) have a slot that matches one of the alignment tools described in Chapter 10. The

FIGURE 12-17
INDUCTORS

inductance of that section of the circuit is adjusted by turning the threaded slug with a plastic tool instead of a metal screwdriver. If a metal tool were used, it would increase the inductance of the coil. Once the metal tool were removed, the inductance would once again change, causing misalignment or improper tuning.

SUMMARY

Magnetism is the ability to attract iron and steel and is a fundamental concept in the study of electronics. Magnets are items that possess magnetism. They can be either temporary or permanent. Temporary magnets have low retentivity and lose most of their magnetism once the magnetizing force has been removed. Permanent magnets have high retentivity and, therefore, high residual magnetism.

The field around a magnet is made up of lines that leave the north pole of a magnet and enter the south pole. These lines take the easiest path. Also, they can bend, but they never cross. The strength of the magnetic field around a magnet depends on the number of lines and their density. A force of attraction exists between unlike poles, while a force of repulsion exists between like poles. The strength of this force is inversely proportional to the square of the distance between the poles.

A temporary magnet can be produced by wrapping a coil around a core and passing current through this coil. This temporary magnet is called an electromagnet. The magnet's strength depends on the amount of current, number of turns, and permeability of the core.

To determine the polarity of the magnetic field around an electromagnet, we use the left-hand rule. If we place our left hand so that our fingers point in the direction of current flow, our thumb will point in the direction of the magnetic lines. Electromagnets are found in a number of products. Doorbells, relays, speakers, and headphones all use electromagnets.

Current passing through a conductor produces a magnetic field. Similarly, a moving magnetic field causes a current to flow. This induced current has a direction based on the direction of the magnet motion and a strength based on the rate of magnet motion. The combination of these two effects is the basis for components called inductors. An inductor is a coil wrapped around a core. A current just beginning to pass through the inductor's coil causes a magnetic field to expand, which then crosses adjacent turns in the coil. These crossings induce a current that moves in a direction opposite to that of the current in the coil. The net result is a slow rather than instant rise of the current to its intended value.

Inductance is the property of a circuit that opposes change in the circuit current. The degree to which an inductor opposes this change is described by the *RL* time constant, which is the time it takes the current to rise to 63% or to drop to 37% of its maximum value. The time constant of an *RL* circuit is directly proportional to the inductance and inversely proportional to the resistance in a circuit.

Inductors come in many sizes and shapes, and they may have air cores or iron cores. The basic purpose of all, though, is filtering. The inductor's opposition to changes in current has a smoothing effect on current flow in a circuit.

REVIEW TERMS

amplification: process by which a small current controls a larger one

back emf: force produced in an inductor that is opposed to the intended current flow

electromagnet: magnet whose field is caused by an electric current

electromagnetism: temporary magnetic field produced by electron flow

flux density: number of magnetic lines per unit area, expressed in teslas

henry: unit of measure for inductance

induced current: current caused by magnetic lines crossing a conductor

inductance: property of a circuit that opposes change in the current

inductor: component that operates on induced current and opposes change in a circuit's current

law of magnets: like poles repel, and unlike poles attract

left-hand rule: description of the relationship between current and magnetic lines

magnet: object that possesses magnetism

magnetic field: force surrounding a magnet produced by magnetic lines

magnetic flux: field around a magnet produced by magnetic lines

magnetic lines: lines (invisible) that form loops around a magnet, leaving the north pole and entering the south pole

magnetism: ability to attract iron and steel

mutual inductance: inductance resulting from the interaction of adjacent inductors

permanent magnet: magnet that retains its magnetism after the magnetizing force has been removed

permeability: ease with which magnetic lines pass through an object

reluctance: opposition to the flow of magnetic lines

residual magnetism: magnetism that remains after a magnetizing force has been removed

retentivity: ability of a material to retain magnetism

right-hand rule: description of the relationship between the directions of flux lines, conductor motion, and induced current

***RL* time constant:** time it takes the current in an *RL* circuit to rise to 63% and fall to 37% of its maximum value

steady state: condition that does not change with time

temporary magnet: magnet that does not retain its magnetism after the magnetizing force has been removed

tesla: unit of measure for the density of a magnetic field

weber: unit of measure for the number of magnetic lines

REVIEW QUESTIONS

1. Define magnetism.

2. Define permanent magnet.

3. Define electromagnet, and describe electromagnetism.

4. Define inductance, and name the unit of inductance.

5. Define permeability.

6. Describe the characteristics of magnetic lines.

7. Define magnetomotive force.

8. Compare the permeability of copper with that of iron.

9. Describe what occurs between the fields of force of like poles and between the fields of unlike poles.

10. Describe and sketch the magnetic field that surrounds a current-carrying conductor.

11. What factors affect the direction and strength of the magnetic field around a conductor?

12. Describe the left-hand rule for conductors.

13. Describe the right-hand rule for induced current.

14. Contrast the effects of resistance and inductance on a DC series circuit.

REVIEW PROBLEMS

1. Calculate the magnetomotive force of a 4000-turn coil having a current of 125 microamperes. Repeat the calculation for a current of 16 milliamperes.

2. Calculate the magnetomotive force for a coil having a current of 60 milliamperes and 800 turns. Repeat the calculation for a current of 1.2 amperes.

3. Calculate the field strength of the coils in Problem 1 if the coil length is 2.5 centimeters. Repeat the calculation for a coil length of 5 centimeters.

4. Calculate the field strength for the coils in Problem 2 if the coil length is 0.4 centimeter. Repeat the calculation for a coil length of 0.3 centimeter.

5. Calculate the induced voltage of an inductor with a current change of 250 microamperes in 200 milliseconds and an inductance of 100 millihenrys.

6. Calculate the induced voltage of an inductor with a current change of 0.5 ampere in 0.5 second and an inductance of 1.5 henrys.

7. Calculate the inductance of 1.5- and 2.5-henry inductors connected in series.

8. Calculate the inductance of a 150-millihenry inductor and a 350-millihenry inductor connected in series.

9. Repeat the calculation of Problem 7, but add the effect of 0.3 henry of aiding mutual inductance. Recalculate with the mutual inductance in opposition.

10. Repeat the calculation of Problem 8, but add the effect of 30 millihenrys of aiding mutual inductance. Recalculate with the mutual inductance in opposition.

11. Calculate the time constant for the circuit in Figure 12–18 with $V_T = 12$ volts, $R = 330$ ohms, and $L = 1.5$ henrys.

12. Calculate the time constant for the circuit in Figure 12–18 with $V_T = 28$ volts, $R = 680$ ohms, and $L = 75$ millihenrys.

FIGURE 12-18

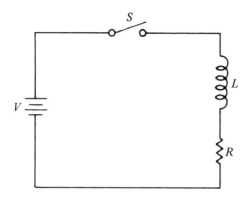

13. Sketch the inductor voltage, current, and resistor voltage waveshapes at 1, 2, 3, 4, and 5 time constants of the circuit described in Problem 11.

14. Sketch the inductor voltage, current, and resistor voltage waveshapes at 1, 2, 3, 4, and 5 time constants of the circuit described in Problem 12.

15. Repeat Problems 11 and 13, but change the inductance to 40 millihenrys.

16. Repeat Problems 12 and 14, but change the resistance to 22 kilohms.

CAPACITANCE

Define capacitance, and describe three factors that affect capacitance.

Determine the total value of capacitors connected in series.

State the safety concerns when working with capacitors.

Define and calculate the time constant of a capacitive circuit.

Measure the time constant of a capacitive circuit.

CAPACITANCE DEFINED

The objective of this chapter is to introduce some properties of capacitance and of capacitors. **Capacitance** is the property of a circuit that opposes a change in the voltage in the circuit. It is not like resistance, which opposes the flow of current. However, in some ways, inductance and capacitance are similar, while in other ways, they are completely unlike each other. The key words to remember about capacitance are *change* and *voltage*. (You recall that the key words for inductance were *change* and *current*.)

A capacitor is an electrical component that takes advantage of the characteristic called capacitance. Basically, a **capacitor** consists of two conductors separated by an insulator. Many capacitors are, in fact, made by rolling two long and narrow aluminum foil sheets, separated by an insulator, into a cylinder. While any two conductors separated by an insulator will have some capacitance, it is generally so small that it is insignificant. However, in later chapters, we will see that even a small amount of capacitance can sometimes be of concern, especially at very high frequencies.

The unit of capacitance is the **farad** (abbreviated F). A capacitance of 1 farad will store a charge of 1 coulomb when 1 volt is placed across it. Recall that 1 coulomb (abbreviated C) is equal to 6.24×10^{18} electrons. Capacitors do not come in values of farads; they come in values of microfarads (μF) and picofarads (pF). One microfarad equals 1×10^{-6} farad, and 1 picofarad equals 1×10^{-12} farad.

Energy Storage

As we saw in the previous chapter, inductors work on the principle of electromagnetism. That is, a field of magnetic flux expands and contracts as the current level varies in an in-

FIGURE 13-1
SIMPLE CAPACITIVE CIRCUIT

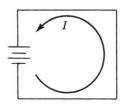

A. Battery and conductor

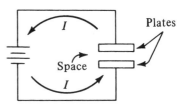

B. Plates connected to circuit

ductor. Once a steady flow of current is established, a magnetic field surrounds the inductor. Any attempt to reduce the flow of current causes the field to collapse. As the magnetic lines contract, they cross adjacent conductors and induce a current that tries to maintain the original level of the current. In other words, an inductor stores energy in a magnetic field. Capacitors have a similar characteristic.

Key point: Capacitors store energy in an electric field.

Let's see how a capacitor works. A battery with both terminals connected by a conductor is illustrated in Figure 13–1A. Electrons move through the conductor from negative to positive. Figure 13–1B shows the conductor cut, the ends of the conductor slightly separated, and metal plates attached to the ends. In this situation, current stops flowing, and the separated ends of the conductor, with

FIGURE 13-2
CHARGING THE PLATES

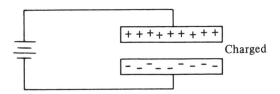

A. Plates charged by applied voltage

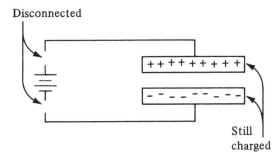

B. Plates still charged with battery disconnected

the plates, form a capacitor. Some electrons will still move, however. Free electrons in the conductor connected to the positive terminal flow toward that terminal of the battery. As shown in Figure 13–2A, the plate connected to the positive terminal then becomes positively charged since it now has a deficiency of electrons. Electrons flow in the other end of the wire also, but from the negative terminal of the battery toward the other plate. That plate has an excess of electrons and, therefore, becomes negatively charged.

An attraction across the space will develop between atoms with electron deficiencies on one side and free electrons on the other side. Current will not flow, though, because the space is an insulator.

If these two plates are disconnected from the battery, as shown in Figure 13–2B, they will remain charged, since one still has an ex-

cess and the other a deficiency of electrons. Suppose that the ends that were connected to the battery are now touched to each other. Current will briefly flow, because the difference in charge is an electromotive force. Thus, electrons will move in the direction that will balance the charges. That is, they will move from where there is an excess to where there is a deficiency — from negative to positive — until the two charges are equal. When the charges are equal, the flow will stop. This process is summarized in Figure 13–3.

A capacitor works in the way just described for the two plates. If a capacitor is connected across a voltage source, a current will briefly flow. However, it will not flow between the plates of the capacitor because they are separated by an insulator. The current in both conductors results from electrons being drawn from one plate by the positive terminal and forced into the other by the negative terminal. Once there are no more electrons that can be removed from the one plate and no more room for electrons in the other, the flow will stop.

If the capacitor plates are disconnected from the source, they will remain charged. They can remain charged for a long time if they are left alone. But if they are connected together, current will flow from one to the other as the charges balance. When the electrons have been evenly distributed throughout both plates, the discharge process is complete. Thus, we have the following rule.

> **Rule:** The quantity of charge (Q), in coulombs, depends on the capacitance of the capacitor (C), in farads, and the amount of force (V), in volts.

In other words, we have the following equation:

$$Q = CV$$

FIGURE 13-3
CHARGING AND DISCHARGING A
CAPACITOR

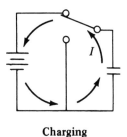

Charging

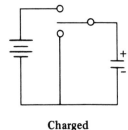

Charged

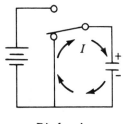

Discharging

Factors Affecting Capacitance

Three factors affect the capacitance of a capacitor:

1. The area of the plates;
2. The distance between the plates:
3. The characteristics of the insulator between them.

Here is the rule for the area of the plates.

Rule: The capacitance of a capacitor is directly proportional to the area *(A)* of the plates.

The situation is depicted in Figure 13–4A. If the area of these plates is doubled, the capacitance will also double. This idea makes sense because the plates are attempting to store electrons. The more area in the plates, the more electrons that can be stored.

The second factor affecting the capacitance of a capacitor is the distance between the plates.

Rule: Capacitance is inversely proportional to the distance *(d)* between the plates, or conductors.

See Figure 13–4B. Recall that an electric attraction occurs across the gap between the two conductors. It is this attraction that draws the positive and negative charges in each conductor toward the gap between them. As for magnetic fields, the attraction between unlike charges gets stronger as the distance between them gets smaller. Bringing the conductors closer together increases the attraction, which causes more charges to move toward the gap. Therefore, the quantity of charge increases.

The third factor affecting the capacitance of a capacitor is the quality of the insulator, or **dielectric,** separating the plates. This factor is described by the **dielectric constant *(K)*,** which allows comparison of one material with another.

Rule: The capacitance of a capacitor is directly proportional to the dielectric constant of the material between the plates.

The dielectric constants for various materials are shown in Table 13–1.

FIGURE 13-4
FACTORS CAUSING INCREASED
CAPACITANCE

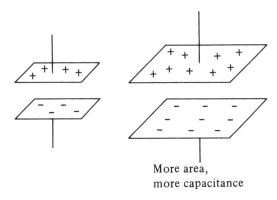

More area,
more capacitance

A. Area increased

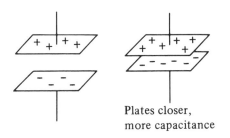

Plates closer,
more capacitance

B. Distance decreased

In summary, we have the following formula:

$$C = \frac{(8.84 \times 10^{-12})KA}{d}$$

This formula describes all the factors affecting the capacitance of a capacitor: C is the capacitance, in microfarads; A is the area of the plates, in square meters; and d is the distance between the plates, in meters.

To see how the formula works, let's consider a mica capacitor with two 1.5-square-

TABLE 13-1
DIELECTRIC CONSTANTS

Air	1.0
Oil	4.0
Mica	5.0
Tantalum	25.0
Ceramic	10,000

centimeter plates 1 millimeter apart. The capacitance is

$$C = \frac{(8.84 \times 10^{-12})KA}{d}$$

$$= \frac{8.84 \times 10^{-12} \times 5 \times 1.5 \times 10^{-2}\ m^2}{1 \times 10^{-3}\ m}$$

$$= 663 \times 10^{-12}\ F$$

$$= \textbf{663 pF}$$

Now, consider a variable capacitor with 14 plates, each 6 square centimeters in area and spaced 2 millimeters apart. The capacitance is

$$C = \frac{8.84 \times 10^{-12} \times 1.0 \times 6 \times 10^{-2}\ m^2 \times 7}{2.0 \times 10^{-3}\ m}$$

$$= \textbf{185.6 pF}$$

The preceding formula indicates that the capacitance of a capacitor can be increased by selecting a high dielectric constant, by making the capacitor larger with an increase in area, or by moving the plates closer together. The degree to which the plates can approach each other is limited by the dielectric strength of the material between the plates. **Dielectric strength** is a number that indicates the breakdown voltage of the dielectric material. Table 13–2 lists the dielectric strengths for some common materials.

TABLE 13-2 DIELECTRIC STRENGTH OF MATERIALS (IN VOLTS/MIL)	
Air	20
Oil	300
Ceramic	1000
Paper	1300
Mica	2000

CAPACITORS

Capacitors are components designed to utilize the effect called capacitance. They are found almost as often as resistors in electronic products. **Fixed capacitors** are the most common. They have one value and are not adjustable. **Variable capacitors** can be adjusted up to a specified total value. Descriptions of various types of capacitors are given in this section.

Fixed Capacitors

Some of the earliest fixed capacitors were called paper capacitors, and many of them are still around. The name *paper* identifies the dielectric material rather than the outside appearance. That is, a paper capacitor consists of two long sheets of conducting foil separated by a special paper insulator.

Mica capacitors were also used for a long period of time for very small values of capacitors. While mica has a high dielectric constant and is durable, it cannot be rolled into compact packages, as paper can be. Figure 13–5 shows a range of fixed capacitors, from a large electrolytic capacitor at the top to a small ceramic disk at the bottom.

Ceramic capacitors have been the most popular capacitors for years because of their high quality and small size. Increasing demands by designers for components that are smaller, more accurate, and less expensive have led to the continued popularity of ceramic as a dielectric. Multilayer ceramic capacitors like those shown in Figure 13–6 have become the most commonly used capacitor today because of their applicability to surface-mounted technology (SMT). Tantalum oxide has also become a very popular dielectric for very small and stable capacitors. In addition, polyester film is quite common because it produces low tolerance values. Other types common today are polystyrene film, dipped silver mica, metallized polyester, and sintered tantalum.

Electrolytic capacitors are fabricated by separating the plates with a liquid or paste dielectric. After assembly, a voltage is applied across the plates. The voltage causes a chemical reaction to take place, which, in turn, causes a thin film of insulating oxide to cover the positive plate. This oxide forms the capacitor's dielectric and gives it a much higher capacitance by volume than regular, or nonelectrolytic, capacitors.

There are precautions to be followed in the use of electrolytic capacitors. The polarity of the forming voltage must be observed in all future uses of those capacitors. Thus, polarity markings are placed on electrolytic capacitors because a polarity reversal will often destroy the capacitor. And if the capacitor is destroyed, a short circuit might develop, which, in turn, could damage other components. Care must also be taken in the soldering or desoldering of tantalum electrolytic capacitors since they can explode when overheated.

Capacitor Ratings

Capacitors are rated in many ways. Capacitance values may be printed on the device, or they may be expressed in a code. Further explanation of capacitor codes is given in the Appendix. We simply note here that there are many varieties of capacitor codes.

Capacitor tolerances of 10% and 20% were quite common in the past. Improvements in the quality of materials and the manufacturing processes have resulted in film capacitors with tolerances down to 0.1%. Such accuracy, of course, increases cost. Tolerance to extremes in temperature is another important factor for capacitors used in a variety of environments. Ceramic capacitors are affected the most by temperature change. However, other capacitors are available for locations that will be colder than −40°C and hotter than +100°C.

Voltage rating is also important. If the maximum rated voltage is exceeded, the dielectric will break down, since even nonconductors will conduct if enough force (voltage) is applied. Once the maximum voltage is reached, most capacitors break down and become defective. However, film capacitors such as metalized polypropylene have self-healing capabilities. Electrolytic capacitors can explode when extreme overvoltages are applied. Therefore, voltage ratings are printed on capacitors so that users can avoid breakdowns or malfunctions.

Capacitors often must withstand 600 or 1000 volts. For these capacitors, oil-treated dielectrics have been used in the past. Today, though, the new ceramic dielectrics are more commonly used. Another way to increase the voltage rating of a capacitor is to increase the distance between the plates, but this process reduces the capacitance. Thus, plates with larger areas are needed to restore the lost capacitance. In other words, an increase in voltage rating or capacitance value generally results in an increase in the physical size of a capacitor.

Variable Capacitors

There are two basic forms of variable capacitors used today. The first type, which is the

FIGURE 13-5
ASSORTED FIXED CAPACITORS

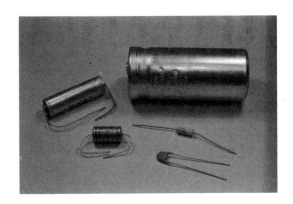

FIGURE 13-6
CERAMIC CAPACITORS

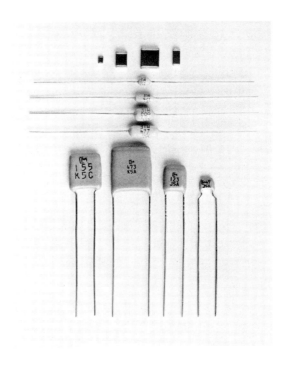

FIGURE 13-7
VARIABLE CAPACITOR

one commonly used as a station selector in radios, is called, simply, a variable capacitor. Figure 13–7 shows a variable capacitor. It consists of a group of stationary plates, called **stators,** and a group of movable plates, called **rotors,** that move when the dial is turned. As the shaft is turned, the amount of rotor that is adjacent to the stator varies from zero to maximum. In other words, the capacitance is varied by changing the area of the plates. This capacitor uses air as a dielectric.

The second type of variable capacitor is a **trimmer capacitor,** which is smaller than the variable capacitor just described. It is usually adjusted by a technician using an alignment tool. Turning a screw causes the small metal plate to move toward or away from the other plate. Thus, capacitance is adjusted by changing the distance between the plates. Some trimmers use a thin insulator such as mica between the plates to keep them from touching. Others use film or Teflon. Rotary ceramic capacitors are also used as trimmers.

Series Connection

Capacitors connected in series act just the opposite of the way resistors in series act; capacitors in series reduce. That is, the total capacitance in the circuit decreases. They reduce because the distance between the plates becomes greater. For example, the upper plate of the upper capacitor in Figure 13–8 becomes one plate of the series capacitors, while the lower plate of the lower capacitor becomes the other plate. The remaining plates just appear as a conductor isolated by a double thickness of dielectric.

For two capacitors, C_1 and C_2, connected in series, the total capacitance C_T of the circuit is calculated by using the following formula:

$$C_T = \frac{C_1 C_2}{C_1 + C_2}$$

For instance, consider the circuit in Figure 13–9. The total capacitance of the circuit is

$$C_T = \frac{C_1 C_2}{C_1 + C_2} = \frac{4\ \mu F \times 6\ \mu F}{4\ \mu F + 6\ \mu F}$$

$$= \frac{4 \times 10^{-6}\ F \times 6 \times 10^{-6}\ F}{4 \times 10^{-6}\ F + 6 \times 10^{-6}\ F}$$

$$= \frac{24 \times 10^{-12}\ F}{10 \times 10^{-6}\ F} = \textbf{2.4}\ \boldsymbol{\mu}\textbf{F}$$

The voltage distribution must also be carefully considered when you connect capacitors in series. In this situation, voltage divides inversely with capacitance. The distribution of the voltage across the capacitors in Figure 13–9 can be determined in this way: First, we find the total charge across the capacitors by using the formula $Q = CV$ introduced earlier. Thus, Q is

$$Q = CV = 2.4 \times 10^{-6}\ F \times 12\ V$$

$$= \textbf{28.8} \times \textbf{10}^{-6}\ \textbf{C}$$

Next, since $Q = CV$, we also have the following formula:

$$V = \frac{Q}{C}$$

We use this formula with each capacitor to calculate the voltage distribution across each one. The calculations are as follows:

$$V_1 = \frac{Q}{C_1} = \frac{28.8 \times 10^{-6}\ C}{4 \times 10^{-6}\ F} = \mathbf{7.2\ V}$$

$$V_2 = \frac{Q}{C_2} = \frac{28.8 \times 10^{-6}\ C}{6 \times 10^{-6}\ F} = \mathbf{4.8\ V}$$

Thus, we see that the smaller capacitor ends up with the larger share of the voltage. This result is also opposite to that of resistance. So, whenever you combine capacitors to achieve a desired value, be sure to consider the final distribution of voltage.

Let's consider the effect of changing the circuit values in Figure 13–9 to these values:

$$V_T = 16\ V \quad C_1 = 240\ pF \quad C_2 = 360\ pF$$

The total capacitance is

$$C_T = \frac{C_1 \times C_2}{C_1 + C_2}$$

$$= \frac{240 \times 10^{-12}\ F \times 360 \times 10^{-12}\ F}{240 \times 10^{-12}\ F + 360 \times 10^{-12}\ F}$$

$$= \frac{8.64 \times 10^{-20}\ F}{6 \times 10^{-10}\ F} = 1.44 \times 10^{-10}\ F$$

$$= \mathbf{144\ pF}$$

The total charge is determined next.

$$Q = C \times V = 144 \times 10^{-12}\ F \times 16\ V$$

$$= 2304 \times 10^{-12}\ C = \mathbf{2.3 \times 10^{-12}\ C}$$

The distribution of charge can now be calculated.

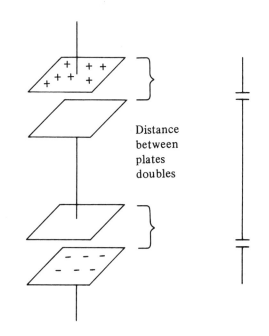

FIGURE 13-8
SERIES CONNECTION OF CAPACITORS

Distance between plates doubles

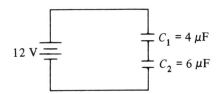

FIGURE 13-9
CAPACITORS IN SERIES

12 V

$C_1 = 4\ \mu F$

$C_2 = 6\ \mu F$

$$V_1 = \frac{Q}{C_1} = \frac{2.3 \times 10^{-9} \text{ C}}{240 \times 10^{-12} \text{ F}} = \textbf{9.6 V}$$

$$V_2 = \frac{Q}{C_2} = \frac{2.3 \times 10^{-9} \text{ C}}{360 \times 10^{-12} \text{ F}} = \textbf{6.4 V}$$

RC TIME CONSTANT

The preceding chapter described how inductance opposes changes in current and creates a delay in its rise and fall. By calculating the *RL* time constant, we were able to determine just how long it would take the current in an inductive circuit to reach its final level. Capacitors also have a time constant. The *RC* **time constant** is the time it takes capacitor voltage to rise to 63% or to drop to 37% of its maximum value. Capacitor voltage, like inductor current, essentially levels off after 5 time constants.

The circuit in Figure 13–10A will help demonstrate the *RC* time constant. Moving the switch to position 1, as shown in Figure 13–10B, places the capacitor, resistor, and battery in series. The battery force moves electrons through the resistor onto the lower plate of the capacitor. Meanwhile, the battery also draws electrons from the upper plate. The rate at which it does both things is affected by the opposition of the resistor and the capacity of the capacitor. An uncharged capacitor acts like a short circuit because it causes high current to flow initially while it charges. Once it is charged, though, it acts like an open circuit. Since the charged capacitor acts like an open circuit, electrons do not cross from one plate to another in any significant amount. Thus, eventually, the capacitor charges up to the same voltage as the battery, and electron flow stops. Disconnecting the capacitor from the battery (Figure 13–10C) will not affect the charge on the capacitor.

Moving the switch to position 2, as shown

in Figure 13–10D, places the resistor and capacitor in series with each other. The charged capacitor now becomes a force that moves electrons from a place where there is an excess to a place where there is a deficiency. The electrons move from the lower plate, up through the resistor, and on to the upper plate of the capacitor. As the capacitor slowly discharges, its voltage drops and eventually reaches zero. At this point, both plates have an equal number of electrons, and there is no current flow. The capacitor is fully discharged, and the switch can be opened, as shown in Figure 13–10E. The battery will once again be the only source of voltage in the circuit.

Calculation

Calculating the time constant for the circuit in Figure 13–10A is quite simple. The equation we use is $t = RC$, where time t is in seconds, resistance R is in ohms, and capacitance C is in farads. The calculation is

$$t = RC = (1.5 \times 10^6 \text{ } \Omega)$$
$$\times (10 \times 10^{-6} \text{ F})$$
$$= \textbf{15 s}$$

Decreasing the capacitance to 0.5 microfarad causes a decrease in the time constant:

$$t = R \times C$$
$$= (1.5 \times 10^6 \text{ } \Omega) \times (0.5 \times 10^{-6} \text{ F})$$
$$= \textbf{0.75 s}$$

While the rise time is less than it was before, the eventual maximum voltage remains unchanged at 10 volts.

A graph of voltage and current in an *RC* circuit is shown in Figure 13–11. Notice that capacitor voltage V_C reaches the same levels that inductor current reached at 1, 2, 3, 4, and 5 time constants (see Chapter 12). The voltage rises slowly and falls slowly. Also, notice that

FIGURE 13-10
CHARGING AND DISCHARGING A
CAPACITOR

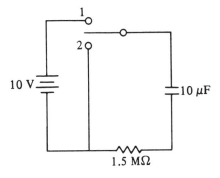

A. Circuit

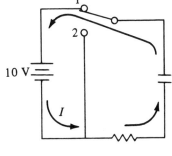

B. Switch in position 1

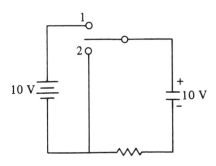

C. Battery disconnected

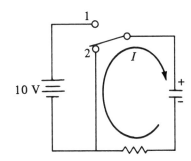

D. Switch in position 2

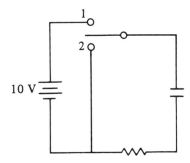

E. Battery disconnected

FIGURE 13-11
VOLTAGE AND CURRENT IN AN *RC* CIRCUIT

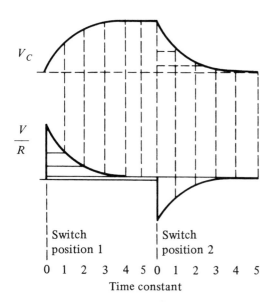

V_C

$\dfrac{V}{R}$

Switch position 1   Switch position 2

0 1 2 3 4 5 0 1 2 3 4 5

Time constant

FIGURE 13-12
EXPERIMENTAL *RC* CIRCUIT

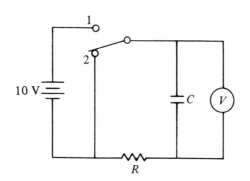

10 V

C V

R

the current (equal to V/R) instantly rises to its maximum level, as determined by V and R. The reversal of current direction during discharge is also shown on the graph. As the graph shows, the circuit is essentially stable after 5 time constants, or 75 seconds.

Measurement

There are two methods for measuring the *RC* time constant of a circuit: You can measure capacitor voltage or circuit current. To make these measurements, you could use a multimeter. But as shown in Figure 13–11, capacitor voltage reaches 63% of its maximum and circuit current falls to 37% of its maximum after only 1 time constant. Thus, measuring capacitor voltage with a meter can be difficult because the meter cannot respond fast enough. Furthermore, a multimeter may load down the capacitor if the meter is not electronic. Although measuring current is not likely to interfere with the circuit, a milliammeter may also be too slow to respond to the quick drop in current. Therefore, an oscilloscope, which does respond quickly, is normally used for these measurements.

However, since you probably are not yet experienced with an oscilloscope, you should try using a meter. Set up a circuit like the one shown in Figure 13–12. The time constant of this circuit is long enough so that a meter should work. Current will not be measured for this circuit because the level is too low. Instead, you will measure capacitor voltage.

Before beginning your measurements, you should determine the voltage level for 1 time constant. When the switch is in position 1 in Figure 13–12, 1 time constant is the time it takes capacitor voltage to rise to 63% of its maximum. So, you need to determine what amount of voltage that is. That amount is calculated as follows:

$$63\% \text{ of } 10 \text{ V} = 10 \text{ V} \times 0.63 = \textbf{6.3 V}$$

TABLE 13-3
VOLTAGE IN AN *RC* CIRCUIT

Time Constant	Seconds	Charging % Maximum	V	Discharging % Maximum	V
1	15	63	6.3 V	37	3.7 V
2	30	87	8.7 V	13	1.3 V
3	45	95	9.5 V	5	0.5 V
4	60	98	9.8 V	2	0.2 V
5	75	99	9.9 V	1	0.1 V

The time it takes the capacitor voltage to rise to 6.3 volts is the time constant of this circuit.

When you move the switch to position 2 (see Figure 13–12), the capacitor voltage should drop to 37% of 10 volts in 1 time constant. The calculation is

$$37\% \text{ of } 10 \text{ V} = 10 \text{ V} \times 0.37 = \textbf{3.7 V}$$

Once you have the circuit connected, you need two more things: an assistant and a stopwatch. Your assistant should start the watch when you place the switch in position 1. Now, you can begin. Watch the meter, and inform your assistant to stop the watch as the meter pointer passes 6.3 volts. Once the capacitor has charged, move the switch to position 2 and measure discharge time. You should try measuring both charging and discharging times for 2, 3, 4, and 5 time constants. The results should be similar to the results shown in Table 13–3.

When you become familiar with triggered oscilloscopes in Chapters 19 and 20, repeat this experiment and compare your results.

SUMMARY

Capacitance is the characteristic of a circuit that opposes changes in the voltage of the circuit. It occurs between any two conductors separated by an insulator, although only in very small amounts. Capacitors are often made by rolling two long and narrow foil sheets separated by an insulator into a cylinder. The unit of capacitance is the farad, although capacitors are usually rated in the picofarad (10^{-12}) or microfarad (10^{-6}) range. Capacitors also have a maximum voltage rating.

Capacitors do not pass DC current since their center is an insulator, or dielectric. Instead, electrons pile up on the plate connected to the negative source, while atoms with electron deficiencies accumulate on the opposite plate. The amount of charge depends on the size of the capacitor and the voltage applied. The charge is measured in coulombs, with 1 coulomb equal to 6.24×10^{18} electrons.

Fixed capacitors have one value of capacitance and a voltage rating. The capacitance of a capacitor is determined by three factors. Capacitance is directly proportional to the area of the plates and inversely proportional to the distance between them. The quality of the dielectric is the third factor that affects the capacitance.

Adjustable capacitors usually operate by varying the distance between the plates or their adjacent areas.

When capacitors are connected in series, the total capacitance decreases, because the effective distance between the outer plates increases. In series connection, the voltage divides, with the smallest capacitor getting the largest voltage.

The *RC* time constant of circuits having resistors and capacitors is the time it takes a capacitor to reach 63% of its maximum charge. It is directly proportional to the resistance in the circuit and the capacity of the capacitor. The time in seconds is equal to the resistance in ohms multiplied by the capacitance in farads — that is, *t = RC*.

The *RC* time constant can be measured by connecting a voltmeter across the capacitor and timing the charging with a stopwatch. Usually, though, this procedure will not work because most time constants are in milliseconds or microseconds, and a voltmeter cannot respond this quickly. Thus, a triggered oscilloscope is needed for these measurements. You will have an opportunity to experiment with this equipment in a later exercise in Chapters 19 and 20.

CHAPTER 13

REVIEW TERMS

capacitance: property of a circuit opposing change in voltage

capacitor: electronic component consisting of two conductors separated by an insulator

ceramic capacitor: small high-quality capacitor commonly used in electronic products

dielectric: insulating material separating plates in a capacitor

dielectric constant: number indicating the relative quality of dielectric materials, with air having a constant of 1

dielectric strength: number, expressed in volts per mil, indicating the breakdown voltage of dielectric materials

electrolytic capacitor: capacitor made with a paste or liquid dielectric that produces a high capacitance per unit volume

farad: unit of measure for capacitance

fixed capacitor: capacitor having only one value

RC time constant: time it takes the capacitor voltage in an *RC* circuit to rise to 63% and fall to 37% of its maximum value

rotor: moving element (plate) in a variable capacitor

stator: fixed element (plate) in a variable capacitor

trimmer capacitor: small variable capacitor with a mica dielectric often used in radios

variable capacitor: capacitor that can be adjusted up to a specified value

REVIEW QUESTIONS

1. Define capacitance, and name the unit of capacitance.

2. Describe what is meant by dielectric.

3. Describe how the plates in a capacitor are charged.

4. Define the *RC* time constant.

5. Describe a trimmer capacitor.

6. Describe two safety concerns when working with capacitors.

7. Name and describe three types of capacitors.

8. Name and describe four types of capacitor dielectrics.

9. Name three factors that affect the capacitance of a capacitor.

10. Describe how current flows in an *RC* circuit when a DC voltage is applied.

11. Describe how a capacitor stores a charge.

12. Describe the results of placing capacitors in series, and explain why these results occur.

13. Explain the relationship between dielectric strength, capacitance, and the voltage rating of a capacitor.

14. Considering dielectric constant and dielectric strength, what material seems to be the best compromise?

REVIEW PROBLEMS

1. Calculate the capacitance of a mica capacitor having 1.2-square-centimeter plates spaced 1.2 millimeters apart.

2. Repeat the calculation of Problem 2 for a ceramic capacitor.

3. Calculate the capacitance of a tantalum capacitor having 20-square-centimeter plates spaced 0.15 millimeter apart.

4. Repeat the calculation of Problem 3 for a mica capacitor.

5. Calculate the capacitance of a variable capacitor having seven pairs of 3.8-square-centimeter plates spaced 1.15 millimeters apart.

6. Repeat the calculation of Problem 5 for plates spaced 2.45 millimeters apart.

7. Calculate the capacitance of two 0.01-microfarad capacitors connected in series. Calculate the capacitance for parallel connection.

8. Calculate the capacitance of two 0.47-microfarad capacitors connected in series. Calculate the capacitance for parallel connection.

9. Calculate the voltage distribution for 10 volts connected across 0.01- and 0.03-microfarad capacitors in series.

10. Calculate the voltage distribution for 14 volts connected across 0.47- and 0.15-microfarad capacitors in series.

FIGURE 13-13

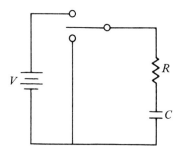

11. Calculate the time constant for the circuit in Figure 13–13 with V_T = 12 volts, R = 1000 ohms, and C = 1.5 microfarads.

12. Calculate the time constant for the circuit in Figure 13–13 with V_T = 18 volts, R = 33 kilohms, and C = 0.01 microfarad.

13. Sketch the current, capacitor voltage, and resistor voltage waveshapes at 1, 2, 3, 4, and 5 time constants for the circuit described in Problem 11.

14. Sketch the current, capacitor voltage, and resistor voltage waveshapes at 1, 2, 3, 4, and 5 time constants for the circuit described in Problem 12.

15. Repeat the calculations of Problems 11 and 13 for a capacitance of 2200 picofarads.

16. Repeat the calculations of Problems 12 and 14 for a resistance of 1.5 megohms.

PARALLEL CIRCUITS

Describe and identify a parallel circuit.

State four basic rules of a parallel circuit.

Calculate the total resistance of a parallel circuit.

Calculate the current in each branch of a parallel circuit and the total current.

Calculate the power dissipation of each resistor and the total power in a parallel circuit.

INTRODUCTION

The objective of this chapter is to introduce parallel connection, which is a continuation of the introduction to basic electrical theory. In Chapter 8, we saw that series connection means that components follow one after another. Consequently, current has to pass through each and every component in the circuit in the order of connection. In parallel connection, though, the components are electrically side by side rather than in order. For example, a three-resistor parallel circuit is like a three-lane highway. Current has a choice of paths to follow, and like an automobile driver, it will take the lane with least opposition.

A single resistor in a circuit provides a certain amount of resistance to the circuit. Connecting a resistor in series with another resistor *increases* the total resistance of the circuit, because the current must pass through more resistance along the way. Connecting a resistor in parallel with another resistor *decreases* the total resistance of the circuit. The total resistance decreases because the current has a choice of paths, as shown in Figure 14–1. The net effect of adding resistors in parallel depends on their values and will be described later in this chapter.

Recognizing Parallel Connection

Parallel circuits, like series circuits, can be identified by tracing the current path from one place to another. Recall that we identified series circuits in Chapter 8 by beginning at the negative terminal of the power supply and tracing the current path to the positive terminal. On a schematic diagram, we can trace a series circuit from negative to positive without the pencil leaving the paper. But in tracing the current of a **parallel circuit,** like the ones shown in Figure 14–1, we find that the pencil

will leave the paper for each new path back to the battery. Both of the circuits in Figure 14–1 are parallel circuits because the current can go through any one of four paths. Note that each path has only one resistor, which makes these circuits simple parallel circuits. More complex versions will be discussed later in Chapters 16 and 18.

Effect of Parallel Connection

Suppose a lamp is connected across a battery. We can treat the lamp as a resistor if we know the resistance of its filament. Let's assume that we have a 12-volt battery and a 24-ohm lamp. The circuit will resemble the one shown in Figure 14–2, and current can be calculated as follows, using Ohm's law:

$$I = \frac{V}{R} = \frac{12 \text{ V}}{24 \text{ }\Omega} = \textbf{0.5 A}$$

Now, suppose that we need more light, so we connect another lamp across the same battery. If it has the same resistance as the first lamp, it will also draw a current of 0.5 ampere from the battery. Assuming the battery is good, the second lamp will have no effect on the first lamp. The circuits are independent and parallel. Figure 14–3 shows how the parallel circuits are actually connected and how they are usually drawn.

The important point to note here is the effect of parallel connection on the total circuit. The resistance for each lamp is, and remains, 24 ohms. And each lamp draws a current of 0.5 ampere. Therefore, the battery must supply 0.5 ampere for each lamp. The current of one lamp does not pass through the other lamp. Thus, the battery sees a circuit that draws 1 ampere. Since the battery is a 12-volt DC source, the effect of the parallel connection on the circuit can be determined with a variation of Ohm's law. That is, since $I =$

V/R, then $R = V/I$. Hence, the total resistance in the circuit is

$$R = \frac{V}{I} = \frac{12 \text{ V}}{1 \text{ A}} = 12 \text{ }\Omega$$

Placing two 24-ohm resistors in parallel has had the effect of producing a total circuit resistance that is less than either of the original resistances. Parallel connection increased the current.

RULES AND CALCULATIONS

Many components in electronic products are connected in parallel with other components. If you are to successfully troubleshoot these circuits, you must understand how components in parallel interact with one another. In addition, since the characteristics of a circuit may not always be given to you, you must be able to determine them yourself. For these reasons, in this section, we examine the basic rules and calculations for a parallel resistive circuit.

Resistance

The preceding problem could have been evaluated another way — by using a special formula. The formula for determining the total circuit resistance of two resistors connected in parallel is

$$R_T = \frac{R_1 R_2}{R_1 + R_2}$$

Thus, for the circuit in Figure 14–3, the total resistance can be calculated as follows:

$$R_T = \frac{R_1 R_2}{R_1 + R_2} = \frac{24 \text{ }\Omega \times 24 \text{ }\Omega}{24 \text{ }\Omega + 24 \text{ }\Omega}$$

$$= \frac{576}{48} \text{ }\Omega = 12 \text{ }\Omega$$

FIGURE 14-1
SIMPLE PARALLEL CIRCUITS

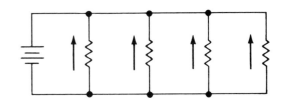

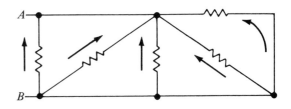

FIGURE 14-2
CURRENT FLOW WITH ONE LAMP

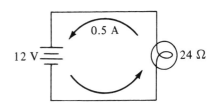

FIGURE 14-3
CURRENT AMOUNTS WITH TWO LAMPS IN PARALLEL

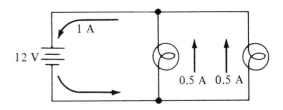

FIGURE 14-4
COMMON-VALUE RESISTORS IN PARALLEL

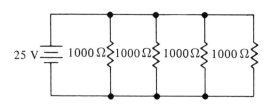

There is another formula for *identical* resistors in parallel. If all the resistors connected in parallel have the same value, divide this value by the number of resistors to find the total resistance. The formula is as follows, where n is the number of resistors:

$$R_T = \frac{R}{n}$$

Thus, the total resistance for the circuit in Figure 14–3 can be calculated as follows:

$$R_T = \frac{R}{n} = \frac{24\ \Omega}{2} = 12\ \Omega$$

Adding a third 24-ohm resistor will further reduce the total resistance. That is,

$$R_T = \frac{R}{n} = \frac{24\ \Omega}{3} = 8\ \Omega$$

Another circuit that can be evaluated in the same manner is shown in Figure 14–4. The total resistance in this circuit is calculated as follows:

$$R_T = \frac{R}{n} = \frac{1000\ \Omega}{4} = 250\ \Omega$$

Now, suppose the circuit has four resistors, each with a different value. The preceding formulas cannot be used directly to find total resistance since we have more than two

resistors and since each resistor has a different value. In this situation, there are two possible ways to proceed. One way is to use a third formula that will be presented here. The second approach is to use the original two-resistor formula and evaluate two resistors at a time. You can use either method. But in this book, we will generally use the second approach since the third formula involves complex fractions. However, that formula is presented here so that you can make your own choice. The formula is

$$\frac{1}{R_T} = \frac{1}{R_1} + \frac{1}{R_2} + \frac{1}{R_3} + \frac{1}{R_4}$$

Another way to arrange this formula is

$$R_T = \frac{1}{\dfrac{1}{R_1} + \dfrac{1}{R_2} + \dfrac{1}{R_3} + \dfrac{1}{R_4}}$$

The circuit shown in Figure 14–5 can be evaluated with this formula. The calculation is

$$R_T = \frac{1}{\dfrac{1}{R_1} + \dfrac{1}{R_2} + \dfrac{1}{R_3} + \dfrac{1}{R_4}}$$

$$= \frac{1}{\dfrac{1}{40}\ \Omega + \dfrac{1}{60}\ \Omega + \dfrac{1}{48}\ \Omega + \dfrac{1}{48}\ \Omega}$$

$$= \frac{1}{0.025\ \Omega + 0.017\ \Omega + 0.021\ \Omega + 0.021\ \Omega}$$

$$= \frac{1}{0.084\ \Omega} = 11.9\ \Omega$$

We can also evaluate this circuit by using the two-resistor formula and evaluating two at a time. When we combine R_1 and R_2, we call our answer R_{12} since it is not R_T. Our next answer is called R_{34}, which is the combination of R_3 and R_4. Then, we find R_T by combining the imaginary R_{12} with the imaginary R_{34}. These calculations are as follows:

$$R_{12} = \frac{R_1 R_2}{R_1 + R_2} = \frac{40\ \Omega \times 60\ \Omega}{40\ \Omega + 60\ \Omega}$$

$$= \frac{2400}{100}\ \Omega = \mathbf{24\ \Omega}$$

$$R_{34} = \frac{R_3 R_4}{R_3 + R_4} = \frac{48\ \Omega \times 48\ \Omega}{48\ \Omega + 48\ \Omega}$$

$$= \frac{2304}{96}\ \Omega = \mathbf{24\ \Omega}$$

$$R_T = \frac{R_{12} R_{34}}{R_{12} + R_{34}} = \frac{24\ \Omega \times 24\ \Omega}{24\ \Omega + 24\ \Omega}$$

$$= \frac{576}{48}\ \Omega = \mathbf{12\ \Omega}$$

These calculations show the following rule.

Rule: The total resistance in a parallel circuit is less than the resistance of the smallest resistor.

Voltage

All components in simple parallel circuits are connected across the voltage source. Therefore, we have the following rule.

Rule: The voltage is the same in all parts of a parallel circuit.

All the regular wall outlets and lamp sockets in a house are connected in parallel. If we measured 117 volts in a living room outlet, we should measure 117 volts in a bedroom outlet. Some outlets, though, such as those for stoves, dryers, and air conditioners, have a higher voltage because they are connected to a different parallel circuit.

Another familiar use of parallel circuits is the connection of automobile accessories. The radio, clock, heater, and lights in an automobile are all connected in parallel with each

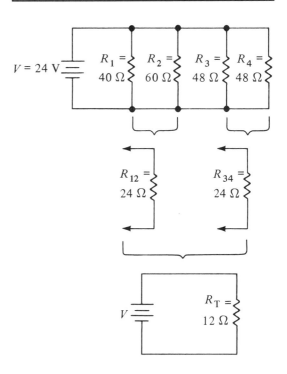

FIGURE 14-5
PARALLEL RESISTORS WITH DIFFERENT VALUES AND A COMBINATION OF PAIRS

other, as illustrated in Figure 14–6. They are 12-volt devices that operate from the 12-volt battery. Home stereo components also use parallel connection.

Current

There are many values of current in a parallel circuit. The current for each branch can be calculated by using Ohm's law, $I = V/R$. Then, the total current is determined by adding the branch currents.

Consider the circuit shown in Figure 14–7. The calculations of the current in this circuit are as follows:

FIGURE 14-6
PARALLEL CIRCUIT FOR AUTOMOBILE
ACCESSORIES

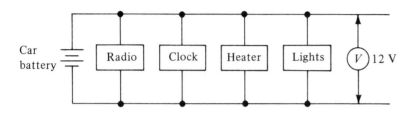

FIGURE 14-7
CURRENT IN A PARALLEL CIRCUIT

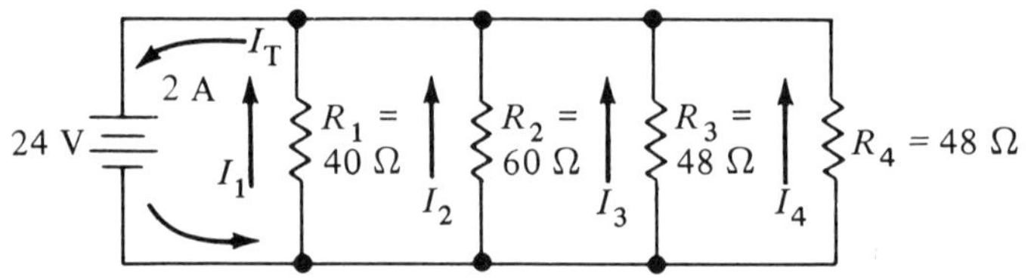

$$I_1 = \frac{V_T}{R_1} = \frac{24 \text{ V}}{40 \text{ }\Omega} = \textbf{0.6 A}$$

$$I_2 = \frac{V_T}{R_2} = \frac{24 \text{ V}}{60 \text{ }\Omega} = \textbf{0.4 A}$$

$$I_3 = \frac{V_T}{R_3} = \frac{24 \text{ V}}{48 \text{ }\Omega} = \textbf{0.5 A}$$

$$I_4 = \frac{V_T}{R_4} = \frac{24 \text{ V}}{48 \text{ }\Omega} = \textbf{0.5 A}$$

$$I_T = I_1 + I_2 + I_3 + I_4$$
$$= 0.6 \text{ A} + 0.4 \text{ A} + 0.5 \text{ A} + 0.5 \text{ A}$$
$$= \textbf{2 A}$$

This answer can be checked by using the value for total resistance determined earlier, as follows:

$$I_T = \frac{V_T}{R_T} = \frac{24 \text{ V}}{12 \text{ }\Omega} = \textbf{2 A}$$

The answer checks.

One rule of parallel circuits is demonstrated in the preceding calculations.

Rule: The total current in a parallel circuit is the sum of the separate branch currents.

The calculations also indicate an important relationship between the size of a resistor and its current.

Key point: Large resistances have smaller currents; small resistances have larger currents.

Power

The total power in a series circuit was determined by adding the power of each resistor. There is a similar rule for determining power in a parallel circuit.

>   **Rule:** The power in a parallel circuit is the sum of the separate branch powers.

Therefore, the total power can be determined in the same two ways it was determined for series circuits. The only difference is that the voltage is the same for all resistors in the parallel circuit, whereas the current was the same for all resistors in the series circuit.

Let's calculate the power for the circuit in Figure 14–7. We first calculate the power in each branch; then, we add these values to find the total power. Here are the calculations.

$$P_1 = V_T I_1 = 24 \text{ V} \times 0.6 \text{ A} = \textbf{14.4 W}$$
$$P_2 = V_T I_2 = 24 \text{ V} \times 0.4 \text{ A} = \textbf{9.6 W}$$
$$P_3 = V_T I_3 = 24 \text{ V} \times 0.5 \text{ A} = \textbf{12 W}$$
$$P_4 = V_T I_4 = 24 \text{ V} \times 0.5 \text{ A} = \textbf{12 W}$$
$$P_T = P_1 + P_2 + P_3 + P_4$$
$$= 14.4 \text{ W} + 9.6 \text{ W} + 12 \text{ W} + 12 \text{ W}$$
$$= \textbf{48 W}$$

This total can be determined another way, as follows:

$$P_T = V_T I_T = 24 \text{ V} \times 2 \text{ A} = 48 \text{ W}$$

Complete Example

The complete solution of a parallel circuit involves determining all values of resistance, current, voltage, and power. We will consider another example here so that you can work through one complete solution before you be-gin solving your own calculations. The circuit consists of four resistors in parallel with a 12-volt power supply. The resistors are color-coded as follows:

— R_1, red, red, red, silver;

— R_2, brown, black, red, silver;

— R_3, brown, green, red, silver;

— R_4, orange, orange, red, silver.

Your first step should always be to draw a schematic of the circuit and indicate all known values. The circuit is illustrated at the top of Figure 14–8.

The next step is to determine the total resistance of the circuit. As mentioned earlier, the two-resistor approach will be used here. Thus, R_{12} will be found first, then R_{34}, and then R_T.

Notice that this example, like most real circuits, does not produce even numbers to work with. Some texts often provide a 10-volt, 5-ohm circuit for calculations so that an easy-to-work-with 2-ampere current results. But real circuits often have values like 12.6 volts and 4.7 ohms, which produce a current of 2.680851 amperes. With numbers like these, it is reasonable to round off answers to three or four significant figures. That is, the current could be rounded off as 2.68 amperes.

The calculations for the resistors are as follows:

$$R_{12} = \frac{R_1 R_2}{R_1 + R_2} = \frac{2200 \ \Omega \times 1000 \ \Omega}{2200 \ \Omega + 1000 \ \Omega}$$
$$= \frac{2,200,000}{3200} \ \Omega = \textbf{687.5 } \Omega$$
$$R_{34} = \frac{R_3 R_4}{R_3 + R_4} = \frac{1500 \ \Omega \times 3300 \ \Omega}{1500 \ \Omega + 3300 \ \Omega}$$
$$= \frac{4,950,000}{4800} \ \Omega = \textbf{1031.25 } \Omega$$

FIGURE 14-8
SAMPLE PARALLEL CIRCUIT AND A
COMBINATION OF PAIRS

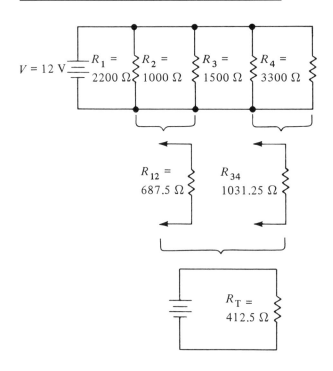

$$R_T = \frac{R_{12}R_{34}}{R_{12} + R_{34}} = \frac{687.5\ \Omega \times 1031.25\ \Omega}{687.5\ \Omega + 1031.25\ \Omega}$$

$$= \frac{708,984.37}{1718.75}\ \Omega = 412.5\ \Omega$$

The final answer has been rounded off from an indefinite decimal. (By this time, you have probably begun to appreciate the value of an electronic calculator.)

Current calculations come next, and they are as follows:

$$I_1 = \frac{V_T}{R_1} = \frac{12\ V}{2200\ \Omega} = 0.00545\ A$$

$$I_2 = \frac{V_T}{R_2} = \frac{12\ V}{1000\ \Omega} = 0.012\ A$$

$$I_3 = \frac{V_T}{R_3} = \frac{12\ V}{1500\ \Omega} = 0.008\ A$$

$$I_4 = \frac{V_T}{R_4} = \frac{12\ V}{3300\ \Omega} = 0.00364\ A$$

$$I_T = I_1 + I_2 + I_3 + I_4$$

$$= 0.00545\ A + 0.012\ A + 0.008\ A$$

$$+ 0.00364\ A$$

$$= 0.029\ A$$

Total current can also be calculated from total resistance. However, rounding off numbers along the way may cause slight differences in final answers. The calculation for total current is

$$I_T = \frac{V_T}{R_T} = \frac{12 \text{ V}}{412.5 \text{ }\Omega} = \mathbf{0.029 \text{ A}}$$

The answer checks.

The final step in the calculations is to determine the circuit power. Notice that the numbers get more difficult to work with as we go on. The power calculations are as follows:

$$P_1 = V_T I_1 = 12 \text{ V} \times 0.00545 \text{ A}$$
$$= \mathbf{0.0654 \text{ W}}$$
$$P_2 = V_T I_2 = 12 \text{ V} \times 0.012 \text{ A}$$
$$= \mathbf{0.144 \text{ W}}$$
$$P_3 = V_T I_3 = 12 \text{ V} \times 0.008 \text{ A}$$
$$= \mathbf{0.096 \text{ W}}$$
$$P_4 = V_T I_4 = 12 \text{ V} \times 0.00364 \text{ A}$$
$$= \mathbf{0.0437 \text{ W}}$$
$$P_T = P_1 + P_2 + P_3 + P_4$$
$$= 0.0654 \text{ W} + 0.144 \text{ W} + 0.096 \text{ W}$$
$$+ \ 0.0437 \text{ W}$$
$$= \mathbf{0.349 \text{ W}}$$

This final value can be checked by another calculation, as follows:

$$P_T = V_T I_T = 12 \text{ V} \times 0.029 \text{ A} = 0.348 \text{ W}$$

Notice that the answers for total power differ because of rounding off. Except for some special cases, this difference is insignificant in all practical applications. In fact, the typical technician will probably report that the total power is between ¼ and ½ watt.

The power calculation completes the solution of the circuit.

It should be noted here that parallel circuits can be drawn many ways. The word *parallel* indicates how they are electrically connected, not physically aligned.

INDUCTORS AND CAPACITORS IN PARALLEL

The rules for inductors in parallel are similar to those for resistors. That is, the total inductance of two inductors in parallel is less than the inductance of the smallest one. The total inductance of the inductors in Figure 14–9A can be found by using the following formula:

$$L_T = \frac{L_1 L_2}{L_1 + L_2}$$

Capacitors in parallel act the opposite of the way resistors and inductors act. The capacitance of two capacitors in parallel increases because the area of the plates increases, as shown in Figure 14–9B. The capacitance of the capacitors in Figure 14–9C can be determined with this formula:

$$C_T = C_1 + C_2$$

USES OF PARALLEL CIRCUITS

Let's take another look at Christmas tree lights; the schematics are shown in Figure 14–10. Recall that they were first discussed in Chapter 8 on series circuits. Now that you are familiar with parallel circuits, both light systems can be compared.

The first difference to notice is one that causes the main problem with series lights, shown in Figure 14–10A. If one bulb burns out, its filament opens the circuit. The broken filament interrupts the current flow, causing all the lamps to go out. Also notice that since the series lamps share the applied voltage, they are 15-volt lamps.

Another difference is that the parallel circuit in Figure 14–10B uses 120-volt lamps instead of the 15-volt ones of the series circuit.

FIGURE 14-9
CAPACITORS VERSUS INDUCTORS

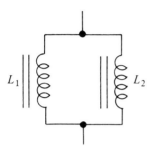

A. Inductors in parallel

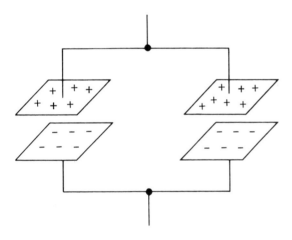

B. Area doubles

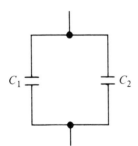

C. Capacitors in parallel

Therefore, manufacturers make their sockets different to prevent people from interchanging them. The number of wires is also different: One wire enters and leaves each series socket, but two enter and leave each parallel socket. Finally, notice that each parallel lamp is independent of all others. If one goes out, it has no effect on any other.

As mentioned earlier, most electric and electronic circuits are connected in parallel with others. The appliances in a home are parallel, as are the accessories in a car. The primary advantage of parallel connection is that one source can power many devices. One battery in a car can operate the starter, lights, radio, and so forth. One set of power lines coming into a home can provide the power for all lights and appliances.

The obvious disadvantage of parallel connection is that if that one source of voltage is lost, all devices connected to that circuit lose their power. You can lose the power in your home by demanding too much current from a source. Now is a good time to consider that problem.

Suppose you are in your room with two 60-watt lamps on. You turn on the television, which uses another 240 watts. The schematic for this situation is shown in Figure 14–11. We can use a variation of the power formula to determine how much current flows through these devices. That is, since $P = VI$, then $I = P/V$. Thus, we have the following calculations:

$$I_{lamp} = \frac{P_{lamp}}{V_T} = \frac{60 \text{ W}}{120 \text{ V}} = \textbf{0.5 A}$$

$$I_{TV} = \frac{P_{TV}}{V_T} = \frac{240 \text{ W}}{120 \text{ V}} = \textbf{2 A}$$

If you switch on a 1200-watt hair dryer, the current demand is increased, as the next calculation shows:

$$I_{dryer} = \frac{P_{dryer}}{V_T} = \frac{1200 \text{ W}}{120 \text{ V}} = \textbf{10 A}$$

FIGURE 14-10
CIRCUIT FOR CHRISTMAS TREE LIGHTS

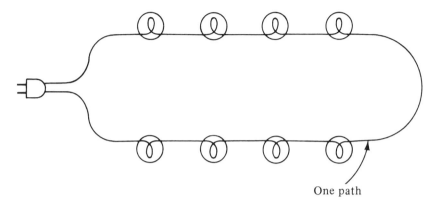

A. Series circuit

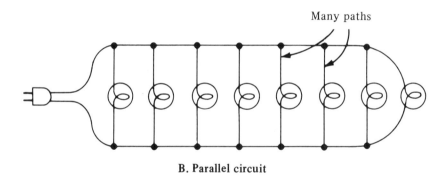

B. Parallel circuit

FIGURE 14-11
PARALLEL CIRCUIT IN A HOME

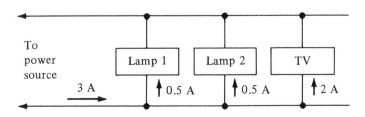

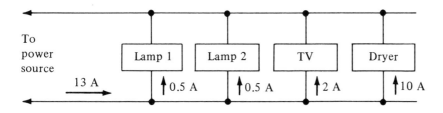

**FIGURE 14-12
TOTAL CURRENT INCREASE WITH ADDED LOADS**

The circuit resembles the one pictured in Figure 14–12. If the power source has a 15-ampere fuse, the circuit is close to an overload. A few more lamps or another appliance will overload the circuit.

There is another point to consider: the total power in your room. What does it cost to operate the items in your room? Suppose you have the television and two lamps on 4 hours (abbreviated h) a day for 30 days (day is not abbreviated). You calculate your power consumption as follows:

$$60 \text{ W lamp} + 60 \text{ W lamp} + 240 \text{ W TV}$$
$$= 360 \text{ W total power}$$

$$360 \text{ W} \times \frac{4 \text{ h}}{1 \text{ day}} \times 30 \text{ days} = 43{,}200 \text{ Wh}$$

The abbreviation Wh stands for watt-hour.

Electric companies determine bills by kilowatt-hours (abbreviated kWh), where 1 kilowatt equals 1000 watts. Therefore, your power consumption for one month is 43.2 kilowatt-hours. Look at the rates on your electric bill and figure out what 1 kilowatt-hour of consumption costs. Now you can determine the cost of operating the television and the 1200-watt hair dryer.

SUMMARY

Parallel circuits have components connected electrically side by side rather than one after another, as in series circuits. One way to think of parallel circuits is by the rules that describe them, which are as follows:

1. The voltage is the same in all parts of a parallel circuit.
2. The total current in a parallel circuit is equal to the sum of the separate branch currents.
3. The total resistance in a parallel circuit is less than the resistance of the smallest resistor.
4. The total power in a parallel circuit is equal to the sum of the separate powers.

Parallel circuits are solved by determining all values of resistance, current, and power, which is done in the order just mentioned. However, there are times when voltage and power are given instead of resistance. In such cases, the basic Ohm's law and power formula can be used by shifting the factors.

It is necessary to be very organized in circuit solution work. Schematics should be drawn, formulas written down, and answers noted. Answers may be rounded off to avoid unmanageable numbers in further calculations.

Parallel circuits, like series circuits, can be found almost everywhere. Therefore, your understanding of the effects of adding one component in parallel with another is absolutely essential. The theory presented in this chapter will be put to practical use in the next chapter, which discusses parallel circuit testing.

CHAPTER 14

REVIEW TERM

parallel circuit: circuit arrangement in which current can flow through alternate paths

REVIEW QUESTIONS

1. Describe a parallel circuit.

2. State the four rules of a parallel circuit.

3. Write three formulas for determining the total resistance of a parallel circuit.

4. What formula can only be used when two resistors are in parallel?

5. What formula can be used when all parallel resistors are of equal value?

6. Give two examples where parallel circuits are used.

7. Describe how current flows in a parallel circuit.

8. Explain an advantage a parallel circuit has over a series circuit.

9. Explain an advantage a series circuit has over a parallel circuit.

10. What effect does doubling the applied voltage have on parallel circuit power?

11. What effect does doubling the total resistance have on parallel circuit power?

12. In a given circuit, which would use the most power, two 1000-ohm resistors in series or two in parallel?

13. What happens in a parallel circuit when another resistor is added?

FIGURE 14-13

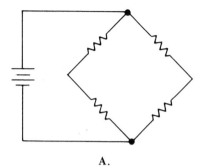

A.

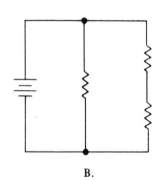

B.

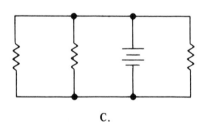

C.

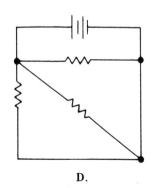

D.

14. Determine which of the circuits shown in Figure 14–13 are parallel.

15. Describe the relationship between resistance and current in the branches of a parallel circuit.

16. Describe the relationship between resistance and power in the branches of a parallel circuit.

REVIEW PROBLEMS

1. Calculate the total resistance of the circuit shown in Figure 14–14 if $R_1 = 40$ ohms and $R_2 = 60$ ohms.

2. Calculate the total resistance of the circuit in Figure 14–14 if $R_1 = 2.5$ kilohms and $R_2 = 3.3$ kilohms.

3. Calculate the total resistance of the circuit pictured in Figure 14–15 if $R_1 = 330$ ohms, $R_2 = 330$ ohms, $R_3 = 220$ ohms, and $R_4 = 220$ ohms. Use two methods of calculation.

FIGURE 14-14

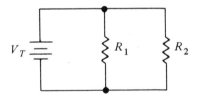

FIGURE 14-15

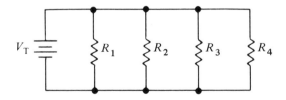

FIGURE 14-16

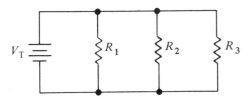

4. Calculate the total resistance of the circuit shown in Figure 14–15 if all resistors are 200 kilohms. Calculate the total resistance if all resistors are 1.2 megohms.

5. Calculate the total resistance of the circuit in Figure 14–15 if R_1 = 330 ohms, R_2 = 330 ohms, R_3 = 470 ohms, and R_4 = 470 ohms.

6. Calculate the total resistance of the circuit in Figure 14–16 if all the resistors are 200 kilohms. Calculate the total resistance if all the resistors are 1.2 megohms.

7. Calculate all values of voltage, current, resistance, and power for the circuit in Figure 14–14 if V_T = 18 volts, R_1 = 600 ohms, and R_2 = 900 ohms.

8. Calculate all values of voltage, current, resistance, and power for the circuit in Figure 14–14 if V_T = 12 volts, R_1 = 1000 ohms, and R_2 = 1200 ohms.

9. Determine all values for the circuit described in Problem 4 if V_T = 48 volts.

10. Calculate all values of voltage, current, resistance, and power for the circuit in Figure 14–15 if V_T = 24 volts, R_1 = 1.2 kilohms, R_2 = 800 ohms, R_3 = 1.2 kilohms, and R_4 = 600 ohms.

11. Determine all values for the circuits described in Problem 6 if V_T = 36 volts.

12. Calculate all values of voltage, current, resistance, and power for the circuit in Figure 14–16 if V_T = 36 volts, R_1 = 900 ohms, R_2 = 1200 ohms, and R_3 = 600 ohms.

13. What effect will doubling V_T have on the values calculated in Problem 7?

14. What changes will occur in the circuit described in Problem 7 if V_T is reduced to half of its original value?

15. How much power is used from a 12-volt automobile battery if the radio requires 1.2 amperes and the lights require 4.8 amperes?

15

PARALLEL CIRCUIT TESTING

OBJECTIVES

Measure the voltage in a parallel circuit.

Measure the current in a parallel circuit.

Measure the resistance in a parallel circuit.

Determine whether a parallel circuit is operating properly.

Locate defects in a parallel circuit.

INTRODUCTION

The objective of this chapter is to introduce the basic techniques of parallel circuit testing. Many of these techniques were used before in testing series circuits, but some new procedures are involved here, too.

Review of Circuit Testing

In Chapter 7, we discussed the use of a multimeter for measuring voltage, current, and resistance. Some essential points from that chapter should be reviewed here. Recall that voltage is measured across sources or components, while current is measured through a circuit. Voltage measurement requires no disconnection, while current measurement does. Both measurements must be made with the power on. Resistance is measured across a component with one of the component's leads disconnected and the power off. Furthermore, primary considerations in all testing are your own safety and the safety of the test equipment and the product under test.

Circuits are generally tested for two basic reasons. First, we test a circuit to determine if it is doing what it should be doing. That is, we compare our measurements with calculated values or manufacturers' specifications. Second, we test a circuit to locate defects and determine their cause. (See Chapter 9 for more detailed discussions of these points.)

All these procedures and reasons for testing apply to parallel circuits. But note that there are two essential differences between series circuits and parallel circuits. Series circuits have many voltages but only one current. Parallel circuits have only one voltage but many currents. Thus, testing parallel circuits involves some new techniques.

To summarize, before you begin testing a

FIGURE 15-1
PARALLEL CIRCUIT TO BE EVALUATED

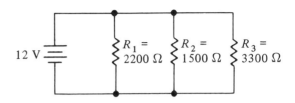

parallel circuit, you must know what it should be doing. You should also continue the practices of writing down formulas, identifying all answers, and placing calculated results in a table for comparison with measured results.

Circuit Calculations

Figure 15–1 shows the parallel circuit we are going to assemble and test in this chapter. We begin by calculating all values and recording them in a table. Recall that the calculations for a parallel resistive circuit were discussed in Chapter 14. Using the formulas presented there, we can calculate the current, resistance, and voltage for the parallel circuit in Figure 15–1. These calculations follow.

$$I_1 = \frac{V_T}{R_1} = \frac{12 \text{ V}}{2200 \text{ } \Omega} = \textbf{5.45 mA}$$

$$I_2 = \frac{V_T}{R_2} = \frac{12 \text{ V}}{1500 \text{ } \Omega} = \textbf{8 mA}$$

$$I_3 = \frac{V_T}{R_3} = \frac{12 \text{ V}}{3300 \text{ } \Omega} = \textbf{3.64 mA}$$

$$I_T = I_1 + I_2 + I_3$$

$$= 5.45 \text{ mA} + 8 \text{ mA} + 3.64 \text{ mA}$$

$$= \textbf{17.1 mA}$$

$$R_T = \frac{V_T}{I_T} = \frac{12 \text{ V}}{0.0171 \text{ A}} = \textbf{701.7 } \Omega$$

Since the voltage is the same in all parts of a parallel circuit, there are no more calculations to do. These results can be entered in a data table similar to Table 15–1. Notice that the calculated values of R_1, R_2, and R_3 are those indicated by color code. In other words, the calculated resistance values are the values specified by the manufacturers.

MEASURING CIRCUIT VALUES

Once the expected (calculated) values for a circuit are known, the actual values must be determined. That is, you must measure the voltage, current, and resistance. Safety is still the first concern here. Do not injure yourself, the meter, or the circuit. Accuracy is the next consideration. If the expected and actual values are not the same, be sure it is because of a real difference, not because of your error. Finally, once again, remember that voltage is measured across and current through a circuit.

Voltage

You begin this phase of evaluation by assembling the circuit according to the schematic in Figure 15–1. The voltage source depends on the quality requirements and the power supplies available. A low-voltage DC power supply can be used here. The power supply should be turned off before you begin. Use clip leads to make all connections. Clip leads allow you to make the breaks needed later for current measurement. Your assembled circuit should look like the one in Figure 15–2.

Before you energize the circuit, connect a VOM across the power supply output to measure the applied voltage. This measurement will provide you with an accurate way to ad-

TABLE 15-1
CALCULATED VALUES OF CIRCUIT PARAMETERS

Parameter	Calculated Value
V_T	12 V
R_1	2200 Ω
R_2	1500 Ω
R_3	3300 Ω
R_T	701.7 Ω
I_1	5.45 mA
I_2	8 mA
I_3	3.64 mA
I_T	17.1 mA

just the supply to 12 volts. Be sure to turn the control all the way down before you turn the supply on. Also, don't set the supply level until the complete circuit is connected. If you set the level first and then connect the circuit, the load on the supply from the circuit will cause the supply voltage to drop.

Measuring the voltage is quite simple. Use the procedures outlined in Chapter 7 to select the range and interpret the indications of the meter. Recall that the voltage is the same in all parts of a parallel circuit (that is, in a properly working parallel circuit).

To test this circuit properly, you must measure the voltage at the terminals of all components. You cannot assume that a clip lead is making a perfect connection. This concern is even more critical with printed circuits. For example, one problem that often occurs in printed circuits is a hairline fracture in the conductor and/or at connections. Thus, a circuit may look perfect, but there may be breaks in it that can only be seen with a magnifying glass. So measure all voltages at the terminals of all the components, and record the results in the data table.

FIGURE 15-2
PARALLEL CIRCUIT TO BE TESTED

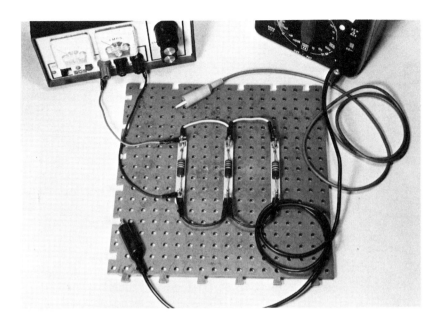

Current

Recall from Chapter 7 that a circuit must be broken in order to measure current and that the meter reconnects the break. The only electrical difference is the addition of the meter in series with the section of the circuit under test. Once again, be sure to turn the power off before you insert the meter.

There are four values of current to be measured in this circuit. So you must know exactly what current you are measuring. Also, be careful not to leave any connections open after you insert the meter. Figure 15–3 shows all of the places you could insert a meter. Some of them are correct and others are not, as we will see next.

As shown in Table 15–2, there are two locations where you can measure I_T. There are also two each for I_1, I_2, and I_3. One mistake

FIGURE 15-3
MEASURING CURRENT IN A PARALLEL CIRCUIT

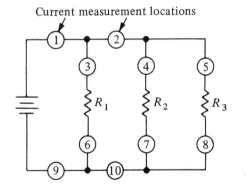

TABLE 15-2

POSITIONS IN CIRCUIT WHERE PARAMETERS CAN BE MEASURED

Position	Parameter
1	I_T
2	$I_2 + I_3$
3	I_1
4	I_2
5	I_3
6	I_1
7	I_2
8	I_3
9	I_T
10	$I_2 + I_3$

people often make is to think they are measuring I_2 at point 2 or 10. But if you look at the schematic in Figure 15–3, you will see that more than I_2 passes point 10; I_3 does also.

A second mistake that inexperienced people make is to try to measure I_1 by connecting a meter from point 3 to point 6. That connection is absolutely wrong. The meter will short out R_1 and possibly be ruined. Remember the following two key points.

Key points: Voltage is measured across a resistor. But current is measured through a circuit by inserting the meter in series.

Now, let's measure some current values. The easiest way to measure I_T is to turn off the supply, disconnect the lead from the negative terminal, connect that lead to the positive lead of the meter, and connect the negative lead of the meter to the negative terminal of the supply. Your connections should be made as shown in Figure 15–4. The meter will be located at position 9 of Figure 15–3.

Before you turn the supply back on, re-check the meter polarity and range. The meter negative lead should go to negative on the supply. And you should have selected the highest current range. Now, turn the supply on, select the best current range, read and record the value, and turn the supply off. You can now remove the meter and reconnect the clip lead to the supply.

Current I_1 can be measured next. Find a place where you can disconnect one side of R_1 without disconnecting any other part. If you select location 6 indicated in Figure 15–3, be sure that points 7, 8, 9, and 10 are still connected with each other after you break the circuit. Once you have broken the circuit, insert the meter. Observe the polarity. Negative on the meter should go toward negative on the supply. Your connections for measuring I_1 should be made as shown in Figure 15–5.

Although you may be tempted to assume that current I_1 will be less than I_T, don't. If you have made an incorrect connection or have shorted some clip leads, the current could be very high.

Again, to measure I_1, start with the highest current range. Energize the circuit, select the best range, read and record the value, and turn the supply off. Now, remove the meter and reconnect the circuit.

This procedure can be repeated for each of the remaining current values. But note that three types of errors are common in this process. First, you may misread the meter. Second, you may mistakenly connect a meter across instead of through the circuit. Remember, voltage across and current through. However, the most likely error you may make is to think you are measuring one current while you are actually measuring something else. As mentioned before, this error occurs when you break the circuit in the wrong place or leave some sections open. So, be careful, and keep these three points in mind before you make a measurement.

FIGURE 15-4
MEASURING TOTAL CURRENT

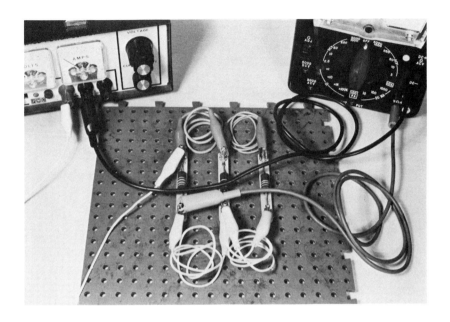

FIGURE 15-5
MEASURING CURRENT I_1

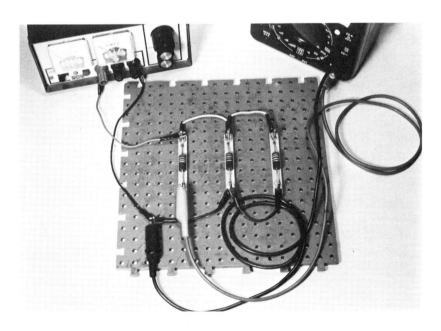

FIGURE 15-6
MEASURING TOTAL RESISTANCE

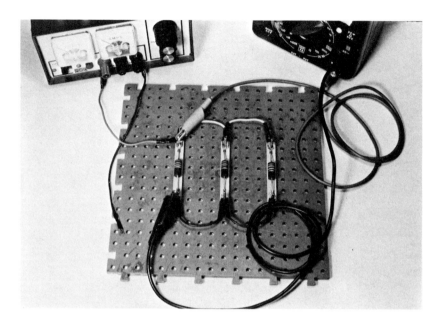

Resistance

The most obvious and logical time to measure resistors is before assembling a circuit. The second choice is to measure them after disassembling the circuit. But there are times when neither of these choices is open — that is, you must measure resistance in an assembled circuit. This third case is the situation you face here. So, before making any measurements, you must de-energize the circuit. Also, disconnect at least one lead of the component or components under test from the remainder of the circuit.

To measure resistance with a meter, you use the procedures outlined in Chapter 7 to zero the meter, select the best range, and interpret the indicated value. To prepare the circuit for measuring a resistance, just remove one lead of the circuit from the power supply

and connect the meter across the circuit leads. You must remove the lead even if the power is off. If you do not do so, there will be another parallel path in the circuit. Thus, for measurement of R_T, the connections should be made as shown in Figure 15–6.

Now, select the best range, read and record the indicated value, and disconnect the meter. You would normally reconnect the circuit to the power supply. However, for the tests here, the power supply must remain off.

The next step is to measure each resistor. Begin by removing one lead of the resistor from the rest of the circuit. Then, measure its resistance by following the steps outlined previously. For example, R_2 can be measured as illustrated in Figure 15–7.

Once R_2 has been measured, reconnect it and disconnect another resistor for measurement. One by one, each resistor should be dis-

**FIGURE 15-7
MEASURING RESISTANCE R_2**

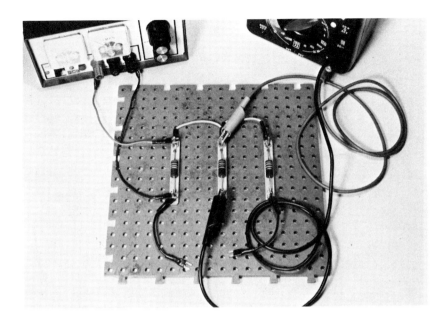

connected, measured, and reconnected as you continue the evaluation of this circuit. When all resistance values have been measured and recorded, the measurement process is complete. Determining the meaning of these results is the next step in circuit evaluation.

MEASUREMENT RESULTS

We have determined what the circuit should be doing and what the circuit is doing. We are now at a point where a decision must be made. Is the circuit working as expected? To answer this question, we must consider another question: Are all measured values of voltage, current, and resistance the same as the calculated values? From these comparisons, we determine whether or not the performance is normal.

Normal Performance

If the circuit you were testing was working exactly as it should, your results would look like the calculated and measured values shown in Table 15–3.

Notice, from the table, that the measured value of V_T should be exactly 12 volts, because you adjusted it. If it measured 11.5 volts, for example, you should have adjusted it to 12 volts. But how close the other measured values are to the calculated values depends on many factors. For instance, as calculated, R_1 should be 2200 ohms. But if it is a 10% resistor, it could be anywhere between 1980 and 2420 ohms and still be normal. Meter error can also affect the measured value. If the resistor was near the high end and the meter error was in the same direction, an indication of 2600 ohms might still be normal. These results can occur for all the resistors.

TABLE 15-3
COMPARISON OF CALCULATED AND
MEASURED VALUES

Parameter	Calculated Value	Measured Value
V_T	12 V	12 V
R_1	2200 Ω	2200 Ω
R_2	1500 Ω	1500 Ω
R_3	3300 Ω	3300 Ω
R_T	701.7 Ω	700 Ω
I_1	5.45 mA	5.5 mA
I_2	8 mA	8 mA
I_3	3.64 mA	3.6 mA
I_T	17.1 mA	17.1 mA

Another problem can occur when measuring current. We calculated I_1 at 5.45 milliamperes. However, if R_1 was acceptably high at 2420 ohms, I_1 would have an expected value of 4.95 milliamperes. A small meter error added to this expected value could produce an even lower measured value of current. This lower value of current would be normal and, therefore, acceptable, though.

So to decide whether operation is normal, you need to know more than the calculated or expected parameter values. You must also know normal variability in all components and the range of error of your test instruments. When more accurate measurements are needed, higher-quality instruments must be used. When measured values must be closer to calculated values, lower-tolerance components must be used.

To judge circuit operation, you must also know how much difference is allowed between measured and calculated values. This difference varies from one circuit to another. Under normal circumstances, however, a difference of up to 10% is probably acceptable.

This much difference can be caused simply by normal variations in parts or normal meter error. If the variation is greater than 10%, there is probably some defect in the circuit.

Short Circuit

A short circuit in a parallel circuit brings immediate disaster. If any of the components becomes shorted, there will be a path of low resistance as an alternate to all other paths. Since that path has low resistance, most current will go that way. Since that path is also an alternate in parallel with others, the total resistance will decrease greatly, causing a large increase in current. As a result, fuses will open.

A new fuse is not the answer, for it will allow the excessive current to flow again, even if only for a moment. This current flow will only cause more damage and will not help in locating the problem. Troubleshooting is the answer here. A measurement of total resistance will confirm that a short exists.

To troubleshoot a short circuit, remove the power, leave the meter connected across the total resistance, and open each branch, one at a time. If no large increase is indicated in R_T as you open a branch, close it and open another. Once one branch appears to cause a significant increase in R_T, it should be checked by itself.

Open Circuit

Open circuits in parallel circuits cause a decrease in current or no current at all, just as in series circuits. If one of the branches in this parallel circuit is open, the total current will be less than it should be. You can sometimes tell which branch is open by the amount of current that should, but does not, flow. For instance, if the open branch contained a lamp, it would be off. If you check the lamp and it is good, then one of the connections is open.

Finding open circuits can be quite difficult. One approach is to draw a schematic and determine what connections are not open — that is, which ones must be good if certain sections are operating as they should. In this approach, you find out what is defective by eliminating everything that is good.

Voltage measurement is an excellent method for locating opens. Recall that the total applied voltage will be dropped across an open circuit. So if you have located an open circuit by measuring branch current and cannot find the cause, measure the voltage across soldered connections. They may not be really soldered. Resistance measurement with the power off is another reliable method for locating opens.

FINDING CAUSES OF PROBLEMS

When you troubleshoot a circuit, locating and replacing a defective component is not all that you must do. You must also determine the probable cause of its failure so that you prevent possible damage from recurring.

If you find a defective resistor in a circuit, you must replace it. But you should not re-energize the circuit until you are sure that you know the cause of the problem, so it is not likely to recur. As mentioned in Chapter 9, shorts cause opens.

A short in a parallel circuit can result in such great current that it can cause part of the circuit to open. If you repair the open without locating and correcting the short, another open — and possibly fire — can result. Thus, in conclusion, always determine the cause of a problem and repair it while correcting the symptoms.

SUMMARY

Parallel circuit parameters are measured for two reasons: (1) to determine if the circuit is doing what it should be doing and (2) to locate defects and determine their cause. Therefore, before you begin to make any measurements, you must know what to expect. Normal operation of circuit components can be determined by reviewing manufacturers' specifications, test manuals, or your own calculations. Also, be sure to follow the proper procedures so that you do not make matters worse.

Voltage measurements in parallel circuits are quite simple. There is only one value of voltage across all components, and it is measured as explained in Chapter 7. However, be certain to measure this voltage across all components, because a component or connection may be open.

Current measurements take more time than voltage measurements because there are so many values to consider. You must measure the value of current for each branch plus the total current. In these measurements, be certain that the meter is connected in series with the branch you are testing. Also, be sure that only the branch under test is involved and that all other branches are connected as they should be.

Parallel circuit resistance measurements can only be done with the power off and with one lead of the component disconnected from the supply. These measurements, especially the one for R_T, are an excellent way to detect short circuits. When you connect an ohmmeter across the total resistance, the presence of a short becomes obvious: The meter will indicate 0 ohms. Branches can then be disconnected, one at a time, until the shorted one is found.

A comparison of measured and calculated values will indicate whether a circuit is working normally. You can usually assume that a variation of less than 20% is acceptable, if some other acceptable variation range has not been specified.

If a short is present, the current will be excessively high and will usually cause fuses to open. When there are no fuses, some component or wire will burn open. In contrast, open circuits will cause less current to flow, and one branch will not perform its intended function. In either case, you must find the cause of the problem.

When a short causes excessive current to flow, an open will result. If all you correct is the open, another open will occur. An aspirin for a toothache stops the pain but does not correct a defective tooth. Likewise, in electronics, correcting symptoms is not enough. Symptoms indicate that a problem exists, and you must find and correct the cause.

CHAPTER 15

REVIEW QUESTIONS

1. Name two reasons for testing a circuit.

2. What effects does an open have on a parallel circuit?

3. What are the symptoms of an open in a parallel circuit?

4. What effect does a short have on a parallel circuit?

5. What are the symptoms of a short in a parallel circuit?

6. Explain how you would proceed in troubleshooting a parallel circuit if you suspected that it had a short.

7. Explain how you would proceed in troubleshooting a parallel circuit if you suspected that it had an open.

8. How can you find a short by measuring current? What are the risks?

9. How can you find a short by measuring voltage?

10. Which method, voltage or current measurement, is the best for finding a short? Why?

11. How can you find an open by measuring current?

12. How can you find an open by measuring voltage?

13. Describe the procedure for measuring the resistance of a resistor that is part of a circuit.

14. Describe how you would determine whether a circuit was working properly.

REVIEW PROBLEMS

1. Describe how to measure the total current in a parallel circuit. Explain all steps and precautions. Draw a sketch of the connections.

2. Describe how to measure the current in the center branch of a three-resistor, parallel DC circuit. Explain all steps and precautions. Draw a sketch of the connections.

3. Describe how to measure the total resistance in a parallel circuit. Explain all steps and precautions. Draw a sketch of the connections.

4. Describe how to measure the resistance of the center resistor in a three-resistor, parallel DC circuit. Explain all steps and precautions. Draw a sketch of the connections.

5. What problems are indicated in the circuit of Figure 15–8 if I_1 = 4.5 milliamperes, I_2 = 1.5 milliamperes, I_3 = 0 milliamperes, and I_T = 6 milliamperes? What is the probable cause of the problems?

6. What problems are indicated in the circuit of Figure 15–8 if I_1 = 4.5 milliamperes, I_2 = 2 milliamperes, I_3 = 3 milliamperes, and I_T = 9.5 milliamperes?

7. What problems are indicated in the circuit of Figure 15–9 if the meter measurement of R shows 500 ohms? What is the probable cause of the problems?

8. What problems are indicated in the circuit of Figure 15–9 if the meter measurement of R shows 0 ohms? What is the probable cause of the problems?

FIGURE 15-8

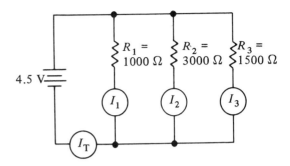

FIGURE 15-9

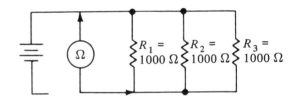

9. What is wrong with the circuit in Figure 15–8 if I_T is 4.5 milliamperes and I_1 is 0 milliamperes?

10. What is wrong with the circuit in Figure 15–8 if I_T is 4.5 milliamperes and I_2 is 0 milliamperes?

11. What can cause the meter in Figure 15–9 to indicate 1000 ohms?

12. What can cause the meter in Figure 15–9 to indicate 0 ohms?

13. What minimum and maximum current would you consider to be acceptable for the circuit in Figure 14–14 if the resistors are all 1000 ohms, 10%? 5%? The power supply voltage is 12 volts.

14. What minimum and maximum current would you consider to be acceptable for the circuit described in Problem 13 if the power supply has an output variation of no more than 2%? 1%?

15. What maximum and minimum current is acceptable for the combined effects described in Problems 13 and 14?

SERIES-PARALLEL CIRCUITS

OBJECTIVES

Define and identify a series-parallel circuit.

State four features of a series-parallel circuit.

Determine the simple equivalent of a series-parallel circuit.

Determine the voltage, current, and power of all resistors in a series-parallel circuit.

INTRODUCTION

The objective of this chapter is to introduce the theory of operation of series-parallel circuits. Series circuits have components connected one after another. They have only one current path. Parallel circuits have components connected electrically side by side. There is more than one current path. If you look at a number of schematics, you will notice that most circuits involve a combination of both series and parallel sections. We have studied these two types of circuits individually first because they are easier to work with. Also, more complex circuits must be separated into series and parallel segments in order to be evaluated.

FIGURE 16-1
SERIES-PARALLEL CIRCUITS

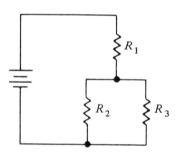

A. R_1 in series with R_2 and R_3 combination

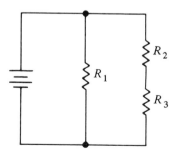

B. R_1 in parallel with R_2 and R_3 combination

A **series-parallel circuit** is one in which some components are connected in series and others are connected in parallel. The circuits in Figure 16–1 are two examples. Some people choose to call the circuit in Figure 16–1A series-parallel and the one in Figure 16–1B parallel-series. That is, the series-parallel circuit has parallel sections that are in series; the **parallel-series circuit** has series sections that are in parallel. To avoid wording problems about "which came first," this text will call all such circuits series-parallel, meaning the circuit has some of each type. It has series sections and parallel sections. What really matters is that you know where to begin in its analysis.

Unlike series or parallel circuits, series-parallel circuits cannot be drawn in just one way. There may be a series group of resistors in parallel with another series group or a few resistors in parallel connected in series with one resistor. The most noticeable characteristic of a series-parallel circuit is that it has some of the features of both types of circuits and must be solved with both approaches.

Recognizing Series-Parallel Connection

The best way to learn to recognize series-parallel circuits is by looking at a few examples. Figure 16–1A is an example of the simplest form of series-parallel circuit. As the figure shows, R_1 is in series with R_2 and R_3. But R_1 is not just in series with R_2, and it is not just in series with R_3. It is in series with the combination of them. Resistors R_2 and R_3 are in parallel with each other. So this circuit is a series-parallel circuit with R_1 in series with the combination of R_2 and R_3.

Another series-parallel circuit can be made from three resistors, as shown in Figure 16–1B. In this case, R_2 and R_3 are in series with each other. Resistor R_1 is in parallel with the combination of R_2 and R_3.

As the number of components increases,

the number of circuits also increases. Then, the simple addition of one conductor can drastically change the operation of a circuit. For example, observe the circuit in Figure 16–2A. It consists of R_1 in series with R_2 and of R_3 in series with R_4. The R_{12} combination is in parallel with the R_{34} combination.

The conductor added to the circuit in Figure 16–2B makes a significant change. Resistor R_1 is no longer in series with R_2; R_1 is now in parallel with R_3. And R_2 is in parallel with R_4. Moreover, the R_{13} combination is in series with the R_{24} combination.

Both circuits in Figure 16–2 are series-parallel. Although their components have the same values, their solution is not the same, and their values of R_T will be different. Recognition of circuits as series-parallel is the first step in their solution. The fact that they have both series and parallel segments is the clue to recognition.

Effects of Series-Parallel Connection

Now that we know what series-parallel circuits look like, we can discuss their behavior. Some of their characteristics are as follows:

1. There is more than one value of voltage in a series-parallel circuit.

2. There is more than one value of current in a series-parallel circuit.

3. The total resistance may be more or less than the largest resistor in a series-parallel circuit.

4. The total resistance may be more or less than the smallest resistor in a series-parallel circuit.

5. The total power is the sum of the separate powers in a series-parallel circuit.

Notice that only one of the characteristics is the same as those for series and parallel cir-

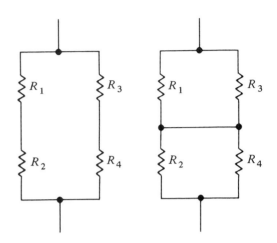

FIGURE 16-2
OTHER SERIES-PARALLEL CIRCUITS

A. R_{12} combination in parallel with R_{34} combination

B. R_{13} combination in series with R_{24} combination

cuits. Furthermore, the evaluation of a series-parallel circuit is more complex than the evaluation of a simple series or parallel circuit. It must be done in a particular order or the results will be wrong. To evaluate a series-parallel circuit, we must first recognize the series parts and the parallel segments.

Although the specific steps to be taken in evaluating these circuits vary from one circuit to another, the general approach remains the same. We begin with the original circuit and work toward one value of R_T. Once this value is determined, I_T is calculated. From there, we work back to the original circuit through a group of intermediate circuits.

RULES AND CALCULATIONS

The procedures for evaluating series and parallel circuits were outlined in earlier chapters. As we have seen, all series circuits are evalu-

FIGURE 16-3
SAMPLE CIRCUIT A

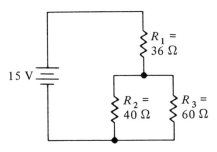

A. Series-parallel circuit

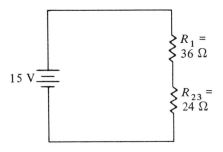

B. Circuit with R_2 and R_3 combined

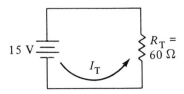

C. Total resistance for circuit

ated in the same manner. The only difference between evaluating one circuit and another is in the number of steps to be taken. Similarly, all parallel circuits are evaluated in the same way. Once again, the only difference in evaluation from one circuit to another is in how many steps are taken.

Series-parallel circuits, though, come in many different configurations. Therefore,

there are many possible ways to calculate the values of series-parallel circuits. However, each circuit has only one correct approach.

Resistance

As mentioned in the previous section, the first step in the evaluation of a series-parallel circuit is to determine the total resistance. To do so, we must first determine a series or parallel segment where we can begin. As we proceed, we will draw intermediate circuits along the way toward R_T to use on our way back. We'll begin with the first series-parallel circuit we looked at in the chapter, redrawn here in Figure 16–3A.

As we noted earlier, in this circuit, R_2 is in parallel with R_3. Therefore, we can represent the circuit as illustrated in Figure 16–3B. The calculation for R_{23} is

$$R_{23} = \frac{R_2 R_3}{R_2 + R_3} = \frac{40\ \Omega \times 60\ \Omega}{40\ \Omega + 60\ \Omega}$$

$$= \frac{2400}{100}\ \Omega = \mathbf{24\ \Omega}$$

The next step is to calculate R_T, as follows:

$$R_T = R_1 + R_{23} = 36\ \Omega + 24\ \Omega = \mathbf{60\ \Omega}$$

The simplified equivalent of the circuit in Figure 16–3A can now be represented as shown in Figure 16–3C.

The next step in the evaluation of this circuit is to calculate all current and voltage values. Those steps will be presented a little later in the chapter. At this time, it may be a good idea to determine the total resistance of a few more series-parallel circuits so that you become familiar with this phase of circuit evaluation.

Figure 16–4A shows another series-parallel circuit. Here, R_1 and R_2 are in series with each other. Resistors R_3 and R_4 are also in series with each other. And R_{12} is in parallel with

FIGURE 16-4
SAMPLE CIRCUIT B

FIGURE 16-5
CIRCUIT B WITH ONE CONDUCTOR
ADDED

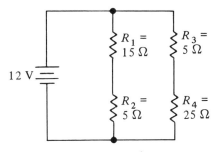

A. Series-parallel circuit

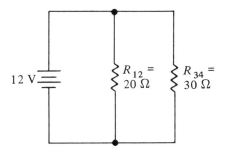

B. Circuit with R_1 and R_2 combined
and R_3 and R_4 combined

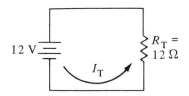

C. Total resistance for circuit

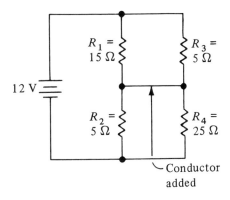

Conductor
added

A. Series-parallel circuit

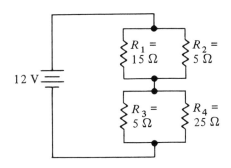

B. Electrically equivalent circuit

R_{34}, as shown in Figure 16–4B. Therefore, we must determine R_{12} and R_{34} before we can determine R_T. The calculations are as follows:

$$R_{12} = R_1 + R_2 = 15\ \Omega + 5\ \Omega = \textbf{20}\ \boldsymbol{\Omega}$$
$$R_{34} = R_3 + R_4 = 5\ \Omega + 25\ \Omega = \textbf{30}\ \boldsymbol{\Omega}$$

Now, we can determine R_T, as follows:

$$R_T = \frac{R_{12}R_{34}}{R_{12} + R_{34}} = \frac{20\ \Omega \times 30\ \Omega}{20\ \Omega + 30\ \Omega}$$

$$= \frac{600}{50}\ \Omega = \textbf{12}\ \boldsymbol{\Omega}$$

The total resistance of the circuit is 12 ohms, as illustrated in Figure 16–4C.

The addition of one conductor to the circuit of Figure 16–4A makes the circuit a much different series-parallel circuit, as shown in Figure 16–5A.

**FIGURE 16-6
SAMPLE CIRCUIT C**

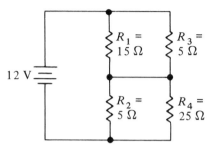

A. Series-parallel circuit

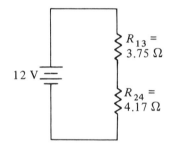

**B. Circuit with R_1 and R_3 combined
and with R_2 and R_4 combined**

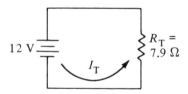

C. Total resistance for circuit

Figure 16–5B is not electrically different from Figure 16–5A. The circuit in Figure 16–5B does, however, make it easier to see the significance of the change caused by an additional conductor. Now, R_1 is in parallel with R_3, and R_2 is in parallel with R_4, as shown in Figure 16–6A. Also, R_{13} is in series with R_{24}, as shown in Figure 16–6B.

While this circuit is also a series-parallel circuit, its evaluation follows two different orders, as shown in the following calculations:

$$R_{13} = \frac{R_1R_3}{R_1 + R_3} = \frac{15\ \Omega \times 5\ \Omega}{15\ \Omega + 5\ \Omega} = \frac{75}{20}\ \Omega$$

$$= \mathbf{3.75\ \Omega}$$

$$R_{24} = \frac{R_2R_4}{R_2 + R_4} = \frac{5\ \Omega \times 25\ \Omega}{5\ \Omega + 25\ \Omega} = \frac{125}{30}\ \Omega$$

$$= \mathbf{4.17\ \Omega}$$

$$R_T = R_{13} + R_{24} = 3.75\ \Omega + 4.17\ \Omega$$

$$= 7.92\ \Omega = \mathbf{7.9\ \Omega}$$

Note that, for ease of later calculations, we round the final answer to 7.9 ohms.

Thus, we see how one added wire changed the circuit. The final equivalent circuit is shown in Figure 16–6C.

Current

Once we have determined the total resistance of a series-parallel circuit, we can determine the total current. From there, we proceed back through the equivalent circuits to determine all voltage values and all other current values. That is, so far, we have worked from the original circuit through a process of equivalent circuits until we reached a circuit having one resistor called R_T. Now, we will work our way back through the equivalent circuits to the original circuit.

The first circuit we looked at in the chapter, circuit A, was presented in Figure 16–3. We have calculated R_T for that circuit. Our next step is to determine total current. See Figure 16–7A. The calculation is

$$I_T = \frac{V_T}{R_T} = \frac{15\ V}{60\ \Omega} = \mathbf{0.25\ A}$$

Before we continue to the next step in circuit evaluation, we should look at the other two circuits. The total current for circuit B of

Figure 16–4 is shown in Figure 16–7B. The total current calculation is

$$I_T = \frac{V_T}{R_T} = \frac{12 \text{ V}}{12 \text{ }\Omega} = \textbf{1 A}$$

The total current for circuit C of Figure 16–6 is shown in Figure 16–7C. The calculation for that total current is

$$I_T = \frac{V_T}{R_T} = \frac{12 \text{ V}}{7.9 \text{ }\Omega} = \textbf{1.5 A}$$

Although the resistors are the same in circuits B and C, that one added wire in circuit C makes the circuits quite different.

Voltage

The next step in circuit evaluation is to determine some of the voltage values. Once again, we will begin with the first of the three circuits, circuit A. We can also note how much we know about the circuit so far. As mentioned earlier, it is a good practice to draw the various forms of a given circuit. We should also continue to mark, on the schematic, how much we know about it. With these observations in mind, we note that circuit A of Figure 16–3 should have the current from Figure 16–7A added to it.

Recall that we began by evaluating the circuits from top to bottom — that is, from part A to part C of the figures. The previous calculations determined all we know about the circuit at the bottom — that is, all values of V, I, and R. (Power will be evaluated later.) Now, we will work on the middle version, part B of the figures. Since we know I_T for part C, we can use that value for the circuits in part B. Therefore, for circuit A in Figure 16–3B, we can calculate the value for V_1 as follows:

$$V_1 = I_T R_1 = 0.25 \text{ A} \times 36 \text{ }\Omega = \textbf{9 V}$$

The other voltage for circuit A is the volt-

FIGURE 16-7
TOTAL CURRENT IN CIRCUITS A, B, AND C

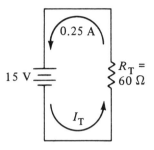

A. Total current in circuit A

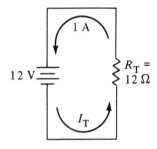

B. Total current in circuit B

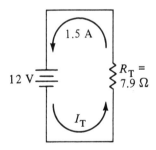

C. Total current in circuit C

age across the two parallel resistors, R_2 and R_3. This value is determined by using the value of their combined resistance and their combined current. While we do not yet know how much current passes through R_2 or R_3, we do know one fact. Since they are in parallel,

FIGURE 16-8
KNOWN VALUES OF CIRCUIT A

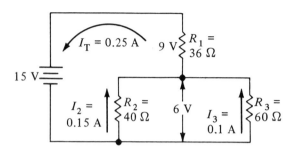

the total current equals the sum of their individual currents, and that sum is I_T. (Their individual currents will be found later.) Thus, we can calculate the voltage across R_{23} as follows:

$$V_{23} = I_T R_{23} = 0.25 \text{ A} \times 24 \text{ } \Omega = \textbf{6 V}$$

This answer can be checked as shown next:

$$V_T = V_1 + V_{23} = 9 \text{ V} + 6 \text{ V} = \textbf{15 V}$$

The answer checks. Now, the schematic is updated by adding these voltage amounts, as shown in Figure 16–8.

Next, we can determine how the current divides through R_2 and R_3. Since we know the voltage across each resistor and their individual resistance values, we can use Ohm's law. Thus, the current values are

$$I_2 = \frac{V_{23}}{R_2} = \frac{6 \text{ V}}{40 \text{ } \Omega} = \textbf{0.15 A}$$

$$I_3 = \frac{V_{23}}{R_3} = \frac{6 \text{ V}}{60 \text{ } \Omega} = \textbf{0.1 A}$$

Since we know that $I_T = 0.25$ ampere, we can check our answer, as shown next:

$$I_T = I_2 + I_3 = 0.15 \text{ A} + 0.1 \text{ A} = \textbf{0.25 A}$$

The answer checks. Finally, we can add all known values to the original version of the schematic for circuit A, as shown in Figure 16–8.

Now, we can pick up where we left off on series-parallel circuit B. First, we note that the circuit of Figure 16–4 should have the current from Figure 16–7B added to it. So, there is nothing left for us to calculate in the version of the circuit shown in Figure 16–4C. Thus, the next step is to determine the left and right branch currents of the circuit in Figure 16–4B. For this calculation, we use the voltage across each branch divided by the resistance of each branch, as follows:

$$I_{12} = \frac{V_T}{R_{12}} = \frac{12 \text{ V}}{20 \text{ } \Omega} = \textbf{0.6 A}$$

$$I_{34} = \frac{V_T}{R_{34}} = \frac{12 \text{ V}}{30 \text{ } \Omega} = \textbf{0.4 A}$$

$$I_T = I_{12} + I_{34} = 0.6 \text{ A} + 0.4 \text{ A} = \textbf{1 A}$$

The answer checks.

Since R_T is known, this answer can also be double-checked, as follows:

$$I_T = \frac{V_T}{R_T} = \frac{12 \text{ V}}{12 \text{ } \Omega} = \textbf{1 A}$$

The answer checks again.

With the middle version, Figure 16–4B, evaluated, we can move to the original circuit. The voltage calculations are the next step, and they are as follows:

$$V_1 = I_{12}R_1 = 0.6 \text{ A} \times 15 \text{ } \Omega = \textbf{9 V}$$

$$V_2 = I_{12}R_2 = 0.6 \text{ A} \times 5 \text{ } \Omega = \textbf{3 V}$$

$$V_T = E_1 + E_2 = 9 \text{ V} + 3 \text{ V}$$
$$= \textbf{12 V} \quad \text{(answer checks)}$$

$$V_3 = I_{34}R_3 = 0.4 \text{ A} \times 5 \text{ } \Omega = \textbf{2 V}$$

$$V_4 = I_{34}R_4 = 0.4 \text{ A} \times 25 \text{ } \Omega = \textbf{10 V}$$

$$V_T = E_3 + E_4 = 2 \text{ V} + 10 \text{ V}$$
$$= \textbf{12 V} \quad \text{(answer checks)}$$

This step completes the V, I, and R calculations for circuit B. The updated schematic with the known values added to it is shown in Figure 16–9.

Finally, we turn to circuit C of Figure 16–6, and we note that part C of that figure should have the current from Figure 16–7C added to it. The voltage calculations for the middle version, Figure 16–6B, can be completed on the way back to the original circuit. These calculations are

$$V_{13} = I_T R_{13} = 1.5 \text{ A} \times 3.75 \text{ } \Omega = \textbf{5.7 V}$$

$$V_{24} = I_T R_{24} = 1.5 \text{ A} \times 4.17 \text{ } \Omega = \textbf{6.3 V}$$

$$V_T = V_{13} + V_{24} = 5.7 \text{ V} + 6.3 \text{ V}$$

$$= \textbf{12 V} \quad \text{(answer checks)}$$

The updated schematic for the middle version of the circuit is shown in Figure 16–10 (bottom).

Now that we know the voltage across known resistance values, the individual current values can be determined, as follows:

$$I_1 = \frac{V_{13}}{R_1} = \frac{5.7 \text{ V}}{15 \text{ } \Omega} = \textbf{0.38 A}$$

$$I_3 = \frac{V_{13}}{R_3} = \frac{5.7 \text{ V}}{5 \text{ } \Omega} = \textbf{1.14 A}$$

Since I_1 and I_3 are in parallel, their sum should equal I_T. We check this answer next:

$$I_T = I_1 + I_3 = 0.38 \text{ A} + 1.14 \text{ A}$$

$$= \textbf{1.52 A}$$

The answer checks with a slight difference caused by rounding. Now, we complete the current calculations, as follows:

$$I_2 = \frac{V_{24}}{R_2} = \frac{6.3 \text{ V}}{5 \text{ } \Omega} = \textbf{1.26 A}$$

$$I_4 = \frac{V_{24}}{R_4} = \frac{6.3 \text{ V}}{25 \text{ } \Omega} = \textbf{0.25 A}$$

FIGURE 16-9
KNOWN VALUES OF CIRCUIT B

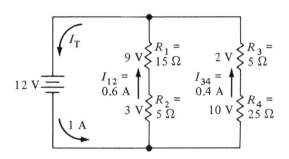

FIGURE 16-10
KNOWN VALUES OF CIRCUIT C

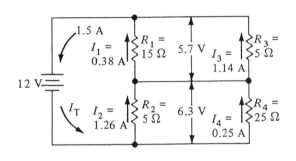

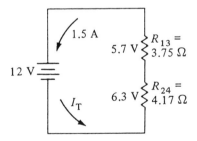

$$I_T = I_2 + I_4 = 1.26 \text{ A} + 0.25 \text{ A}$$

$$= \textbf{1.51 A}$$

Again, the answer checks with a slight difference caused by rounding off.

This step completes the V, I, and R calculations for series-parallel circuit C. The updated schematic is shown in Figure 16–10 (top).

Power

The final step in the evaluation of a series-parallel circuit is the calculation of individual resistor power values and the total power value. Since all voltage and current values are known at this point, we simply select the appropriate values and multiply them. We will again consider circuits A, B, and C, beginning where we left off with them.

For circuit A, refer to Figure 16–8. The power calculations for circuit A are as follows:

$$P_1 = V_1 I_1 = 9 \text{ V} \times 0.25 \text{ A} = \textbf{2.25 W}$$

$$P_2 = V_{23} I_2 = 6 \text{ V} \times 0.15 \text{ A} = \textbf{0.9 W}$$

$$P_3 = V_{23} I_3 = 6 \text{ V} \times 0.1 \text{ A} = \textbf{0.6 W}$$

$$P_T = P_1 + P_2 + P_3$$

$$= 2.25 \text{ W} + 0.9 \text{ W} + 0.6 \text{ W}$$

$$= \textbf{3.75 W}$$

$$P_T = V_T I_T = 15 \text{ V} \times 0.25 \text{ A}$$

$$= \textbf{3.75 W} \quad \text{(answer checks)}$$

Next, we complete the calculations for circuit B, which is shown in Figure 16–9. The power calculations for circuit B are as follows:

$$P_1 = V_1 I_{12} = 9 \text{ V} \times 0.6 \text{ A} = \textbf{5.4 W}$$

$$P_2 = V_2 I_{12} = 3 \text{ V} \times 0.6 \text{ A} = \textbf{1.8 W}$$

$$P_3 = V_3 I_{34} = 2 \text{ V} \times 0.4 \text{ A} = \textbf{0.8 W}$$

$$P_4 = V_4 I_{34} = 10 \text{ V} \times 0.4 \text{ A} = \textbf{4 W}$$

$$P_T = P_1 + P_2 + P_3 + P_4$$

$$= 5.4 \text{ W} + 1.8 \text{ W} + 0.8 \text{ W} + 4 \text{ W}$$

$$= \textbf{12 W}$$

$$P_T = V_T I_T = 12 \text{ V} \times 1 \text{ A}$$

$$= \textbf{12 W} \quad \text{(answer checks)}$$

Finally, we complete the calculations for circuit C, which is shown in Figure 16–10. The power calculations for circuit C are as follows:

$$P_1 = V_{13} I_1 = 5.7 \text{ V} \times 0.38 \text{ A} = \textbf{2.2 W}$$

$$P_2 = V_{24} I_2 = 6.3 \text{ V} \times 1.26 \text{ A} = \textbf{7.9 W}$$

$$P_3 = V_{13} I_3 = 5.7 \text{ V} \times 1.14 \text{ A} = \textbf{6.5 W}$$

$$P_4 = V_{24} I_4 = 6.3 \text{ V} \times 0.25 \text{ A} = \textbf{1.6 W}$$

$$P_T = P_1 + P_2 + P_3 + P_4$$

$$= 2.2 \text{ W} + 7.9 \text{ W} + 6.5 \text{ W} + 1.6 \text{ W}$$

$$= \textbf{18.2 W}$$

$$P_T = V_T I_T = 12 \text{ V} \times 1.5 \text{ A}$$

$$= \textbf{18 W} \quad \text{(answer checks)}$$

Complete Examples

Now that we have discussed each of the steps in a series-parallel evaluation, we can do a complete example. The circuit for this example is shown in Figure 16–11A. Since R_1 is in parallel with the combination of R_2 and R_3, R_{23} must be found first. The calculation is

$$R_{23} = R_2 + R_3 = 600 \ \Omega + 1000 \ \Omega$$

$$= \textbf{1600} \ \boldsymbol{\Omega}$$

The revised circuit is illustrated in Figure 16–11B.

The total resistance of the simplified equivalent circuit in Figure 16–11C can be calculated next:

$$R_T = \frac{R_1 R_{23}}{R_1 + R_{23}} = \frac{1200 \ \Omega \times 1600 \ \Omega}{1200 \ \Omega + 1600 \ \Omega}$$

$$= \frac{1,920,000}{2800} \ \Omega = \textbf{686} \ \boldsymbol{\Omega}$$

Since we know total voltage, we can calculate total current:

$$I_T = \frac{V_T}{R_T} = \frac{22 \text{ V}}{686 \ \Omega} = \textbf{0.032 A}$$

FIGURE 16-11
SAMPLE CIRCUIT D

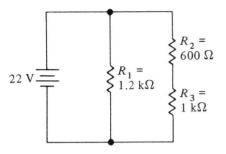

A. Series-parallel circuit

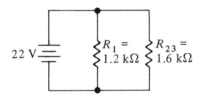

B. Circuit with R_2 and R_3 combined

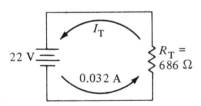

C. Total resistance and current

The branch currents, which are shown in Figure 16–12, can be computed next, as follows:

$$I_1 = \frac{V_T}{R_1} = \frac{22\ \Omega}{1200\ \Omega} = \textbf{0.018 A}$$

$$I_{23} = \frac{V_T}{R_{23}} = \frac{22\ \text{V}}{1600\ \Omega} = \textbf{0.014 A}$$

Now, let's consider voltage. The voltage across the left branch is dropped totally across

FIGURE 16-12
BRANCH CURRENTS AND VOLTAGE DROPS
FOR CIRCUIT D

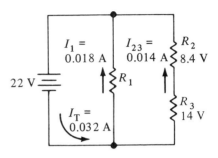

R_1. The voltage drops across R_2 and R_3 are determined as follows:

$$V_2 = I_{23}R_2 = 0.014\ \text{A} \times 600\ \Omega = \textbf{8.4 V}$$
$$V_3 = I_{23}R_3 = 0.014\ \text{A} \times 1000\ \Omega = \textbf{14 V}$$
$$V_T = E_2 + E_3 = 8.4\ \text{V} + 14\ \text{V} = \textbf{22.4 V}$$

This answer is slightly high because I_2 was rounded off from 0.01375 ampere to 0.014 ampere. Whether or not this amount of error is acceptable depends on the accuracy required by the circuit.

Calculating power is the final step. These calculations follow:

$$P_1 = V_T I_1 = 22\ \text{V} \times 0.018\ \text{A} = \textbf{0.39 W}$$
$$P_2 = V_2 I_{23} = 8.4\ \text{V} \times 0.014\ \text{A} = \textbf{0.12 W}$$
$$P_3 = V_3 I_{23} = 14\ \text{V} \times 0.014\ \text{A} = \textbf{0.19 W}$$
$$P_T = V_T I_T = 22\ \text{V} \times 0.032\ \text{A} = \textbf{0.7 W}$$

Let's do one more example. In this example, we will just show the calculations, without explanations. The circuit for this example is shown in Figure 16–13A.

The circuit in Figure 16–13A can be re-

FIGURE 16-13
SAMPLE CIRCUIT E

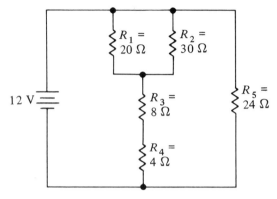

A. Series-parallel circuit

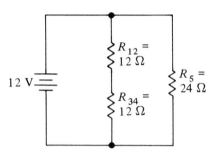

B. Circuit with R_1 and R_2 combined and R_3 and R_4 combined

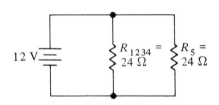

C. Circuit with R_{12} and R_{34} combined

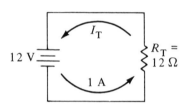

D. Total resistance and current

duced to the simpler form shown in Figure 16–13B. The resistance calculations are

$$R_{12} = \frac{R_1 R_2}{R_1 + R_2} = \frac{20\ \Omega \times 30\ \Omega}{20\ \Omega + 30\ \Omega} = \frac{600}{50}\ \Omega$$

$$= 12\ \Omega$$

$$R_{34} = R_3 + R_4 = 8\ \Omega + 4\ \Omega = 12\ \Omega$$

Another reduction in the circuit can be made, as shown in Figure 16–13C. The resistance calculation is

$$R_{1234} = R_{12} + R_{34} = 12\ \Omega + 12\ \Omega$$

$$= 24\ \Omega$$

The final reduction, as shown in Figure 16–13D, can now be made. The calculations for this version follow:

$$R_T = \frac{R_{1234} R_5}{R_{1234} + R_5} = \frac{24\ \Omega \times 24\ \Omega}{24\ \Omega + 24\ \Omega}$$

$$= \frac{576}{48}\ \Omega = 12\ \Omega$$

$$I_T = \frac{V_T}{R_T} = \frac{12\ V}{12\ \Omega} = 1\ A$$

Now, we can work back and calculate the branch currents shown in Figure 16–14A, as follows:

$$I_5 = \frac{V_T}{R_5} = \frac{12 \text{ V}}{24 \text{ }\Omega} = \textbf{0.5 A}$$

$$I_{1234} = \frac{V_T}{R_{1234}} = \frac{12 \text{ V}}{24 \text{ }\Omega} = \textbf{0.5 A}$$

The branch voltages shown in Figure 16–14B are then calculated, as follows:

$$V_{12} = I_{1234}R_{12} = 0.5 \text{ A} \times 12 \text{ }\Omega = \textbf{6 V}$$
$$V_{34} = I_{1234}R_{34} = 0.5 \text{ A} \times 12 \text{ }\Omega = \textbf{6 V}$$

Finally, the remaining voltage and current values, shown in Figure 16–14C, and the branch and total power values can be calculated, as shown next:

$$I_1 = \frac{V_{12}}{R_1} = \frac{6 \text{ V}}{20 \text{ }\Omega} = \textbf{0.3 A}$$

$$I_2 = \frac{V_{12}}{R_2} = \frac{6 \text{ V}}{30 \text{ }\Omega} = \textbf{0.2 A}$$

$$I_{12} = I_1 + I_2 = 0.3 \text{ A} + 0.2 \text{ A}$$
$$= \textbf{0.5 A} \quad \text{(answer checks)}$$
$$V_3 = I_{1234}R_3 = 0.5 \text{ A} \times 8 \text{ }\Omega = \textbf{4 V}$$
$$V_4 = I_{1234}R_4 = 0.5 \text{ A} \times 4 \text{ }\Omega = \textbf{2 V}$$
$$V_T = V_{12} + V_3 + V_4$$
$$= 6 \text{ V} + 4 \text{ V} + 2 \text{ V}$$
$$= \textbf{12 V} \quad \text{(answer checks)}$$
$$P_1 = V_{12}I_1 = 6 \text{ V} \times 0.3 \text{ A} = \textbf{1.8 W}$$
$$P_2 = V_{12}I_2 = 6 \text{ V} \times 0.2 \text{ A} = \textbf{1.2 W}$$
$$P_3 = V_3I_{1234} = 4 \text{ V} \times 0.5 \text{ A} = \textbf{2 W}$$
$$P_4 = V_4I_{1234} = 2 \text{ V} \times 0.5 \text{ A} = \textbf{1 W}$$
$$P_5 = V_TI_5 = 12 \text{ V} \times 0.5 \text{ A} = \textbf{6 W}$$
$$P_T = P_1 + P_2 + P_3 + P_4 + P_5$$
$$= 1.8 \text{ W} + 1.2 \text{ W} + 2 \text{ W} + 1 \text{ W}$$
$$\quad + 6 \text{ W}$$
$$= \textbf{12 W}$$
$$P_T = V_TI_T = 12 \text{ V} \times 1 \text{ A}$$
$$= \textbf{12 W} \quad \text{(answer checks)}$$

FIGURE 16-14
FINAL VALUES FOR CIRCUIT E

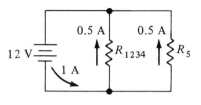

A. Branch currents

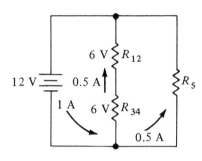

B. Branch voltages

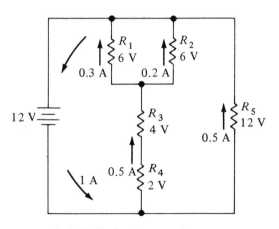

C. Individual voltages and currents

USES OF SERIES-PARALLEL CIRCUITS

In Chapter 8, we saw that circuits were designed in series form to gain some convenience. Series Christmas lights, for example, require fewer conductors than parallel circuits and allow the use of lower-voltage bulbs. We also saw that circuits were designed in a parallel form to gain other advantages. For instance, parallel circuits allow the use of one power source for many circuits. And when electronic products are completely assembled, many series-parallel circuits usually result.

The voltage divider described in Chapter 8 was a simple series circuit. However, when the circuits are added that will utilize those different voltage levels, a complex series-parallel circuit is produced, as shown in Figure 16–15. The evaluation of a loaded voltage divider like this one requires a series-parallel approach.

So circuits are not designed as series-parallel for convenience, as series circuits or parallel circuits are. Instead, series-parallel circuits are usually a result of connecting simple circuits together to produce a system.

SUMMARY

Circuits can be designed as series or parallel, depending on the needs of the designer. Eventually, simple series and parallel circuits become part of a larger system, forming a series-parallel circuit. A series-parallel circuit is one in which some components are connected in series and other components are connected in parallel.

Since series-parallel circuits can come in many forms, their rules and characteristics are a bit complex. The basic rules are as follows:

1. There is more than one value of voltage in a series-parallel circuit.

2. There is more than one value of current in a series-parallel circuit.

3. The total resistance may be more or less than the largest resistor in a series-parallel circuit.

4. The total resistance may be more or less than the smallest resistor in a series-parallel circuit.

5. The total power is the sum of the separate powers in a series-parallel circuit.

The evaluation of a series-parallel circuit proceeds with a series of equivalent-circuit schematics. Each of the two to four schematics should involve a step in combining series or parallel resistors into one equivalent resistor. The final circuit should have only one resistor, R_T. Resistance values for each equivalent circuit are calculated along the way toward the final circuit. Once the value of R_T is calculated, I_T is determined.

Step by step, all voltage and current values are calculated on the way back from the

---

FIGURE 16-15
VOLTAGE DIVIDER

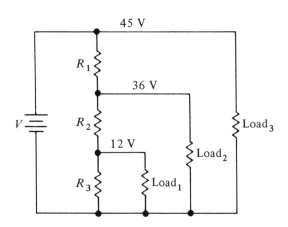

simplest equivalent circuit with R_T to the original circuit as drawn. After all values are known for an equivalent circuit, the next and more complex circuit is evaluated. Writing the values on the schematic as they are calculated is the best way to keep organized in this time-consuming task. Once all voltages and currents are known, the power values are calculated.

The key to evaluating series-parallel circuits is organization. Keep track of each voltage as it is calculated. Label the voltages V_1, V_2, V_{12}, and so on, since these values will be needed later for other calculations. One wrong label or calculation early in the work can produce many wrong answers. The ability to correctly evaluate complex circuits is an important skill for success as a technician.

CHAPTER 16

REVIEW TERMS

parallel-series circuit: circuit arrangement in which some series sections are in parallel

series-parallel circuit: circuit arrangement in which some parallel sections are in series; the term is often used as the name for circuits with both series and parallel sections

REVIEW QUESTIONS

1. Define series-parallel circuit.

2. State four characteristics of a series-parallel circuit.

3. Determine which of the circuits shown in Figure 16–16 are series-parallel.

4. Name two uses for series-parallel circuits.

5. Explain how current flows in a series-parallel circuit.

6. Describe the steps taken when determining all values in a series-parallel circuit.

7. Explain why it is important to draw additional schematics when determining all values in a series-parallel circuit.

FIGURE 16-16

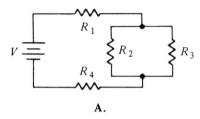

A.

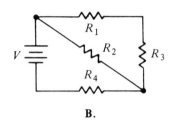

B.

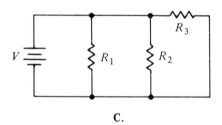

C.

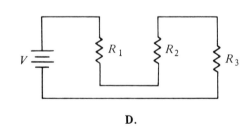

D.

8. Which resistors would you combine first when evaluating the characteristics of the circuit in Figure 16–17? Why?

9. Which resistors would you combine first when evaluating the circuit in Figure 16–16A? Why?

FIGURE 16-17

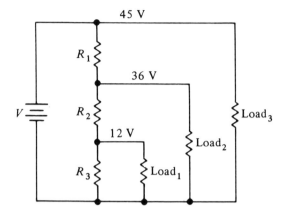

10. Which resistors would you combine first when evaluating the circuit in Figure 16–16B? Why?

11. What is the effect on circuit performance if the applied emf in a series-parallel circuit is doubled?

12. What is the effect on circuit power if the total resistance in a series-parallel circuit is doubled?

13. Explain why R_1 and R_2 are added for the circuit in Figure 16–2A but not for the circuit in Figure 16–2B.

14. Under what circumstances would the presence or absence of the jumper wire in Figure 16–2B make no difference?

15. What would have the greatest impact on the total power of the circuit in Figure 16–4A, doubling R_1 or doubling R_2? Why?

REVIEW PROBLEMS

1. Determine the total resistance of the circuit pictured in Figure 16–18 if $R_1 = 24$ ohms, $R_2 = 80$ ohms, and $R_3 = 120$ ohms.

2. Determine the total resistance of the circuit in Figure 16–18 if R_1 = 1.2 kilohms, R_2 = 1.5 kilohms, and R_3 = 2.2 kilohms.

3. Determine the total resistance of the circuit pictured in Figure 16–19 if R_1 = 60 ohms, R_2 = 35 ohms, and R_3 = 25 ohms.

4. Determine the total resistance of the circuit in Figure 16–19 if R_1 = 3.3 megohms, R_2 = 1.5 megohms, and R_3 = 2.2 megohms.

5. Calculate the total resistance of the circuit shown in Figure 16–20 if each resistance is 1000 ohms.

6. Calculate the total resistance of the circuit in Figure 16–20 if R_1 = 240 ohms, R_2 = 400 ohms, R_3 = 600 ohms, and R_4 = 180 ohms.

7. Calculate the total resistance of the circuit in Figure 16–20 if R_1 = 2.2 megohms, R_2 = 4.7 megohms, R_3 = 1.5 megohms, and R_4 = 1 megohm.

8. Determine all values of resistance, current, voltage, and power for the circuit in Figure 16–18 if V_T = 24 volts, R_1 = 120 ohms, R_2 = 240 ohms, and R_3 = 180 ohms.

9. Determine all values of resistance, current, voltage, and power for the circuit in Figure 16–18 if V_T = 12 volts, R_1 = 2.4 kilohms, R_2 = 3.6 kilohms, and R_3 = 1.8 kilohms.

10. What will happen to the circuit values calculated for Problem 9 if the applied voltage is reduced to 10 volts?

11. Determine all values of resistance, current, voltage, and power for the circuit in Figure 16–19 if V_T = 18 volts, R_1 = 33 ohms, R_2 = 12 ohms, and R_3 = 24 ohms.

12. Determine all values of resistance, current, voltage, and power for the circuit in Figure 16–19 if V_T = 9 volts, R_1 = 1.5

FIGURE 16-18

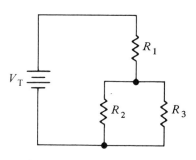

FIGURE 16-19

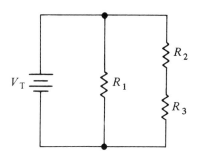

FIGURE 16-20

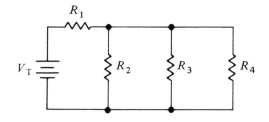

megohms, R_2 = 470 kilohms, and R_3 = 630 kilohms.

13. What will happen to the circuit values calculated for Problem 12 if the applied voltage is increased to 12 volts?

14. Using Figure 16–20, determine all values of resistance, current, voltage, and power

for this circuit if V_T = 48 volts, R_1 = 2.2 kilohms, R_2 = 4.7 kilohms, R_3 = 3.3 kilohms, and R_4 = 5.6 kilohms.

15. Determine all values of resistance, current, voltage, and power for the circuit in Figure 16–20 if V_T = 12 volts, R_1 = 48 ohms, R_2 = 82 ohms, R_3 = 76 ohms, and R_4 = 68 ohms.

16. What will happen to the circuit values calculated for Problem 15 if the applied voltage is changed to 15 volts?

SERIES-PARALLEL CIRCUIT TESTING

Measure all voltages, currents, and resistances in a series-parallel circuit.

Compare all measured values with calculated values, and identify any discrepancies.

Describe what is meant by an acceptable range of variation.

Identify the cause of any discrepancies in the operation of a series-parallel circuit.

CALCULATING CIRCUIT VALUES

The objective of this chapter is to expand your testing and troubleshooting abilities to more complex circuits. Now that you have seen how to evaluate series-parallel circuits and have learned how to troubleshoot both series and parallel circuits, you are ready for testing series-parallel circuits.

Recall from earlier chapters that you must know what a circuit *should* be doing before you determine what it *is* doing. Manufacturers' specifications are one way of determining what a circuit should be doing. But in the absence of specifications, you must produce your own calculations of circuit values. So in this section, we will find calculated values for two experimental circuits that you will build and test in this chapter.

Note that this chapter will present the topics in a parallel fashion rather than in series. That is, we will work on similar parts of both circuits at the same time. Thus, both sets of specifications will be developed, tests for similar segments of both circuits will be conducted, and corresponding data will be analyzed.

Experimental Circuit A

The first circuit we consider is the three-resistor circuit shown in Figure 17–1A. It is as simple a series-parallel circuit as we can have. The combined resistance, shown in Figure 17–1B, is

$$R_{23} = \frac{R_2 R_3}{R_2 + R_3} = \frac{3300\ \Omega \times 4700\ \Omega}{3300\ \Omega + 4700\ \Omega}$$

$$= \frac{15{,}510{,}000}{8000}\ \Omega = \mathbf{1939\ \Omega}$$

The total resistance and total current for the equivalent circuit in Figure 17–1C are then calculated as follows:

$$R_T = R_1 + R_{23} = 1800\ \Omega + 1939\ \Omega$$

$$= \mathbf{3739\ \Omega}$$

$$I_T = \frac{V_T}{R_T} = \frac{18\ V}{3739\ \Omega} = \mathbf{0.0048\ A}$$

The voltage calculations are done next:

$$V_1 = I_T R_1 = 0.0048\ A \times 1800\ \Omega = \mathbf{8.64\ V}$$

$$V_{23} = I_T R_{23} = 0.0048\ A \times 1939\ \Omega = \mathbf{9.3\ V}$$

$$V_T = V_1 + V_{23} = 8.6\ V + 9.3\ V = \mathbf{17.9\ V}$$

This answer for V_T is slightly low because of the rounding off of R_{23} and I_T earlier.

The final calculations are for branch currents, as shown next:

$$I_2 = \frac{V_{23}}{R_2} = \frac{9.3\ V}{3300\ \Omega} = \mathbf{0.0028\ A}$$

$$I_3 = \frac{V_{23}}{R_3} = \frac{9.3\ V}{4700\ \Omega} = \mathbf{0.0020\ A}$$

$$I_T = I_2 + I_3 = 0.0028\ A + 0.0020\ A$$

$$= \mathbf{0.0048\ A} \quad \text{(answer checks)}$$

Once the final values are calculated, the schematic can be updated by including all calculated values on it, as shown in Figure 17–1D.

A table of specifications, similar to Table 17–1, can now be prepared. This table indicates all values we expect according to our calculations and which parameters to measure. Note that R_{23} is an intermediate value used to calculate R_T, and hence it will not be measured. A column titled Measured Value should be added to the table so that you can enter values as they are measured. This technique makes it easier to compare the measured values with the calculated ones.

FIGURE 17-1
EXPERIMENTAL CIRCUIT A

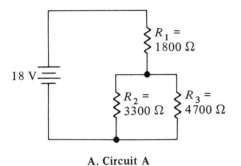

A. Circuit A

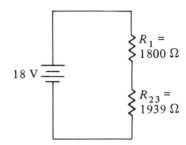

B. Circuit A with R_2 and R_3 combined

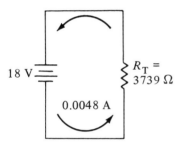

C. Equivalent circuit

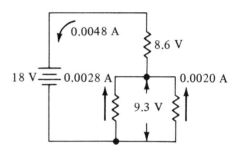

D. Final values for circuit A

TABLE 17-1
CALCULATED VALUES OF EXPERIMENTAL
CIRCUIT A

Parameter	Calculated Value
V_T	18 V
V_1	8.6 V
V_{23}	9.3 V
R_1	1800 Ω
R_2	3300 Ω
R_3	4700 Ω
R_{23}	1939 Ω
R_T	3739 Ω
I_T	0.0048 A
I_2	0.0028 A
I_3	0.002 A

Experimental Circuit B

Circuit B is shown in Figure 17–2A. Notice that this circuit is more complex than circuit A.

The first resistance calculation for circuit B is

$$R_{45} = \frac{R_4 R_5}{R_4 + R_5} = \frac{470 \ \Omega \times 560 \ \Omega}{470 \ \Omega + 560 \ \Omega}$$

$$= \frac{263,200}{1030} \ \Omega = \mathbf{255 \ \Omega}$$

The circuit can now be drawn as in Figure 17–2B. Next, we calculate R_{2345}:

$$R_{2345} = R_2 + R_3 + R_{45}$$

FIGURE 17-2
EXPERIMENTAL CIRCUIT B

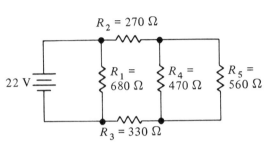

A. Circuit B

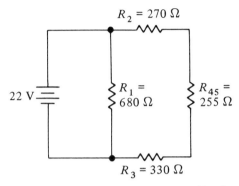

B. Circuit B with R_4 and R_5 combined

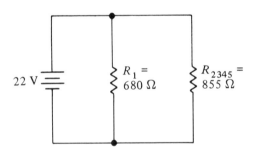

·C. Circuit B reduced to a parallel circuit

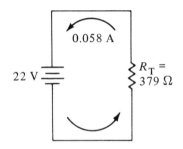

D. Equivalent circuit

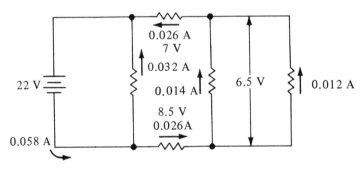

E. Final values for circuit B

$$= 270 \ \Omega \ + \ 330 \ \Omega \ + \ 255 \ \Omega$$

$$= \mathbf{855 \ \Omega}$$

The circuit is now reduced to a simple parallel circuit, as shown in Figure 17–2C, from which we can calculate the total resistance and current, as follows:

$$R_T = \frac{R_1 R_{2345}}{R_1 + R_{2345}} = \frac{680 \ \Omega \ \times \ 855 \ \Omega}{680 \ \Omega \ + \ 855 \ \Omega}$$

$$= \frac{581,400}{1535} \ \Omega = \mathbf{379 \ \Omega}$$

$$I_T = \frac{V_T}{R_T} = \frac{22 \ V}{379 \ \Omega} = \mathbf{0.058 \ A}$$

The final simplified circuit is shown in Figure 17–2D.

Now that total resistance and total current have been calculated, the steps back to the original circuit can begin. Branch current and voltage calculations are as follows:

$$I_1 = \frac{V_T}{R_1} = \frac{22 \ V}{680 \ \Omega} = \mathbf{0.032 \ A}$$

$$I_{2345} = \frac{V_T}{R_{2345}} = \frac{22 \ V}{855 \ \Omega} = \mathbf{0.026 \ A}$$

$$V_2 = I_{2345} R_2 = 0.026 \ A \ \times \ 270 \ \Omega$$

$$= \mathbf{7 \ V}$$

$$V_3 = I_{2345} R_3 = 0.026 \ A \ \times \ 330 \ \Omega$$

$$= \mathbf{8.5 \ V}$$

$$V_{45} = I_{2345} R_{45} = 0.026 \ A \ \times \ 255 \ \Omega$$

$$= \mathbf{6.5 \ V}$$

$$I_4 = \frac{V_{45}}{R_4} = \frac{6.5 \ V}{470 \ \Omega} = \mathbf{0.014 \ A}$$

$$I_5 = \frac{V_{45}}{R_5} = \frac{6.5 \ V}{560 \ \Omega} = \mathbf{0.012 \ A}$$

These final values are then recorded on the schematic, as shown in Figure 17–2E.

The specifications for this circuit can now be listed in a specifications table, like Table

TABLE 17-2
CALCULATED VALUES OF EXPERIMENTAL CIRCUIT B

Parameter	Calculated Value
V_T	22 V
V_2	7 V
V_3	8.5 V
V_{45}	6.5 V
R_1	680 Ω
R_2	270 Ω
R_3	330 Ω
R_4	470 Ω
R_5	560 Ω
R_T	379 Ω
I_T	0.058 A
I_1	0.032 A
I_4	0.014 A
I_5	0.012 A

17–2. Note that some intermediate values are left out. Only those parameters to be measured are included in the table.

MEASURING CIRCUIT VALUES

You should now get the eight resistors, power supply, multimeter, and assorted leads necessary for building and testing circuits A and B. Since the circuits are not yet assembled, it makes sense to measure all resistors now.

Resistance

As you measure each resistor, record its values in a table similar to Table 17–1 or Table 17–2. Check each one to see whether it falls within the range indicated by its color code. Keep in mind, however, the range of possible error of the multimeter. Any resistor far out of range should be replaced.

At this point in circuit evaluation, other steps are often taken. For instance, if more accurate measurements are needed, a resistance bridge may be used instead of a multimeter. If more accurate values are needed, 1% resistors are used rather than 10% ones. Finally, the calculations should all be repeated by using measured resistance values rather than color-coded values. In this way, a more realistic indication of what to expect from the circuits is obtained.

After you have measured and recorded the value of each resistor, assemble both circuits. However, you should measure and record R_T for each circuit before connecting the resistor network to a power supply. Once R_T is measured, connect the first circuit to a power supply and adjust for a V_T of 18 volts.

Voltage

The first voltage measurement you should always make is V_T. If it is not 18 volts for circuit A, then you must adjust the power supply. Be sure you adjust it with the circuit connected in order to account for the loading effect.

After measuring and recording the three voltages for circuit A, remove that circuit and connect circuit B. Readjust the power supply to 22 volts. Measure and record the four voltages for this circuit. This step completes the normal voltage measurements. However, there is one additional experiment to try.

Add two columns to your copy of Table 17–2. One of them should be labeled R_4 Shorted, and the other should be labeled R_4 Opened. Now, place a clip lead across R_4 to simulate a defect in the resistor. Measure and record all voltages. Remove the short and R_4, making sure not to open R_5 also. Once again, measure and record all voltages. This step completes the voltage measurements.

Current

Once again, measuring current is the most difficult step. It is difficult because there are so many ways to make mistakes in measuring current. For example, you might connect the meter the wrong way or in the wrong location. Before you begin, review the procedures outlined in Chapter 15.

Now, connect circuit A to the power supply and adjust the supply for 18 volts. Then, remove the lead from the negative terminal so that you can place the multimeter in series. This measurement for I_T is also the same as that for I_1, so one location is enough. But the meter must be removed and the lead replaced before you can measure I_2 and I_3. Remember to turn the supply off (not down) when you connect or disconnect the meter for current measurement.

Branch currents I_2 and I_3 are measured next. Be sure you interrupt the circuit in the proper location so that you measure the desired current and not any other. Figure 17–3 shows the correct locations for circuit A. After you have measured and recorded the currents for the first circuit, turn off the supply, disconnect the circuit, and connect circuit B.

Once circuit B is connected, adjust the power supply for 22 volts. Then, turn off the supply, connect the multimeter to measure I_T, and turn the supply on again. Measure and record I_T. Now, you begin the process of measuring all the remaining currents. Follow this procedure: Turn off the supply, insert the meter, turn the supply on, measure and record the current, turn off the supply, and reconnect the circuit.

The locations for measuring all remaining current values of circuit B are indicated in Figure 17–4. Current I_3 does not need to be measured since it should be the same as I_2. When all currents have been measured and recorded, you can reintroduce the first defect described earlier. That is, short R_4 with a clip lead. Measure and record all values of current, including I_4. These values should be recorded in your copy of Table 17–2 in the column R_4 Shorted.

Remove the short across R_4 and remove R_4. Measure and record all values of current (note that there is no I_4). These values should be recorded in the column R_4 Opened. Once this step is done, disassemble the circuit and prepare your data for analysis.

MEASUREMENT RESULTS

As in Chapter 15, you now have two sets of data. Your calculations tell you what the circuit should be doing; they are the expected results. These values are also called the circuit specifications. Your second set of data consists of the measured or actual results; they tell you what the circuit is doing. The next step is to determine whether or not circuit performance is normal. The type of analysis and decision making involved in this step is a key part of a technician's job.

Normal Performance

Your copies of Tables 17–1 and 17–2 provide convenient ways for you to compare the expected values with the measured values of all circuit parameters. Look at each measured value and ask, Does this result seem reasonable? Is this result within the normal range of error that could be caused by component tolerance or meter accuracy? As noted before, if the actual value is within 10% of the calculated value, it is probably acceptable. Your results for circuit A should be that close. If they are not, you should find out why.

The second circuit presents additional concerns. With more components that can vary, the differences between calculated and

FIGURE 17-3
MEASURING CURRENT IN CIRCUIT A

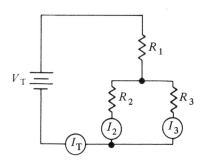

FIGURE 17-4
MEASURING CURRENT IN CIRCUIT B

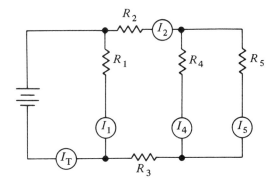

measured values can be even greater than they are in a simpler circuit. Still, some results are predictable. For example, the sum of the voltage drops should equal the total source voltage. While one resistor may be out of tolerance and have a higher share of the voltage, one in series with it will have less. Branch currents should add to the total even though one may be more or less than expected. If the voltages or currents do not add properly, check your measurements.

FIGURE 17-5
EFFECTS OF A SHORT IN CIRCUIT A

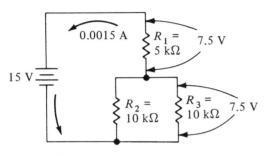

A. Normal operation for circuit A

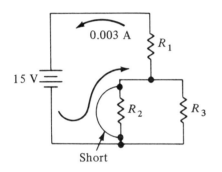

B. Shorting R_2

FIGURE 17-6
EFFECTS OF A SHORT IN CIRCUIT B

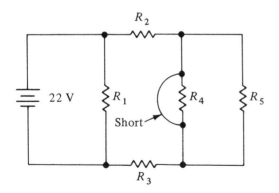

Short Circuit

Regardless of its location, a short will cause an increase in current. To see why, we will examine the circuit shown in Figure 17–5A. This circuit is similar to circuit A, but different values are used. As you know, R_2 and R_3 in parallel should produce an R_{23} of 5000 ohms. That resistance in series with the 5000 ohms of R_1 should cause equal voltage drops of 7.5 volts. And I_T will be 0.0015 ampere.

If R_2 is shorted, as shown in the circuit of Figure 17–5B, R_3 will also be shorted, since they are in parallel. In this situation, the total resistance will be 5000 ohms, and I_T will increase to 0.003 ampere. The voltage across R_1 will increase to 15 volts, and V_{23} will drop to zero. Thus, the circuit will be completely different.

Now, consider the results for shorting R_4 in circuit B, as shown in Figure 17–6. In this situation, R_4 and R_5 no longer enter the calculations, for they are both shorted. The revised circuit, shown in Figure 17–7A, can be analyzed in this way: First, we calculate I_1 and the combined resistance R_{23}:

$$I_1 = \frac{V_T}{R_1} = \frac{22 \text{ V}}{680 \text{ }\Omega} = \textbf{0.032 A}$$

$$R_{23} = R_2 + R_3 = 270 \text{ }\Omega + 330 \text{ }\Omega$$
$$= \textbf{600 }\boldsymbol{\Omega}$$

Now, the circuit is reduced to the simple parallel circuit shown in Figure 17–7B, which can be analyzed as follows:

$$I_{23} = \frac{V_T}{R_{23}} = \frac{22 \text{ V}}{600 \text{ }\Omega} = \textbf{0.037 A}$$

$$V_2 = I_{23}R_2 = 0.037 \text{ A} \times 270 \text{ }\Omega = \textbf{10 V}$$

$$V_3 = I_{23}R_3 = 0.037 \text{ A} \times 330 \text{ }\Omega = \textbf{12 V}$$

$$I_T = I_1 + I_{23} = 0.032 \text{ A} + 0.037 \text{ A}$$
$$= \textbf{0.069 A}$$

FIGURE 17-7
EVALUATING THE EFFECTS OF A SHORT

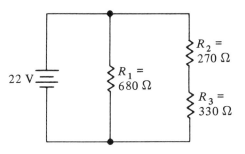

A. Circuit B with R_4 and R_5 shorted

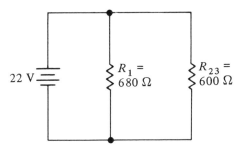

B. R_2 and R_3 combined

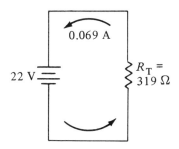

C. Equivalent circuit

$$R_T = \frac{V_T}{I_T} = \frac{22 \text{ V}}{0.069 \text{ A}} = \textbf{319 } \Omega$$

Figure 17–7C shows the total values and the equivalent circuit.

Table 17–3 tabulates the effect that short-

TABLE 17-3
EFFECTS OF A SHORT IN EXPERIMENTAL
CIRCUIT B

Parameter	Calculated Normal Circuit Values	R_4 Shorted
V_T	22 V	22 V
V_2	7 V	10 V
V_3	8.5 V	12 V
V_{45}	6.5 V	0 V
R_T	379 Ω	319 Ω
I_T	0.058 A	0.069 A
I_1	0.032 A	0.032 A
I_5	0.012 A	0 A

ing R_4 has on circuit operations. Calculated values without a short have been taken from Table 17–2. A comparison of the two data columns in Table 17–3 indicates some significant facts about partially shorted circuits. As the table shows, the total resistance has decreased and the total current has increased. There is no current through R_5 because that component is in parallel with a short. So, the first conclusion is that this circuit is no longer working as it is supposed to be.

However, there is another problem: R_3 is a 330-ohm resistor that we expect will have 8.5 volts across and 0.026 ampere through. Thus, we might install a ¼-watt resistor for R_3 because of the estimated power, which is calculated as follows:

$$P_3 = V_3 I_3 = 8.5 \text{ V} \times 0.069 \text{ A} = \textbf{0.22 W}$$

But notice what happens to this resistor when R_4 becomes shorted. The new calculation is

$$P_3 = V_3 I_3 = 12 \text{ V} \times 0.037 \text{ A} = \textbf{0.44 W}$$

There is a 100% increase in the power dissipation of R_3. If R_3 is a ¼-watt resistor, it will probably open. The total current will go to zero, and you will suddenly notice something wrong. A brief look at the circuit will show a

FIGURE 17-8
OPEN IN CIRCUIT A

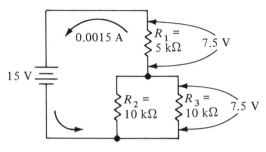

A. Normal characteristics of circuit A

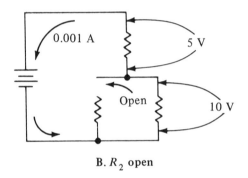

B. R_2 open

burned-out R_3. If the short is not found, a new R_3 will also burn out.

A good troubleshooter will notice the above-normal and below-normal voltage drops across the resistors. The most significant one, zero voltage across a component, indicates a short across the component being tested or no voltage from the source.

Open Circuit

An open will always reduce the current and increase the total resistance in a parallel section of a circuit. Consider a simple series-parallel circuit like circuit A but with one resistor open. Under normal conditions, as shown in Figure 17–8A, R_T will be 10,000 ohms, I_T will

be 0.0015 ampere, and V_1 and V_{23} will each be 7.5 volts.

Now, let's see what happens if R_2 opens up, as shown in Figure 17–8B. To begin with, the parallel option at the bottom of the circuit has been reduced to only one path. With R_2 open, R_1 will be in series with 10,000 ohms, causing R_T to rise to 15,000 ohms and I_T to fall to 0.001 ampere. The voltage drops will also readjust, with V_1 dropping to 5 volts and V_2 rising to 10 volts. The circuit will still be functioning, but it will be operating differently from what was intended.

Now, consider circuit B with an open, as shown in Figure 17–9A. Resistor R_4 is opened, so it can just be ignored. The resulting circuit is shown in Figure 17–9B, and it can be analyzed in this way:

$$I_1 = \frac{V_T}{R_1} = \frac{22 \text{ V}}{680 \text{ }\Omega} = \textbf{0.032 A}$$

$$R_{235} = R_2 + R_3 + R_5$$
$$= 270 \text{ }\Omega + 330 \text{ }\Omega + 560 \text{ }\Omega$$
$$= \textbf{1160 }\Omega$$

At this point, the circuit is reduced to a parallel circuit, as shown in Figure 17–9C, which can be analyzed as follows:

$$I_{235} = \frac{V_T}{R_{235}} = \frac{22 \text{ V}}{1160 \text{ }\Omega} = \textbf{0.019 A}$$

$$V_2 = I_{235}R_2 = 0.019 \text{ A} \times 270 \text{ }\Omega$$
$$= \textbf{5.13 V}$$

$$V_3 = I_{235}R_3 = 0.019 \text{ A} \times 330 \text{ }\Omega$$
$$= \textbf{6.27 V}$$

$$V_5 = I_{235}R_5 = 0.019 \text{ A} \times 560 \text{ }\Omega$$
$$= \textbf{10.6 V}$$

$$I_T = I_1 + I_{235} = 0.032 \text{ A} + 0.019 \text{ A}$$
$$= \textbf{0.051 A}$$

$$R_T = \frac{V_T}{I_T} = \frac{22 \text{ V}}{0.051 \text{ A}} = \textbf{431 }\Omega$$

FIGURE 17-9
OPEN IN CIRCUIT B AND EFFECTS OF THE
OPEN

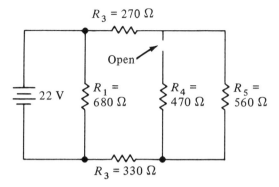

A. Circuit B with R_4 open

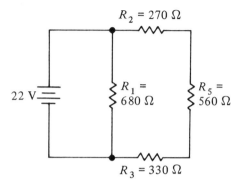

B. Resulting circuit B

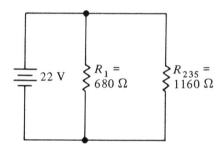

C. Circuit B reduced to
a parallel circuit

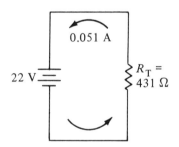

D. Equivalent circuit

The final equivalent circuit is shown in Figure 17–9D.

Table 17–4 lists the effect that opening R_4 has on circuit operations. Calculated values without an open have been taken from Table 17–2. A comparison of the two data columns in Table 17–4 indicates some significant facts about partially opened circuits. As the table shows, the total resistance has increased and total current has decreased. The voltage share across R_5 has increased because the resistance of that branch has increased. This increase, in turn, has caused a reduction in the share of R_2 and R_3.

A circuit with opened sections is not likely to create more damage, as one with shorts is likely to do, because the defect lowers total current. However, some problems could arise in the component left across the open. Consider R_5. Before the open, P_5 is

$$P_5 = V_5 I_5 = 6.5 \text{ V} \times 0.012 \text{ A}$$
$$= \textbf{0.078 W}$$

After the open, P_5 has the following value:

$$P_5 = V_5 I_5 = 10.6 \text{ V} \times 0.019 \text{ A}$$
$$= \textbf{0.2014 W}$$

TABLE 17-4
EFFECTS OF AN OPEN IN EXPERIMENTAL
CIRCUIT B

Parameter	Calculated Normal Circuit Values	R_4 Opened
V_T	22 V	22 V
V_2	7 V	5.13 V
V_3	8.5 V	6.27 V
V_5	6.5 V	10.6 V
R_T	379 Ω	431 Ω
I_T	0.058 A	0.051 A
I_1	0.032 A	0.032 A
I_{235}	0.012 A	0.019 A

FIGURE 17-10
EFFECTS OF RESISTOR TOLERANCE ON
CURRENT

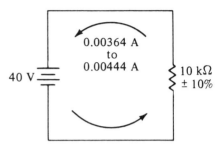

While the power level is still quite low, there is a significant increase after the open. Once again, one defective component can cause the destruction of other components.

ACCEPTABLE RANGE OF VARIATION

The data in Tables 17–3 and 17–4 indicate the variations that can occur when there are defects present in a circuit. For example, I_T, which should be 58 milliamperes, increased to 69 milliamperes when there was a short. If you were checking I_T, you would have probably noticed that there was a problem. However, the opened resistor only caused I_T to drop from 58 to 51 milliamperes. You might not have noticed that drop for it is only a minor difference. The open would have been detected when V_5 was found to be 10.6 volts instead of the expected 6.5 volts.

The acceptable range of variation should be defined by the person who designed the circuit or the person who will use the circuit. But when specifications are not given, you must calculate the range of normal variation. This

task can be rather involved, as the following example shows.

Consider the possible range of normal current values for the simple series circuit shown in Figure 17–10. The total current is

$$I_T = \frac{V_T}{R_T} = \frac{40\ V}{10,000\ \Omega} = \textbf{0.004 A}$$

Since the resistor has a 10% tolerance, it may actually have a value between 9000 ohms and 11,000 ohms. This range can cause some variation in the current between these two limits, as the following calculations show:

$$I_T = \frac{V_T}{R_T} = \frac{40\ V}{11,000\ \Omega} = \textbf{0.00364 A}$$

$$I_T = \frac{V_T}{R_T} = \frac{40\ V}{9000\ \Omega} = \textbf{0.00444 A}$$

Thus, normal variations in the resistor can cause the current to actually be anywhere between 0.00364 and 0.00444 ampere and still be normal.

The next problem in variation arises when you check the 40-volt supply. If you use a voltmeter to check it, which you must, then you have to consider the meter error. Suppose you use a voltmeter with 3% error on the 100-volt range to adjust the 40-volt supply. Your actual setting can be within 3% of 100 volts, or somewhere between 37 and 43 volts. This possible error will also be within normal operation.

Finally, variation may result from current measurement. You are expecting a current of 4 milliamperes, which you would probably measure on the 10-milliampere range. If the meter has 3% accuracy, this measurement can reasonably be off by 3% of 10 milliamperes, or 0.3 milliampere. Thus, the meter may indicate a current of 4 milliamperes, but because of meter error, the actual current may be somewhere between 3.7 and 4.3 milliamperes.

Errors can add up, or compound. Consider the circuit in Figure 17-11. You intended to adjust the voltage to 40 volts; but because of meter error, it actually was adjusted to 43 volts, as the figure indicates. The color-coded value of the resistor is 10,000 ohms, but the actual value is 9000 ohms. Thus, according to your calculations, the current should be 4 milliamperes. However, because of the actual values of voltage and resistance, the current is really about 4.8 milliamperes.

Even though the indicated current is far from what you expected, it is the true value. Thus, the circuit is normal, because the voltage and current were measured as closely as possible, given the meter capabilities. The resistor is also normal because it is within tolerance. As mentioned earlier, if you need results closer to the calculated values, you must use lower-tolerance components and more accurate equipment.

The preceding example has indicated how values can vary within normal reasonable limits. Acceptable variation is an important aspect of electronics. As a technician, you must be able to decide quickly when values are acceptable and when they are not. For example, you must know when it is acceptable to have a specified 10 volts rising to 11 volts but unacceptable to have the 10 volts rising to 10.2 volts. Further study of circuit theory will help you make such decisions.

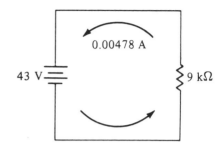

SUMMARY

The testing of series-parallel circuits requires more care than the testing of individual series or parallel circuits. For instance, shorts and opens in series or parallel circuits are almost immediately noticeable. But they can pass unnoticed in a series-parallel circuit. As with series or parallel circuits, though, evaluation of a series-parallel circuit begins with the knowledge of what to expect — that is, knowing what a circuit should be doing.

If published specifications for the circuit are not available, values must be calculated. The final step before beginning testing is to determine the reasonable range of normal operation. Resistor tolerance and meter accuracy will indicate the reasonable ranges.

Voltage measurements are the easiest, quickest, and safest tests to make since no circuit interruptions are needed. In fact, a good troubleshooter will use voltage and resistance measurements to locate most problems. A knowledge of circuit theory helps here. That is, you can use the fact that the sum of the voltage drops in a series circuit should add to the total source voltage. Also, the voltage drops should be in proportion to resistor size. Thus, extreme variations in the voltage proportions are an indication of a problem. Furthermore, a shorted component will indicate a zero voltage drop even though there is still source voltage. An open component will indicate a higher-than-normal voltage drop, sometimes as high as the source voltage.

Current measurements in a series-parallel circuit can provide useful information, but they should be avoided in assembled circuits whenever possible. If, however, you must make them, proceed with caution. These measurements are the most difficult to make because it is not always easy to locate a specific branch and not interfere with any others. Therefore, open only one circuit at a time and only with the power off. Label leads before you remove them so that there will be no error in replacing them. Place the meter in series with the branch being tested, making sure you

have not shorted or opened any other branch.

Resistance measurements in a series-parallel circuit follow the same rules as those for other circuits, and they may be the most useful. If possible, test components before you install them. If not, turn the power off and disconnect one lead of the component being tested. If a circuit is suspect, measure its total resistance. A short will tend to lower the total resistance; an open will tend to increase it. One of the most useful resistance tests in any complex circuit is a test for opens. Questionable solder joints and circuit breaks can be quickly checked for their resistance. A good connection should indicate zero ohms.

In conclusion, you can learn to troubleshoot with only a multimeter. Although it cannot measure every parameter, it can be used to locate most defects.

CHAPTER 17

REVIEW QUESTIONS

1. Examine Figure 17–12. What symptoms will you expect to find if R_1 is open?

2. What symptoms will you expect to find if R_1 in Figure 17–12 is shorted?

3. What measurements would you perform if you believed that R_1 in the circuit of Figure 17–12 was open? Why?

4. What measurements would you perform if you believed that R_1 in the circuit of Figure 17–12 was shorted? Why?

FIGURE 17-12

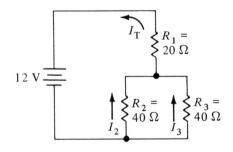

5. What will happen in the circuit of Figure 17–12 if R_3 is shorted?

6. What will happen in the circuit of Figure 17–12 if R_3 is open?

7. What measurements would you perform if you believed that R_3 in the circuit of Figure 17–12 was open? Why?

8. What measurements would you perform if you believed that R_3 in the circuit of Figure 17–12 was shorted? Why?

9. Explain why voltage measurement is preferred over current and resistance measurement in troubleshooting a circuit.

10. What are the risks when you are measuring current? Resistance?

11. Name three advantages of using a manufacturer's published specifications in troubleshooting a circuit.

12. How are normal operational characteristics determined when the manufacturer's specifications are not available?

13. Name three factors that allow an acceptable range of variation rather than require absolute values.

14. Explain how you would set the cutoff points between acceptable and unacceptable circuit performance.

REVIEW PROBLEMS

1. What is wrong with the circuit in Figure 17–13 if I_T measures 0 milliamperes, V_1 measures 33 volts, and V_2 and V_3 measure 0 volts?

2. What is wrong with the circuit in Figure 17–13 if I_T measures 10 milliamperes, V_1 measures 33 volts, and V_2 and V_3 measure 0 volts?

FIGURE 17-13

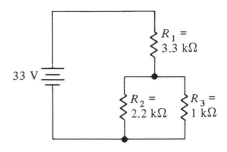

FIGURE 17-14

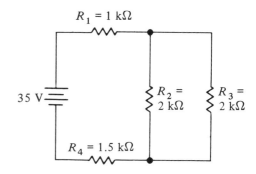

3. What is wrong with the circuit in Figure 17–13 if V_1 measures 19.8 volts and I_T measures 6 milliamperes?

4. What is wrong with the circuit in Figure 17–13 if V_1 measures 0 volts and I_T measures 48 milliamperes?

5. What is wrong with the circuit in Figure 17–14 if V_1 measures 7.8 volts, V_3 measures 15.5 volts, and V_4 measures 11.7 volts?

6. What is wrong with the circuit in Figure 17–14 if I_T measures 14 milliamperes, I_2 measures 0 milliamperes, and I_4 measures 14 milliamperes?

7. What is wrong with the circuit in Figure 17–14 if V_1 measures 14 volts and V_4 measures 21 volts?

8. How will a short in R_2 in the circuit of Figure 16–3A affect V_1 and V_3?

9. How will an open in R_3 in the circuit of Figure 16–3A affect V_1 and V_2?

10. What will an open in R_2 in the circuit of Figure 16–4A do to V_1, V_3, and V_4?

11. What will a short in R_2 in the circuit of Figure 16–4A do to V_1, V_3, and V_4?

12. How will a short in R_1 change the operation of the circuit in Figure 16–6A?

13. How will an open in R_4 change the operation of the circuit in Figure 16–6A?

14. What will cause the circuit in Figure 16–11A to indicate $V_1 = 22$ volts, $V_2 = 0$ volts, and $V_3 = 22$ volts?

15. What steps can you take to determine a specific cause for the defect you found in the circuit described in Problem 14?

ADVANCED CIRCUIT THEORY

Determine the parameters of a circuit by using Kirchhoff's laws.

Determine the parameters of a circuit by using the superposition theorem.

Determine the parameters of a circuit by using a Thevenin equivalent circuit.

FIGURE 18-1
CIRCUIT WITH UNKNOWN RESISTANCES

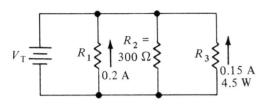

INTRODUCTION

The objective of this chapter is to expand circuit-analyzing skills to advanced circuits. Most circuits can be evaluated by simple series, parallel, or series-parallel techniques. There are three situations, however, when these approaches are not enough. In the first situation, not all voltage and resistance values are given. Instead, power and current may be given. The method of evaluation in this case is not much different from the methods presented earlier. The order of evaluation is a little different, though, and requires some imagination.

In the second situation, the circuit is not a series, parallel, or series-parallel circuit but something else. In the third situation, there is more than one voltage for the circuit. The second and third situations are the main topics of this chapter.

The first situation is easy to resolve since it simply involves a different order of steps from the ones discussed previously. We can evaluate that type of circuit quite quickly right now.

Consider the circuit in Figure 18–1. We cannot begin with the usual approach to evaluation since we do not know V_T or the three resistor values. All we can do is find a place where we know two factors for a component and can calculate the third. Then, with the given information, we see whether we can use Ohm's law or the power formula to determine

something else. Step by step, pieces of the puzzle begin to fit. The formulas we use are those we have been using all along. All that is different is the order in which we use them. Let's evaluate the circuit.

We know two factors for R_3, current and power. So we can use the power formula to find the third factor. Since $P_3 = V_T I_3$, then $V_T = P_3/I_3$. Therefore, V_T is

$$V_T = \frac{P_3}{I_3} = \frac{4.5 \text{ W}}{0.15 \text{ A}} = \textbf{30 V}$$

Since we know R_2, we can now determine I_2, as follows:

$$I_2 = \frac{V_T}{R_2} = \frac{30 \text{ V}}{300 \text{ Ω}} = \textbf{0.1 A}$$

And since we know I_1 and V_T, we can determine R_1, as follows:

$$R_1 = \frac{V_T}{I_1} = \frac{30 \text{ V}}{0.2 \text{ A}} = \textbf{150 Ω}$$

The rest of the calculations are quite simple. They are as follows:

$$I_T = I_1 + I_2 + I_3$$
$$= 0.2 \text{ A} + 0.1 \text{ A} + 0.15 \text{ A}$$
$$= \textbf{0.45 A}$$
$$R_T = \frac{V_T}{I_T} = \frac{30 \text{ V}}{0.45 \text{ A}} = \textbf{66.7 Ω}$$
$$P_1 = V_T I_1 = 30 \text{ V} \times 0.2 \text{ A} = \textbf{6 W}$$
$$P_2 = V_T I_2 = 30 \text{ V} \times 0.1 \text{ A} = \textbf{3 W}$$
$$P_T = P_1 + P_2 + P_3$$
$$= 6 \text{ W} + 3 \text{ W} + 4.5 \text{ W}$$
$$= \textbf{13.5 W}$$
$$P_T = V_T I_T = 30 \text{ V} \times 0.45 \text{ A}$$
$$= \textbf{13.5 W} \quad \text{(answer checks)}$$

Thus, we see that the approach for this situation is not more complicated than those we have used before. All that is necessary is a little imagination and a place to begin the circuit evaluation.

KIRCHHOFF'S LAWS

So far, most of the methods we have used for circuit calculations began with resistance. That is, we applied rules about the sum of resistances in a circuit and reduced the circuit to one total equivalent resistance. However, Kirchhoff's first law, which is used to evaluate complex circuits, is based on the relationship among *voltages* in a circuit.

Voltage Law

Circuits having more than one source of voltage cannot be evaluated by using the techniques already discussed in this text. One of the techniques that will work, though, is the application of the following law.

> **Kirchhoff's voltage law:** The algebraic sum of all voltage sources and drops around a complete circuit is zero.

The term *algebraic sum* means that we must consider the polarity of the voltage as well as its magnitude.

Consider the circuit in Figure 18–2. If we start at point A and add the voltages, we begin with the source and we go from $-$ to $+24$ volts, arriving at point B. We have had a voltage change of $+24$ volts. As we go from point B to point C, the voltage is from $+$ to $-$, and the magnitude is 8 volts. Algebraically, this voltage change is -8 volts. Note that this "minus" voltage change is the reason for the common expression *voltage drop* used when we speak about the voltage across passive components like resistors. Continuing around the loop, the voltage from point C to point D is -10 volts. Finally, from point D back to the start at point A, we have a voltage change

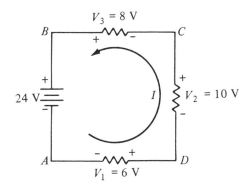

FIGURE 18-2
KIRCHHOFF'S VOLTAGE LAW

from $+$ to $-$ of 6 volts. Adding the voltages from point A, around B, C, and D, and back to A, we have

$$+24\text{ V} - 8\text{ V} - 10\text{ V} - 6\text{ V} = \mathbf{0\ V}$$

Current Law

The use of voltage and/or resistance relationships in complex circuit analysis was demonstrated in the previous section. Current relationships can also be used, according to the following law.

> **Kirchhoff's current law:** The algebraic sum of all currents at any junction, or common point, is zero.

Consider the junction at point A in Figure 18–3, the place where the three resistors join. We call the two currents coming into the junction plus or positive because they are adding to the number of electrons at that point. Then, we must call the one current leaving a minus, since it represents the depletion (or loss) of electrons that have minus polarity. As with the voltage law, it does not matter how we apply the polarity as long as we are consistent. If we call one entering current positive, then all en-

FIGURE 18-3
KIRCHHOFF'S CURRENT LAW

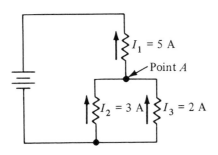

FIGURE 18-4
CIRCUIT EVALUATED WITH KIRCHHOFF'S
LAWS

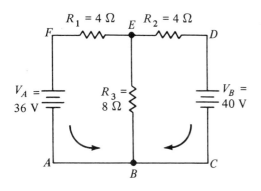

tering currents are positive. All currents leaving must then be called negative.

In Figure 18–3, three currents can be identified at point A: I_2 and I_3 are entering, and I_1 is leaving. This situation can be described by Kirchhoff's current law in the following way:

$$I_2 - I_3 - I_1 = 0 \text{ A}$$

or

$$3 \text{ A} - 2 \text{ A} - 5 \text{ A} = 0 \text{ A}$$

So if you know some of the currents in a circuit and do not know others, this approach will help in evaluating the circuit.

Applying Kirchhoff's Laws

Both of Kirchhoff's laws work well together in the evaluation of a complex circuit. Consider the circuit in Figure 18–4. There are two voltage sources, so previous techniques will not work. To evaluate this circuit with Kirchhoff's laws, we will define some unknown parameters in terms of others. The mathematics gets a little complex, so we will outline each step here.

The current at point E can be described in the following way:

$$+I_3 - I_1 - I_2 = 0$$

Therefore, the expression for I_3 is

$$I_3 = I_1 + I_2$$

We will use this formula later.

The voltage loop $ABCDEFA$ can be described as follows:

$$+V_B - V_2 + V_1 - V_A = 0$$

Now, we will use Ohm's law and known values to find an equation involving only the terms I_1 and I_2, as follows:

$$+V_B - I_2R_2 + I_1R_1 - V_A = 0$$
$$40 - 4I_2 + 4I_1 - 36 = 0$$
$$4I_1 + 4 = 4I_2$$
$$I_1 + 1 = I_2$$

This last equation also will be used later.

The voltage for loop $ABEFA$ can be described as follows:

$$V_3 + V_1 - V_A = 0$$

Again, we use Ohm's law and known values to find an equation involving only I_3 and I_1:

$$8I_3 + 4I_1 - 36 = 0$$

Substituting the equation previously derived for I_3 into this last equation, we have

$$8(I_1 + I_2) + 4I_1 - 36 = 0$$
$$8I_1 + 8I_2 + 4I_1 - 36 = 0$$

Now, we substitute the equation previously derived for I_2 into this equation and solve the resulting equation for I_1, as follows:

$$8I_1 + 8(I_1 + 1) + 4I_1 - 36 = 0$$
$$8I_1 + 8I_1 + 8 + 4I_1 - 36 = 0$$
$$20I_1 - 28 = 0$$
$$20I_1 = 28$$
$$I_1 = \textbf{1.4 A}$$

Next, we work backward to find other current values. Substituting this value for I_1 into the equation for I_2 that we derived earlier, we can find I_2:

$$I_1 + 1 = I_2$$
$$1.4 + 1 = I_2$$
$$I_2 = \textbf{2.4 A}$$

Substituting these values into the equation for I_3 that we derived earlier, we can find I_3:

$$I_3 = I_1 + I_2 = 1.4 \text{ A} + 2.4 \text{ A} = \textbf{3.8 A}$$

Now, we continue with the calculations, using Ohm's law to determine voltages, as follows:

$$V_3 = I_3R_3 = 3.8 \text{ A} \times 8 \text{ }\Omega = \textbf{30.4 V}$$
$$V_2 = I_2R_2 = 2.4 \text{ A} \times 4 \text{ }\Omega = \textbf{9.6 V}$$
$$V_1 = I_1R_1 = 1.4 \text{ A} \times 4 \text{ }\Omega = \textbf{5.6 V}$$

The accuracy of these values can be checked with Kirchhoff's voltage law. Around loop *ABEFA*, the voltages are

$$+V_3 + V_1 - V_A = 0$$
$$30.4 + 5.6 - 36 = \textbf{0} \qquad \text{(answer checks)}$$

Around loop *BCDEB*, the voltages are

$$+V_B - V_2 - V_3 = 0$$
$$40 - 9.6 - 30.4 = \textbf{0} \qquad \text{(answer checks)}$$

Power can be determined in the usual manner, as follows:

$$P_3 = V_3I_3 = 30.4 \text{ V} \times 3.8 \text{ A}$$
$$= \textbf{115.52 W}$$
$$P_2 = V_2I_2 = 9.6 \text{ V} \times 2.4 \text{ A} = \textbf{23.04 W}$$
$$P_1 = V_1I_1 = 5.6 \text{ V} \times 1.4 \text{ A} = \textbf{7.84 W}$$
$$P_T = P_3 + P_2 + P_1$$
$$= 115.52 \text{ W} + 23.04 \text{ W} + 7.84 \text{ W}$$
$$= \textbf{146.40 W}$$

SUPERPOSITION THEOREM

Another process for evaluating circuits with more than one voltage source is called the superposition method. The steps in this procedure are described in the following sections.

Introduction

The superposition method of circuit analysis involves a multiple use of the series-parallel procedures described in previous chapters. We begin with a statement of the superposition theorem.

Superposition theorem: The total current in a circuit with more than one source is the algebraic sum of the currents caused by each source.

That is, we can determine the current caused by each source and then algebraically add it to the current caused by the other source or sources. For example, if one source is causing a current of 1 ampere toward the left and another source is causing 2 amperes toward the left, the net current is 3 amperes to-

260

FIGURE 18-5
CIRCUIT WITH TWO VOLTAGE SOURCES

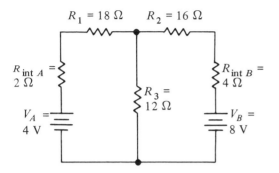

$R_1 = 18\ \Omega$ $R_2 = 16\ \Omega$

$R_{\text{int } A} = 2\ \Omega$ $R_{\text{int } B} = 4\ \Omega$

$R_3 = 12\ \Omega$

$V_A = 4\ V$ $V_B = 8\ V$

ward the left. If one source causes 2 amperes toward the left and another causes 3 amperes toward the right, the net current, or algebraic sum, is 1 ampere toward the right.

Before we continue, we must introduce another concern. While we have assumed so far that all voltage sources are ideal, they are not. All voltage sources have some losses in what is called internal resistance. **Internal resistance** is the resistance that appears along with voltage between the terminals of a source such as a battery. Generally, internal resistance (R_{int}) is small compared to load resistance (R_L), and it can be ignored. Consider the relationship among the resistances in such a circuit:

$$R_T = R_{\text{int}} + R_L$$

As this battery ages, its internal resistance increases. Thus, while the chemical processes inside can produce energy, losses due to internal resistance increase, causing a significant reduction at the terminals.

The internal resistance of voltage sources was introduced here because it must be considered when applying some advanced circuit evaluation techniques. An understanding of this feature and of algebraic sums is necessary for an understanding of the superposition theorem.

Applying the Superposition Theorem

Consider the circuit in Figure 18–5. It has two voltage sources, so we cannot use the usual series or parallel approach. Instead, we consider the impact of one source with all others shorted. Then, we find the algebraic sum of all currents caused by each source. So the first step is to short all sources except V_A and evaluate the circuit shown in Figure 18–6A in the usual manner.

By now, we have evaluated enough circuits so that you should not need a detailed explanation of each step. Thus, we will outline the evaluation here. As shown in Figure 18–6A, we now have a series-parallel circuit. It must be reduced as shown in Figures 18–6B and 18–6C. The calculations are as follows:

$$R_2 + R_{\text{int } B} = 16\ \Omega + 4\ \Omega = \textbf{20}\ \boldsymbol{\Omega}$$

$$R_{32\text{ int } B} = \frac{R_3 R_{2\text{ int } B}}{R_3 + R_{2\text{ int } B}}$$

$$= \frac{12\ \Omega \times 20\ \Omega}{12\ \Omega + 20\ \Omega} = \frac{240}{32}\ \Omega$$

$$= \textbf{7.5}\ \boldsymbol{\Omega}$$

The total resistance, as seen by this source, will be called R_{TA} since it is not the total resistance for the circuit. Because of the two voltage sources, there is not one value of total resistance for the circuit. The calculations are as follows:

$$R_{TA} = R_{\text{int } A} + R_1 + R_{32\text{ int } B}$$

$$= 2\ \Omega + 18\ \Omega + 7.5\ \Omega = \textbf{27.5}\ \boldsymbol{\Omega}$$

$$I_{TA} = \frac{V_A}{R_{TA}} = \frac{4\ V}{27.5\ \Omega} = \textbf{0.145 A}$$

FIGURE 18-6
CIRCUIT WITH RIGHT VOLTAGE SOURCE
REMOVED

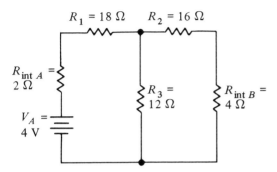

A. Circuit with V_B shorted

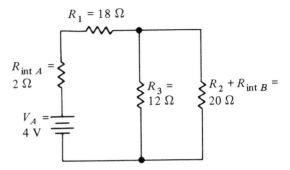

B. Reducing the circuit: Step 1

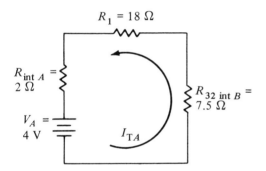

C. Reducing the circuit: Step 2

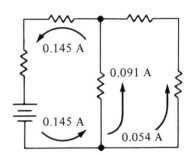

D. Currents caused by left source only

Current I_{TA} will flow through $R_{\text{int }A}$, R_1, and the combination $R_{32 \text{ int }B}$, as shown in Figure 18–6C. To find out how it divides through R_3 and $R_{2 \text{ int }B}$, we must determine the voltage drop and branch currents. The calculations are as follows:

$$V_{32 \text{ int }B} = I_{TA}R_{32 \text{ int }B} = 0.145 \text{ A} \times 7.5 \text{ }\Omega$$

$$= \mathbf{1.09 \text{ V}}$$

$$I_3 = \frac{V_{32 \text{ int }B}}{R_3} = \frac{1.09 \text{ V}}{12 \text{ }\Omega} = \mathbf{0.091 \text{ A}}$$

$$I_{2 \text{ int }B} = \frac{V_{32 \text{ int }B}}{R_{2 \text{ int }B}} = \frac{1.09 \text{ V}}{20 \text{ }\Omega} = \mathbf{0.054 \text{ A}}$$

All these current values can be added to the original circuit with source A only, as shown in Figure 18–6D.

Now, the procedure can be done over again, this time using only source V_B. All other sources must be shorted. The second evaluation begins with the circuit shown in Figure 18–7A. Then, the circuit is reduced as shown in Figures 18–7B and 18–7C. The calculations are as follows:

$$R_1 + R_{\text{int }A} = 18 \text{ }\Omega + 2 \text{ }\Omega = \mathbf{20 \text{ }\Omega}$$

$$R_{31 \text{ int }A} = \frac{R_3 R_{1 \text{ int }A}}{R_3 + R_{1 \text{ int }A}}$$

FIGURE 18-7
**CIRCUIT WITH LEFT VOLTAGE SOURCE
REMOVED**

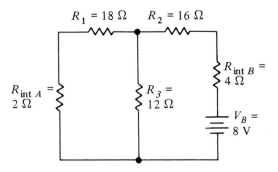

A. Circuit with V_A shorted

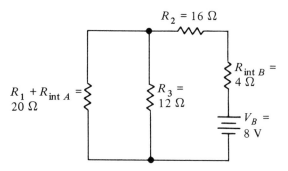

B. Reducing the circuit: Step 1

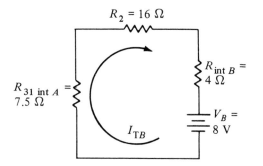

C. Reducing the circuit: Step 2

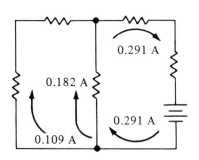

D. Currents caused by right source only

$$= \frac{12 \ \Omega \times 20 \ \Omega}{12 \ \Omega + 20 \ \Omega} = \frac{240}{32} \ \Omega$$

$$= \textbf{7.5} \ \Omega$$

The total resistance seen by this source will be called R_{TB}. The calculations follow:

$$R_{TB} = R_{int \ B} + R_2 + R_{31 \ int \ A}$$

$$= 4 \ \Omega + 16 \ \Omega + 7.5 \ \Omega = \textbf{27.5} \ \Omega$$

$$I_{TB} = \frac{V_B}{R_{TB}} = \frac{8 \ V}{27.5 \ \Omega} = \textbf{0.291 A}$$

This amount of current will flow through $R_{int \ B}$, R_2, and the combination $R_{31 \ int \ A}$, as

shown in Figure 18–7C. To find out how it divides through R_3 and $R_{1 \ int \ A}$, we must determine the voltage drop and branch currents, as follows:

$$V_{31 \ int \ A} = I_{TB}R_{31 \ int \ A} = 0.291 \ A \times 7.5 \ \Omega$$

$$= \textbf{2.18 V}$$

$$I_3 = \frac{V_{31 \ int \ A}}{R_3} = \frac{2.18 \ V}{12 \ \Omega} = \textbf{0.182 A}$$

$$I_{1 \ int \ A} = \frac{V_{31 \ int \ A}}{R_{1 \ int \ A}} = \frac{2.18 \ V}{20 \ \Omega} = \textbf{0.109 A}$$

All these current values can be added to

the original circuit with source *B* only, as shown in Figure 18–7D.

Figure 18–7D shows the amounts of current that source *B* alone is attempting to cause. Figure 18–6D shows the amounts and directions of current that source *A* is attempting to cause. Finding their algebraic sums is quite simple. If two currents are in the same direction, we add them. If their directions are opposite, we find their difference. The direction of the resulting current is the same as the direction of the larger one. The calculations are

$$I_A = 0.145 \text{ A} - 0.109 \text{ A} = \mathbf{0.036 \text{ A}}$$
$$I_B = 0.291 \text{ A} - 0.054 \text{ A} = \mathbf{0.237 \text{ A}}$$
$$I_3 = 0.182 \text{ A} + 0.091 \text{ A} = \mathbf{0.273 \text{ A}}$$

The results are pictured in Figure 18–8.

The voltage drops are calculated next, using these composite current values:

$$V_{\text{int } A} = I_A R_{\text{int } A} = 0.036 \text{ A} \times 2 \text{ }\Omega$$
$$= \mathbf{0.072 \text{ V}}$$
$$V_1 = I_A R_1 = 0.036 \text{ A} \times 18 \text{ }\Omega$$
$$= \mathbf{0.648 \text{ V}}$$
$$V_3 = I_3 R_3 = 0.273 \text{ A} \times 12 \text{ }\Omega$$
$$= \mathbf{3.276 \text{ V}}$$
$$V_A = V_{\text{int } A} + V_1 + V_3$$
$$= 0.072 \text{ V} + 0.648 \text{ V} + 3.276 \text{ V}$$
$$= \mathbf{4 \text{ V}} \quad \text{(answer checks)}$$
$$V_{\text{int } B} = I_B R_{\text{int } B} = 0.237 \text{ A} \times 4 \text{ }\Omega$$
$$= \mathbf{0.948 \text{ V}}$$
$$V_2 = I_B R_2 = 0.237 \text{ A} \times 16 \text{ }\Omega$$
$$= \mathbf{3.792 \text{ V}}$$
$$V_B = V_{\text{int } B} + V_2 + V_3$$
$$= 0.948 \text{ V} + 3.792 \text{ V} + 3.276 \text{ V}$$
$$= \mathbf{8 \text{ V}} \quad \text{(answer checks)}$$

The final solution is shown in Figure 18–9.

FIGURE 18-8
VECTOR SUM OF ALL CURRENTS

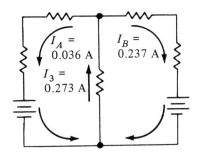

FIGURE 18-9
VOLTAGE DROPS IN COMPLETE CIRCUIT

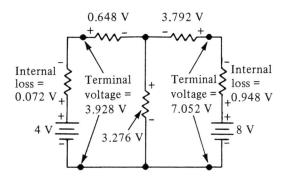

THEVENIN'S THEOREM

The Thevenin approach to circuit analysis has two points of value. As you will see here, it allows you to calculate the parameters of circuits when other methods presented so far will not work. In addition, you can use this approach to analyze transistor amplifier circuits in future studies.

Introduction

Circuits such as the five-resistor bridge shown in Figure 18–10 are difficult to evaluate be-

FIGURE 18-10
FIVE-RESISTOR BRIDGE CIRCUIT

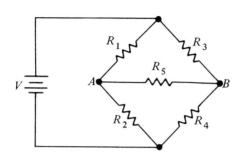

FIGURE 18-11
CIRCUIT EVALUATED WITH THEVENIN'S THEOREM

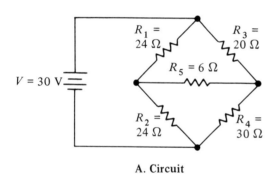

A. Circuit

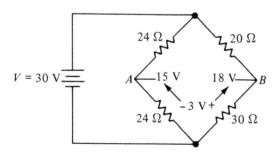

B. Determining equivalent voltage

cause they are not series, parallel, or series-parallel circuits. And they cannot be simplified in the usual manner. Other techniques must be used.

Thevenin's theorem provides a way to evaluate such a circuit. With Thevenin's theorem, we must select a place of concern. That is, we must decide which specific current or voltage we are interested in. Then, we determine what the remainder of the circuit looks like from that particular location. Suppose we want to determine I_5 and V_5 of Figure 18–10. Thevenin's theorem tells us that if we find resistance and voltage values from points A to B with R_5 out, then we can place R_5 in series with those equivalent values and calculate I_5 and V_5.

The general ideas of Thevenin's theorem can be divided into two essential parts.

Thevenin's theorem: First, a complex circuit can be reduced to the voltage and resistance seen at the terminals from which you remove the resistor of interest. That modified circuit is called the **Thevenin equivalent circuit.** Second, the resulting voltage and current of the resistor of interest can be determined by placing that resistor in series with the equivalent circuit.

Thevenin Equivalent Circuit

The best way to understand this approach is to work through an example. Let's consider the five-resistor bridge circuit in Figure 18–11A. We want to know the values of I_5 and V_5.

We begin by removing R_5 and determining the voltage from point A to point B; see Figure 18–11B. With R_5 removed, the circuit is simply two series circuits in parallel, as shown in Figure 18–12A. The calculations follow:

$$R_{12} = R_1 + R_2 = 24\ \Omega + 24\ \Omega = \mathbf{48\ \Omega}$$

FIGURE 18-12
DETERMINING EQUIVALENT RESISTANCE

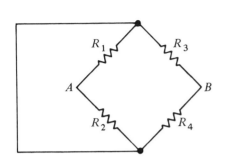

A. Circuit with R_5 removed

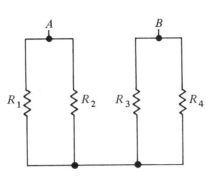

B. Circuit as seen from A to B

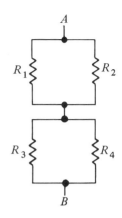

C. Circuit with R_3 and R_4 moved into traditional positions

$$I_{12} = \frac{V_T}{R_{12}} = \frac{30 \text{ V}}{48 \text{ }\Omega} = \textbf{0.625 A}$$

$$V_1 = I_{12}R_1 = 0.625 \text{ A} \times 24 \text{ }\Omega = \textbf{15 V}$$

$$V_2 = I_{12}R_2 = 0.625 \text{ A} \times 24 \text{ }\Omega = \textbf{15 V}$$

Thus, the voltage at point A is 15 volts.

The remainder of the calculations are as follows:

$$R_{34} = R_3 + R_4 = 20 \text{ }\Omega + 30 \text{ }\Omega = \textbf{50 }\Omega$$

$$I_{34} = \frac{V_T}{R_{34}} = \frac{30 \text{ V}}{50 \text{ }\Omega} = \textbf{0.6 A}$$

$$V_3 = I_{34}R_3 = 0.6 \text{ A} \times 20 \text{ }\Omega = \textbf{12 V}$$

$$V_4 = I_{34}R_4 = 0.6 \text{ A} \times 30 \text{ }\Omega = \textbf{18 V}$$

Thus, the voltage at point B is 18 volts.

The voltage from A to B with R_5 removed is the difference between 15 volts and 18 volts, which is 3 volts. This situation is shown in Figure 18–11B. In addition, B is more positive and A is less positive. So from point A to point B with R_5 removed, the circuit looks like 3 volts with B positive and A negative.

The next step is to determine the resistance seen by looking into the circuit at points A and B. Before we do so, however, let's redraw the circuit. Recall that Figure 18–12B shows the circuit as it looks from A to B. If we turn resistors R_3 and R_4, the circuit can be redrawn in a more recognizable and traditional form. Thus, as shown in Figure 18–12C, the circuit from A to B is simply two pairs of parallel resistors connected in series with each other.

With the circuit redrawn, we can proceed. All sources must be shorted for this step. Notice that the three views of the circuit in Figure 18–12 show that it is another form of a series-parallel circuit. Once we have reduced the circuit to this form, the total equivalent resistance is easy to determine. The calculations are as follows:

$$R_{12} = \frac{R_1 R_2}{R_1 + R_2} = \frac{24 \text{ }\Omega \times 24 \text{ }\Omega}{24 \text{ }\Omega + 24 \text{ }\Omega}$$

$$= \frac{576}{48} \text{ }\Omega = \textbf{12 }\Omega$$

FIGURE 18-13
FINAL THEVENIN EQUIVALENT CIRCUIT

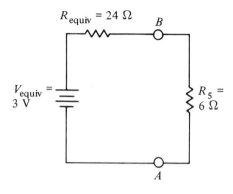

$$R_{34} = \frac{R_3 R_4}{R_3 + R_4} = \frac{20\,\Omega \times 30\ \Omega}{20\ \Omega + 30\ \Omega}$$

$$= \frac{600}{50}\ \Omega = \mathbf{12\ \Omega}$$

$$R_T = R_{12} + R_{34} = 12\ \Omega + 12\ \Omega$$

$$= \mathbf{24\ \Omega}$$

The total equivalent resistance from point A to point B with R_5 removed is 24 ohms.

Now, we place R_5 in series with the equivalent of the remainder of the circuit and treat it as a simple series circuit, as shown in Figure 18–13. The calculations are as follows:

$$I_5 = \frac{V_T}{R_T} = \frac{3\ V}{30\ \Omega} = \mathbf{0.1\ A}$$

$$V_5 = I_5 R_5 = 0.1\ A \times 6\ \Omega = \mathbf{0.6\ V}$$

SUMMARY

Most circuits can be evaluated by using simple series, parallel, or series-parallel techniques. Occasionally, though, circuits cannot be evaluated this way. Bridge circuits and circuits with more than one voltage source are two examples. Application of techniques using Kirchhoff's laws, the superposition theorem, and Thevenin's equivalent circuit is necessary in these circuits.

Kirchhoff's voltage law states that the sum of all voltages around a closed loop is zero. Thus, if we can determine some of the voltages in the loop, we can use that information to determine other voltage values. In applying Kirchhoff's law, we often must describe one voltage in terms of another. This process uses algebra and algebraic sums.

Kirchhoff's current law states that the algebraic sum of all currents at a junction is zero. In applying this law, we assign negative values to current leaving the junction and positive values to current entering the junction.

The combination of the current law and the voltage law provides technicians with a useful mathematical tool. Here are some points to remember:

1. The sum of voltages around a closed loop is zero.

2. The sum of currents at a junction is zero.

3. Unknown values can be written in terms of others in order to reduce the number of unknowns.

4. The number of equations necessary to evaluate a circuit is equal to the number of unknowns.

The superposition theorem presents another way to evaluate circuits having more than one voltage source. When we replace all sources except one by their internal resistances, the effect of that one source can be determined. We then repeat this step for each of the other sources. In this way, we can describe how each source is attempting to cause the current to flow. Adding currents in the same direction and finding the difference of those in opposing directions provide us with values for the actual currents that result.

Once we know the current values and the

given resistance values, we can calculate voltage and power. To do so, we use the same formulas — Ohm's law and the power formula — that we used in earlier chapters, only in a different order.

The Thevenin equivalent circuit provides us with a way to evaluate what is happening to a given resistor in a complex circuit. The steps are as follows:

1. The resistor is removed from the circuit.

2. The open-circuit voltage across those terminals is determined.

3. The open-circuit resistance across those terminals, with all supplies shorted, is determined.

4. An equivalent circuit of that voltage and resistance is connected in series with the resistor of interest, and Ohm's law is applied.

The processes outlined in this chapter are the most complex processes you will see. This chapter completes your introduction to DC circuits.

CHAPTER 18

REVIEW TERMS

internal resistance: resistance that appears along with voltage between the terminals of a source

Kirchhoff's current law: relationship for determining currents at a junction in the analysis of complex circuits — the algebraic sum of currents at a junction equals zero

Kirchhoff's voltage law: relationship for determining voltages in the analysis of complex circuits — the algebraic sum of voltage sources and drops around a complete circuit equals zero

superposition theorem: relationship for determining the current from each of several sources in the analysis of complex circuits — the total current in a circuit with more than one source is the algebraic sum of the currents caused by each source

Thevenin equivalent circuit: representation showing how a complex circuit appears to one component, expressed as a series voltage and resistance

Thevenin's theorem: procedure for determining the parameters of a complex circuit by starting at a specific point in the circuit — a complex circuit, as seen at one component, can be reduced to an equivalent voltage and an equivalent resistance in series with that component

REVIEW QUESTIONS

1. State Kirchhoff's voltage law.

2. State Kirchhoff's current law.

3. State the superposition theorem.

4. State both parts of Thevenin's theorem.

5. Describe what is meant by an equivalent voltage.

6. Describe what is meant by an equivalent resistance.

7. State the relationship between the number of unknown values in a circuit and the number of equations required to determine them.

8. What is meant by algebraic sum?

9. Under what circumstances might you use Kirchhoff's current law?

10. Under what circumstances might you use the superposition theorem?

11. Why are series-parallel techniques not appropriate when you are evaluating five-resistor bridge circuits?

12. How is circuit power determined with the use of Kirchhoff's laws or the superposition theorem?

REVIEW PROBLEMS

1. Determine all values of voltage, current, resistance, and power for the circuit in Figure 18–1 if you are given the following values:

$I_1 = 0.02$ ampere

$R_2 = 1000$ ohms

$I_2 = 0.05$ ampere

FIGURE 18-14

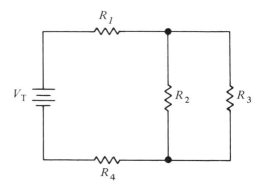

$I_3 = 0.01$ ampere

$P_3 = 0.5$ watt

2. Determine all values of voltage, current, resistance, and power for the circuit in Figure 18–3 if you are given the following values:

$R_1 = 260$ ohms

$I_1 = 0.05$ ampere

$R_2 = 600$ ohms

$V_2 = 12$ volts

3. Determine all values of voltage, current, resistance, and power for the circuit shown in Figure 18–14 if $R_1 = 220$ ohms, $I_1 = 0.02$ ampere, $I_3 = 0.01$ ampere, $V_4 = 6.6$ volts, and $V_T = 24$ volts.

4. Determine all values of voltage and current for the circuit in Figure 18–4 if you are given the following values:

$R_1 = 18$ ohms

$R_2 = 36$ ohms

$R_3 = 18$ ohms

$V_A = 9$ volts

$V_B = 18$ volts

Use Kirchhoff's voltage law.

5. Determine all values of voltage and current for the circuit described in Problem 4 if R_3 is changed to 26 ohms.

6. Determine all values of voltage and current in the circuit shown in Figure 18–15 if R_1 = 220 ohms, R_2 = 180 ohms, R_3 = 240 ohms, R_4 = 200 ohms, R_5 = 160 ohms, V_A = 9 volts, and V_B = 11 volts. Use Kirchhoff's voltage law.

7. Determine all values of voltage and current for the circuit described in Problem 6 if R_3 is changed to 180 ohms.

8. Determine all values of voltage and current for the circuit in Figure 18–5 if you are given the following values:

R_1 = 16 ohms

R_2 = 20 ohms

R_3 = 12 ohms

$R_{int\ A}$ = 8 ohms

$R_{int\ B}$ = 6 ohms

V_A = 12 volts

V_B = 8 volts

Use the superposition theorem.

9. Determine all values of voltage and current for the circuit described in Problem 8 if R_1 is changed to 20 ohms and R_2 is changed to 24 ohms.

10. Determine all values of voltage and current in the circuit shown in Figure 18–16 if R_1 = 240 ohms, R_2 = 330 ohms, R_3 = 220 ohms, R_4 = 560 ohms, R_5 = 470 ohms, $R_{int\ A}$ = 14 ohms, $R_{int\ B}$ = 12 ohms, V_A = 24 volts, and V_B = 18 volts. Use the superposition theorem.

11. Determine all values of voltage and current for the circuit described in Problem 10 if R_5 is changed to 180 ohms and R_4 is changed to 150 ohms.

FIGURE 18-15

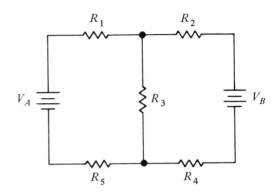

FIGURE 18-16

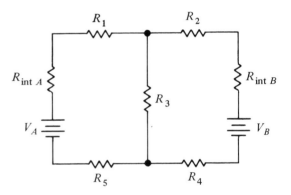

12. Determine the voltage and current of R_5 of Figure 18–10 if you are given the following values:

R_1 = 20 ohms

R_2 = 20 ohms

R_3 = 30 ohms

R_4 = 40 ohms

R_5 = 50 ohms

V = 18 volts

Use a Thevenin equivalent circuit.

FIGURE 18-17

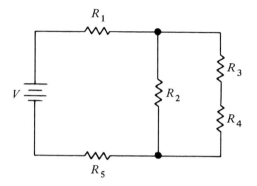

13. Determine the voltage and current of R_5 in the circuit described in Problem 12 if R_1 is changed to 60 ohms.

14. Determine the voltage and current of R_2 shown in the circuit of Figure 18–17 if R_1 = 24 ohms, R_2 = 16 ohms, R_3 = 12 ohms, R_4 = 4 ohms, R_5 = 12 ohms, and V = 48 volts. Use a Thevenin equivalent circuit.

15. Determine the voltage and current of R_2 in the circuit described in Problem 14 if R_5 is changed to 18 ohms and R_3 is changed to 8 ohms.

OSCILLOSCOPES

State the functions of an oscilloscope.

Name the major circuits or sections of an oscilloscope, and explain the basic operation of each.

Name and describe the common controls of an oscilloscope.

Determine the magnitude and polarity of a DC voltage with an oscilloscope.

FIGURE 19-1
PORTABLE OSCILLOSCOPE

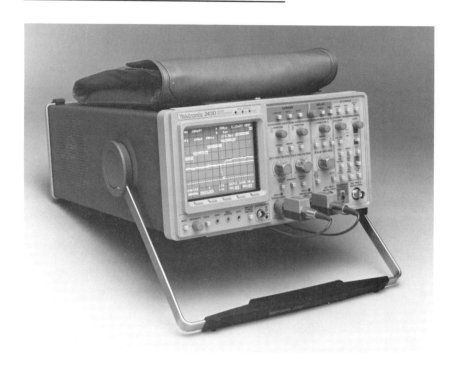

INTRODUCTION

The objective of this chapter is to introduce the **oscilloscope** and its use. You have probably seen oscilloscopes in your school lab already. They are easily recognized by their screen and their many knobs and switches, as shown in Figure 19–1. They range from very simple to quite complex, but they all work basically in the same manner.

Key point: Oscilloscopes measure voltage and time.

The voltages that they measure can be very small and can last for a short duration. Also, more than one voltage can be measured at a time. Finally, the shape as well as the magnitude of a voltage are indicated.

To perform its tasks, an oscilloscope uses a cathode ray tube (CRT). The CRT is similar in operation to a television picture tube, although it is much smaller. An oscilloscope CRT is usually round or rectangular and 4 or 5 inches across. The face has horizontal and vertical grid lines for reference points in measurement. These lines are generally 1 centimeter apart.

Knobs and switches cover the remaining space on the front of the cabinet. They are used for making operational adjustments, selecting ranges, and checking calibration. Voltages to be measured enter the oscilloscope through connections on the front. Leads with special probes provide the connections between the oscilloscope and the equipment under test.

Purpose

As mentioned earlier, the primary purpose of an oscilloscope is to view and measure voltage waveforms. Your first use of an oscilloscope will be to measure the voltage of batteries, although that is not a common use. Voltages that are constantly changing are more often measured. For example, the *RL* time constant can best be measured with an oscilloscope. An oscilloscope can respond to rapidly changing voltages. It shows the shape of the voltage, so you can see the voltage as it changes with time. Thus, you can measure both the voltage and the time.

A typical example of the use of an oscilloscope is in troubleshooting a stereo system. When a stereo system distorts a signal, the shape of the output signal is not the same as the shape of the input. The tape is good, but the sound from the speakers is poor. To find the problem, a troubleshooter plays a test tape. With the aid of an oscilloscope, the technician can check the progress of this test signal as it moves through the stages of the amplifier. He or she can find the location where the incoming signal changes shape and becomes distorted. The location of this defect can be easily found with an oscilloscope because it shows shape, not just magnitude.

Advantages and Disadvantages

The primary advantage of oscilloscopes has just been described: It shows the shape of a voltage. They have many other advantages of equal importance. Oscilloscopes are very sensitive; they can measure voltages down to the microvolt (millionth of a volt) level. Oscillo-

scopes can show multiple traces so that a number of signals can be observed at one time. They have a high input impedance, which means they have little effect on the circuits they are measuring. (Recall that circuit loading was mentioned earlier as a possible problem with some multimeters.)

Oscilloscopes can also measure the time for an event, including events that occur in microseconds. For very detailed observations, oscilloscopes can produce a single image of events that repeat themselves, such as AC voltages.

With all their advantages, you might expect oscilloscopes to be used for all tests. They are not, though, because they do have some disadvantages. For instance, oscilloscopes can measure voltage but not current and resistance. They are often less portable than meters, and they need line voltage to operate. However, small battery-operated models are available. An oscilloscope may yield less accurate results than a multimeter. Loss of accuracy is often caused by technician error in deciding exactly what voltage is indicated. If an absolute value is needed, a meter should be used. However, there are times when a voltage is so small that only an oscilloscope will measure it.

An oscilloscope's advantages generally outweigh its disadvantages. Consequently, oscilloscopes are as common as multimeters in electronics shops and labs.

Range of Options

The cost of an oscilloscope can range from a few hundred dollars to thousands of dollars. As with most test equipment, increased cost means increased quality.

Three factors usually improve as the cost of oscilloscopes increases. The first factor is sensitivity. Recall (from Chapter 7) that sensitivity is also an important feature for multimeters, but it has another meaning for oscilloscopes. **Sensitivity** for an oscilloscope indicates its lowest voltage range. A typical value is 1 millivolt per centimeter. Low-cost oscilloscopes can measure voltages in the millivolt (thousandth of a volt) range, and more expensive ones can measure voltages in the microvolt (millionth of a volt) range.

The second factor is **frequency response** — that is, how quickly the instrument can respond to an event. Low-cost oscilloscopes can usually respond to events that occur in the millisecond range; more expensive ones see events that occur in the nanosecond (billionth of a second) range. Oscilloscopes with improved frequency response and sensitivity are more expensive to produce — and, therefore, cost more — than general-purpose ones.

The third factor affecting oscilloscope cost is the availability of plug-in units. These are devices that plug into the front of a mainframe oscilloscope to adapt it to a special purpose. Transistor curve tracers, differential amplifiers, and integrated circuit testers are just a few of the units available. For a more detailed picture of the range of options available, examine the catalogs of some of the well-known, higher-cost oscilloscope manufacturers. See what type of plug-in units they offer, and note their cost.

THEORY OF OPERATION

Familiarity with the outside of an oscilloscope is not enough. You also should understand a bit about what is going on inside. As with meters, knowing the theory of operation helps in understanding the capabilities and limitations of an instrument. Of course, oscilloscopes are more complicated than meters, but we can discuss the basic operational sections of oscilloscopes. These basic sections are the vertical, horizontal, and sweep circuits and the cathode ray tube.

Cathode Ray Tube

The most important part of an oscilloscope is its **cathode ray tube** (CRT), shown in Figure 19–2. This glass tube holds many sections, or elements, sealed in a vacuum. The heater, or the filament, glows a reddish color and gets very hot as current passes through. This element is the one that you can see glowing inside radio and television tubes. Another element, the cathode, is quite close to the filament. When it heats up, its electrons become agitated and move rapidly in their orbits. If enough heat is applied, electrons will boil off, or leave, the cathode, similar to the way bubbles leave a glass of ginger ale. This action will cause a field of electrons, or space charge, to surround the cathode.

Partway down the tube neck are cylinder-shaped elements, called anodes. They attract electrons with positive voltage and focus them into a beam aimed toward the tube face. One has a positive voltage level in the thousands of volts, which causes electrons to accelerate, or speed up. Since this element is also a hollow tube, electrons speed right through it and collide with the face of the tube.

The inside of the tube face is coated with a phosphor compound that glows when bombarded with electrons. Most oscilloscopes use a compound that glows green, although compounds for other colors are available. Since the brightness or intensity of the glow depends on the electron beam, a control grid is located between the cathode and anodes to control the intensity. The length of time the phosphor glows is called the persistence of a CRT, and various times are available. As electrons begin to gather on the tube face, they are drained off along a carbon-based coating called the aquadag. This coating lines the inside of the CRT neck and allows electrons to return through the power supply for recycling to the cathode.

The final elements of the CRT are the hor-

FIGURE 19-2
CATHODE RAY TUBE

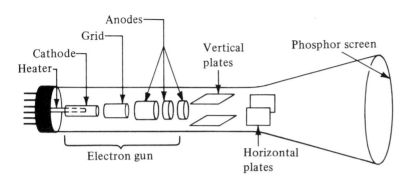

izontal and vertical **deflection plates.** Their purpose is to change the direction of the electron beam as it approaches the CRT face. As previously mentioned, the combination of cathode, grid, and anodes produces a concentrated electron beam that creates a green spot on the screen of the CRT. If a DC voltage source is connected to the vertical deflection plates, it will affect the direction of the beam.

The vertical deflection plates work this way: A positive voltage on the upper plate will attract the passing electrons, and a corresponding negative voltage on the lower plate will repel them. Although these forces are not strong enough to attract electrons onto the plate, they do cause the beam, pictured in Figure 19–3A, to change direction. As shown in Figure 19–3B, the beam deflects upward. A voltage of 1 volt moves the beam up 1 centimeter; 2 volts moves the beam up 2 centimeters. That is, the deflection is linear. Once you know the ratio between volts and deflection, you can place a calibrated grid, or graticule, on the face of the CRT and measure deflection and calculate voltage.

You can also check polarity, because, as shown in Figure 19–3C, reversing the battery polarity changes the direction of beam deflection. Negative voltage on the upper plate and positive voltage on the bottom plate move the

FIGURE 19-3
ELECTROSTATIC DEFLECTION

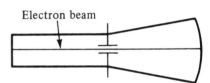

A. Beam undeflected

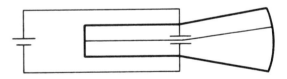

B. Beam deflects up

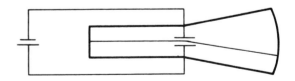

C. Beam deflects down

FIGURE 19-4
OSCILLOSCOPE SCREEN

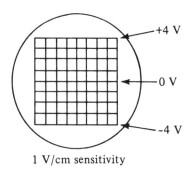

+4 V

0 V

-4 V

1 V/cm sensitivity

beam downward. Of course, the beam continues to the screen undeflected if no plate voltage is applied, as indicated in Figure 19–3A.

The horizontal deflection plates work in a similar manner. Various levels of DC voltages applied to them will move the beam to the right or left of center. These plates are generally not used for voltage measurement; rather, they are used for horizontal sweep circuits. These plates will be described in more detail later.

Attenuators and Amplifiers

There are some limitations in the voltage-measuring system just described. It has a very small voltage range. Assume that 1 volt will cause a deflection of 1 centimeter. Stated in the more common way, the CRT has a sensitivity of 1 volt per centimeter (1 V/cm). A typical 5-inch-diameter CRT has a graticule marked with about 8 centimeters. If we consider the center as 0 volts, then we can measure up to ± 4 centimeters from the center, which corresponds to a maximum voltage of ± 4 volts, as shown in Figure 19–4. Thus, if the deflection is 1.6 centimeters, the voltage is 1.6 volts. But 6 volts will cause the beam to deflect above the top of the graticule, while 0.06 volt will cause no noticeable deflection.

In other words, the voltage to be measured could be too large or too small.

These problems can be resolved with more circuitry. The incoming signal can be passed through circuits consisting of attenuators and amplifiers. These devices are added to the circuitry that deflects the beam. See the block diagram shown in Figure 19–5.

Suppose we wish to measure a voltage that is rather large. The unknown voltage is connected to the vertical input and common terminals of the oscilloscope. Since the voltage is too large to be measured, it must first be reduced. The reduction is done in a manner similar to that of multimeters. A voltage divider circuit called an attenuator is used. In its simplest form, an **attenuator** consists of several series resistors, and it reduces the size of a voltage. For example, a 10 : 1 attenuator reduces an incoming voltage by a factor of 10. Thus, if a 40-volt source is connected to the input terminals, the attenuator will reduce it to 4 volts by the time the voltage gets to the vertical deflection plates. Therefore, the sensitivity of the oscilloscope has been changed to 10 volts per centimeter.

So an oscilloscope measures large voltages by reducing them, with an attenuator, to manageable sizes. But an oscilloscope must also measure very small voltages. As noted earlier, a voltage of 1 volt produces a 1-centimeter deflection. But a voltage of 0.1 volt causes a deflection too small to measure. Thus, the voltage must be increased, and the increase is accomplished by an amplifier. A voltage **amplifier** is a circuit that increases the size of a voltage. For example, a 1 : 10 amplifier in the path of the signal increases the voltage by a factor of 10. Thus, if the signal enters the amplifier as 0.1 volt, it leaves as 1 volt. When it reaches the CRT, it causes a 1-centimeter deflection, which is in the measurable range. The amplifier changes the sensitivity of the oscilloscope to 0.1 volt per centimeter.

The relationship between the sensitivity of

FIGURE 19-5
BLOCK DIAGRAM OF THE DEFLECTION
CIRCUIT

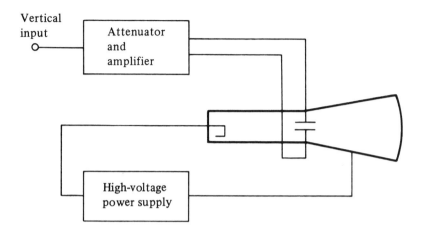

a CRT and the desired per-centimeter range of an oscilloscope determines the amount of gain or attenuation needed. **Gain** is a proportional increase in the voltage or current in a circuit and is produced by an amplifying circuit. Consider an application where a CRT with a sensitivity of 1 volt per centimeter must be used to measure a 10-millivolt signal. The incoming signal is too small to produce a measurable deflection and, therefore, needs amplification. The common symbol for voltage gain, or **amplification,** is A_V. The equation for calculating A_V is

$$A_V = \frac{\text{sensitivity}}{\text{range}} = \frac{1 \text{ V/cm}}{0.01 \text{ V/cm}} = \mathbf{100}$$

The vertical amplifier has to provide a voltage gain of 100.

Now, consider a situation where the incoming signal is such that a range of 50 volts per centimeter is desired. The gain is

$$A_V = \frac{\text{sensitivity}}{\text{range}} = \frac{1 \text{ V/cm}}{50 \text{ V/cm}} = \mathbf{0.02}$$

The vertical amplifier must provide a gain of 0.02. Since this gain is less than 1, it is called an attenuation. **Attenuation,** then, can be defined as a proportional reduction of circuit voltage.

In summary, the voltage-measuring range of a CRT is extended in both directions with the aid of attenuators and amplifiers. Large voltages are reduced and small voltages increased to the normal deflection range of the CRT. This built-in circuitry is selected by using the range switch, marked volts per centimeter. Ranges vary from one oscilloscope to another, but most have ranges of 0.01, 0.1, 1, 10, and 100 volts per centimeter. The range switch allows voltage measurements from millivolts to hundreds of volts.

Sweep Circuits

So far, we have considered a dot on the CRT that can be moved up and down with voltages applied to the vertical input. Actually, having just a dot on the screen is undesirable because

FIGURE 19-6
EFFECTS OF DEFLECTION PLATE VOLTAGE ON THE BEAM

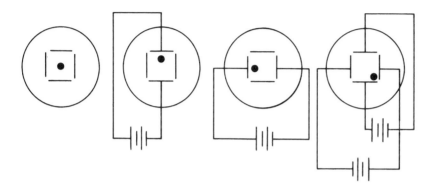

FIGURE 19-7
HORIZONTAL SWEEP VOLTAGE

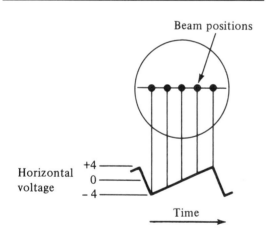

concentrating this energy in one location can damage the phosphor. In fact, technicians and engineers always avoid displaying a dot. Instead, they adjust the controls so that a curve or a line is displayed.

As you will recall, voltages applied to the horizontal plates move the beam left or right, while voltages applied to the vertical plates

move the beam up or down. These ideas are pictured in Figures 19–6. The figure also shows that voltages applied to both sets of plates independently move the beam both horizontally and vertically. That is, horizontal and vertical circuits can command the CRT beam at the same time, independent of one another.

Generally, voltages to be measured are connected to the vertical input of the oscilloscope. These voltages will cause the beam to deflect up or down. And as mentioned earlier, if the unknown voltage is the only input to the CRT, the screen will display only a single spot. However, a shape can be produced — and voltage measurement made easier — with the addition of an internally produced voltage. The horizontal **sweep circuit** produces a ramp or sawtooth voltage that can be applied to the horizontal deflection circuit. As shown in Figure 19–7, this voltage steadily rises from −4 volts to +4 volts, drops quickly to −4 volts, and begins to rise again. If you imagine this waveshape repeated over and over again, you see why it is called a sawtooth voltage. This voltage is then applied to the horizontal deflection plates, causing the beam to sweep across the screen. The figure also indicates beam positions during selected parts of the sweep signal.

The sweep circuit works this way. At the beginning of the ramp, 4 volts are applied across the horizontal deflection plates, with the left plate positive and the right plate negative. The positive on the left attracts the beam; the negative on the right repels it. Since the beam deflects 1 centimeter for each volt applied to its plates, the beam begins at a position 4 centimeters to the left of center, indicating −4 volts. As the ramp progresses toward 0 volts, the attraction and repulsion decline. The beam approaches its normal position in the center. Then, the ramp passes through zero, changes polarity, and rises up to +4 volts. At this time, the beam moves over to the right side, because the right plate is positive and the left plate is negative.

Once the beam reaches the right side of the screen, the sawtooth quickly returns to its −4 volt starting point. This action causes the beam to quickly return to its starting point, the left side of the screen, ready for another sweep. You cannot see the spot move back to the left because the beam is turned off by the sawtooth pulse as it quickly returns to its starting point. This operation is called retrace blanking.

The rate at which the beam sweeps across the screen is called the sweep rate. It can be adjusted over a wide range, from very slow to very fast. A slow rate of 1 second per centimeter will allow you to watch the spot move across the screen. A fast rate of 1 microsecond per centimeter will produce a straight horizontal line. The use of the sweep rate controls will be described shortly.

USING AN OSCILLOSCOPE

Now that you are familiar with the internal workings of an oscilloscope, you can become familiar with using the instrument. In this section, we discuss the names of the common controls, what they do, and how to adjust them. After you've learned to adjust the controls, you will be ready to use the oscilloscope for one of its typical applications, measuring a DC voltage. Keep in mind, though, just what the purpose of an oscilloscope is. Generally, it is used to draw a graph with voltage on the vertical axis and time on the horizontal axis.

Common Controls

There are more controls on an oscilloscope than there are on any other common test instrument. The range is from a minimum of about 10 to a maximum of about 30. Furthermore, the names of controls vary from one brand to another. However, if you learn how a basic oscilloscope works, you should be able to operate almost any one you encounter.

There are some controls that are common to almost all oscilloscopes. These controls are shown in Figure 19–8. Their functions are as follows:

— Off-on turns the oscilloscope off or on.
— Intensity controls the brightness of the trace.
— Focus adjusts the sharpness or fineness of the trace.
— Sweep range and variable allow for sweep rates between the fixed speeds of the sweep range control.
— Horizontal position moves the trace left or right.
— Trigger level and mode select the level at which triggering occurs.
— Trigger source selects the signal that begins the sweep.
— Trigger input provides connections for external triggering sources.

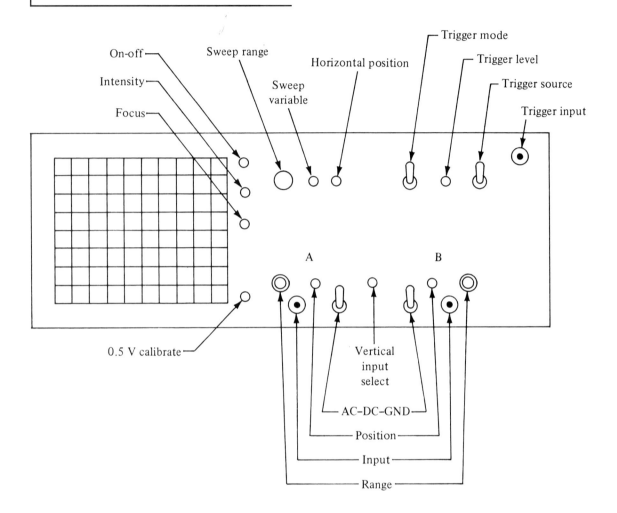

FIGURE 19-8
OSCILLOSCOPE CONTROLS

— Vertical input select determines which input channel (A or B or both) is to be viewed at a given time.

— AC–DC–GND informs the vertical circuitry whether the voltage being measured is AC or DC or whether the input is to be grounded.

— Vertical position moves the trace up or down.

— Vertical input provides a connection for the voltage to be measured.

— Vertical range selects the voltage range and is usually marked in volts per centimeter (V/cm).

— 0.5 V calibrate provides an accurate voltage for checking the gain of the oscilloscope circuitry.

Basic Adjustments

It is a good idea to have the use of an oscilloscope (and its manuals) as you read about the basic adjustments. These adjustments are the typical ones you make before measuring a voltage. Begin by turning the instrument on and allowing it a minute or so to warm up. There is no value in turning any control yet, for all the present positions may be correct. If, after 1 minute, there is no trace on the screen, make the following adjustments: Place the focus, intensity, and position controls in the middle of their ranges. Then, place the vertical range switch on 1 V/cm and the trigger on internal +. If you do not have a trace now, consult your instructor.

The first adjustment to make is for intensity. The intensity should be just bright enough for you to see the trace clearly without causing the trace to bloom or smear. The focus and astigmatism adjustments come next. If there is no astigmatism control, it may be inside; so you can ignore it. Adjust these controls at the same time. The purpose of each is to make the trace as sharp or as fine a line as possible. So as you turn one and the other, observe the effects. Once these controls are adjusted, you can leave them alone.

Move the position controls and observe what occurs. One control moves the trace up and down, while the other moves the trace left and right. The normal vertical position, as shown in Figure 19–9, places the trace on the line that goes across the center of the screen. Since no voltage has been applied yet for measurement, this line represents 0 volts. The normal horizontal position shows the trace going from the left end of the grid to the right end. Adjustment of the horizontal gain control must also be made at this time in order to get the correct size and position of the trace.

The trigger controls (sometimes called sync) are more complicated and are best understood by experimenting with them while

FIGURE 19-9
OSCILLOSCOPE TRACE WITH ZERO AT CENTER

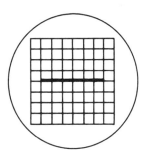

following the directions of the operating manual. However, they must be in proper positions if any trace is to appear on the oscilloscope. For most measurements, you can place the trigger source controls on internal and +. The trigger level should be placed on preset or at 0. The sweep controls should not be set until the need arises, such as when you want to analyze an AC voltage. For now, all you need are settings that produce a solid horizontal line. You are now ready to calibrate the vertical circuits.

Calibration

Oscilloscopes are unlike meters in that they must be calibrated before voltage can be measured. The process is quite simple, and your measurements will be incorrect if you do not follow it. Simply put the vertical variable control in its calibrated position.

An oscilloscope must be calibrated before you can accurately measure with it. However, an uncalibrated oscilloscope can be used to look at the shape of a waveform when you are not interested in its voltage value.

FIGURE 19-10
OSCILLOSCOPE TRACES

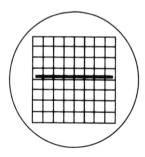

**A. Trace for 1 1/2 volts on the
10-volt-per-centimeter range**

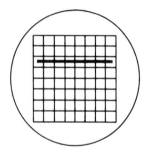

**B. Trace for 1 1/2 volts on the
1-volt-per-centimeter range**

MEASURING DC VOLTAGE

The first steps in measuring an unknown DC voltage are to turn the instrument on, make the basic adjustments, and calibrate the vertical circuits. Then, place the vertical controls on DC and the 10-volt-per-centimeter range. Connect the scope leads to vertical input (sometimes called Y) and ground or common. Check the probe at the end of the scope lead to see if it is marked 10×. If it is, the probe will reduce the voltage by a factor of 10 before the oscilloscope sees it. In this case, 50 volts will be shown as 5 volts, so you must multiply your results by 10. Let's assume you have a 1× probe so that the results are as they appear.

Voltage is measured across the source in the same manner it was measured with a multimeter. So connect the leads across a 1½-volt battery. If you connect ground to minus and vertical input to plus, the line will move up. Since you are on the 10-volt-per-centimeter range, the line will not move very far, as shown in Figure 19–10A. You must change the range to 1 volt per centimeter, as shown in Figure 19–10B.

Each time you change ranges, you must rezero the oscilloscope. To rezero, disconnect the leads from the battery and connect them together. By doing so, you are sure that the oscilloscope is looking at 0 volts. As mentioned earlier, you can place zero wherever you choose. However, zero is usually placed in the center. Now, reconnect the battery.

A voltage of 1½ volts on the 1-volt-per-centimeter range should produce a beam deflection of 1½ centimeters. Keep in mind that you multiply the deflection in centimeters by the range in volts per centimeter. Unlike a multimeter, which can measure only up to 1 volt on the 1-volt range, an oscilloscope can measure up to 6, 8, or 10 volts on the 1-volt range, depending on the number of centimeters from the bottom to the top of the screen.

Once you have measured the battery voltage, reverse the leads. You should notice the trace move down. It moves down because the voltage entering the vertical input is negative compared to the one entering ground. From ground, a positive voltage will make the trace move up, and a negative voltage will make the trace move down.

There are two more points to consider in voltage measurement with an oscilloscope. First, you cannot determine the *exact* voltage of the battery. That is, you cannot find a specific value closer than, say, 1.5 to 1.6 volts or

1.6 to 1.7 volts. An oscilloscope is not as precise as a multimeter. So you cannot determine a value to as many significant figures. Three significant figures, such as 1.62 volts, is reasonable with a meter. But you can only measure to two significant figures with an oscilloscope.

The second point involves the DC–AC–GND switch. Place this switch on AC instead of DC, and try to measure the voltage. It should read 0. Most measurements made with oscilloscopes are made at locations where both AC and DC voltages appear together. This switch allows you to ignore one so that you can measure the other. Therefore, you cannot measure DC with the switch on AC.

SUMMARY

Oscilloscopes are second only to multimeters in popularity as electronic test instruments. They are easily recognized by their cathode ray tubes and dozens of knobs and switches. Technicians use them to measure voltages and show waveshapes. With them, you can *see* voltages.

The main advantages of oscilloscopes are their ability to measure events that occur in microseconds and their high input impedance. This last advantage means that an oscilloscope has little effect on the circuit under test. Disadvantages are large size and weight and the need for line voltage, which reduce portability. The inability to measure current and resistance also limits their versatility.

The center of operation of an oscilloscope is its cathode ray tube. The CRT is a large, vacuum-filled glass component with elements that direct a ray of electrons from cathode to tube face. A phosphor coating inside the face glows when bombarded with the electron beam. Voltages to be measured find their way

to deflection plates and establish an electrostatic field. This field, in turn, causes the beam to move up or down. The direction of motion is determined by voltage polarity, while the distance of deflection is determined by voltage magnitude. Since it is a beam rather than a pointer that is moved, an oscilloscope is much faster and more sensitive than a multimeter.

If a voltage to be measured is too large for the CRT, it is reduced in the vertical circuits by an attenuator. Voltages that are too small are increased with amplifiers. Most oscilloscopes can measure microvolts as well as hundreds of volts. Internal sweep circuits produce a voltage that varies at a constant rate as a function of time. By the internal connection of this circuit to the horizontal deflection plates, a graph is produced. The horizontal axis of the graph is an appropriate unit of time, while the vertical axis is the unknown voltage.

The dozen or so controls on an oscilloscope must be properly adjusted before any voltage can be measured. Controls such as focus, intensity, and position are common to most oscilloscopes and can usually be left alone once set. Vertical range and gain controls require calibration. The procedure varies greatly from one instrument to another, so you must read the manual. These controls are found in a variety of locations and can have many different names. However, as with multimeters, you must select AC or DC and a range.

Once an oscilloscope is calibrated, it is ready for use. To measure voltage, connect the leads across the unknown voltage. If the beam goes up, the voltage at the vertical input terminal is positive. If the beam moves down, the voltage is negative. Now, count the number of centimeters the beam moves and multiply that number by the range. As before, always select the lowest possible range for the measurement.

CHAPTER 19

REVIEW TERMS

amplification: see *gain*

amplifier: oscilloscope circuit that increases the size of a voltage

attenuation: proportional reduction of circuit voltage

attenuator: oscilloscope circuit that decreases the size of the voltage being measured to a workable magnitude

cathode ray tube (CRT): vacuum tube used in oscilloscopes for displaying measurement results

deflection plates: elements in a cathode ray tube used to electrically control the beam position

frequency response: instrument specification that indicates frequency range for acceptable performance or indicates how quickly the instrument will respond

gain: proportional increase in the voltage or current in a circuit through the use of electronic circuitry

oscilloscope: instrument with a cathode ray display and used to measure voltage and time

sensitivity: oscilloscope specification that indicates its lowest voltage range

sweep circuit: oscilloscope circuit producing a voltage that varies at a constant rate as a function of time

REVIEW QUESTIONS

1. Describe the deflection plates of a cathode ray tube.

2. Describe what is meant by oscilloscope sensitivity.

3. Describe frequency response.

4. What is meant by the persistence of a cathode ray tube?

5. Describe sweep circuit.

6. Describe what is meant by calibration.

7. Name two types of measurements that can be made with an oscilloscope.

8. Explain what causes the screen of an oscilloscope to illuminate.

9. Explain how a line is produced on an oscilloscope screen.

10. Name three advantages of an oscilloscope.

11. Name three disadvantages of an oscilloscope.

12. Describe how to adjust the focus and intensity controls of an oscilloscope.

13. Explain the purpose of the sweep range and sweep variable controls of an oscilloscope.

14. Explain the purpose of oscilloscope trigger controls.

15. Describe the process by which DC voltage is measured with an oscilloscope.

REVIEW PROBLEMS

FIGURE 19-11

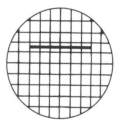

1. What is the largest voltage that can be measured by an oscilloscope with an 8-centimeter grid if its lowest range is 1 millivolt per centimeter and its highest range is 10 volts per centimeter?

2. What is the smallest voltage that the oscilloscope described in Problem 1 can reasonably measure? Why?

3. What is the largest voltage that can be measured by an oscilloscope with an 8-centimeter grid if its lowest range is 5 microvolts per centimeter and its highest range is 50 volts per centimeter?

FIGURE 19-12

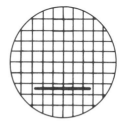

4. What is the smallest voltage that the oscilloscope described in Problem 3 can reasonably measure? Why?

5. Determine the gain or attenuation ratios required by the oscilloscope described in Problem 1 if the CRT has a sensitivity of 1 volt per centimeter.

6. Determine the gain or attenuation ratios required by the oscilloscope described in Problem 3 if the CRT has a sensitivity of 0.2 volt per centimeter.

7. What voltage is indicated in Figure 19–11 if the range is 0.1 volt per division and 0 volts is in the center?

8. What voltage is indicated in Figure 19–11 if the range is 50 volts per division and 0 volts is at the bottom line?

9. What voltage is indicated in Figure 19–12 if the range is 10 microvolts per division and 0 volts is in the center?

10. What voltage is indicated in Figure 19–12 if the range is 0.5 volt per division and 0 volts is at the top line?

ALTERNATING CURRENT AND VOLTAGE

Define alternating current, and describe alternating voltage.

Name and describe the four types of AC voltage values.

Name several AC voltage sources and their voltage level.

Determine the magnitude of an AC voltage by using an oscilloscope and a multimeter.

ALTERNATING CURRENT

The objective of this chapter is to introduce the concepts of alternating current. These concepts are best understood if we briefly review the theory behind induced voltage and current.

Induced Voltage and Current

In Chapter 12, we saw that current flowing in a conductor produces a magnetic field around the conductor. The direction of the field is determined by the direction of the current, and the strength of the field is determined by the strength of the current. As this current changes — that is, as it increases or decreases — the magnetic field expands and contracts. Now, another rule enters the picture. Just as a changing current produces a changing magnetic field around a conductor, a magnetic field crossing a conductor produces a force to move electrons. In Chapter 12 on inductance, we saw that this induced force is in opposition to the original current flow. Its net effect is to slow the rise of the intended current to its expected value.

Attempts to reduce the rate of current flow in a circuit with an inductor find a similar opposition. This opposition to change occurs when a decreasing current allows the magnetic field to start to collapse. Once again, the magnetic lines cross adjacent conductors and induce a force in the opposite direction. This force is called a counter emf (electromotive force) or a back emf, and the current it causes is an induced current. The net effect of this induced current is to slow the decline of the falling current, just as the earlier induced current slowed the rise of the current.

An understanding of inductor theory is a basis for understanding the concepts of alternating current generation. One characteristic of inductors provides a starting point for theory of AC generation.

> **Key point:** Magnetic lines crossing a conductor produce a force (voltage) that can cause a current to flow in that conductor.

The direction of the force depends on the direction of the crossing. This force can occur by moving the inductor through a magnetic field or by moving the magnetic field across the conductor.

AC Generator

Alternating current (AC) is a current that alternates or continually changes in direction. The alternating current we will be considering here continually changes in amplitude, or size, as well. As Ohm's law tells us, current is a result of voltage. So, an alternating current is the result of an *alternating voltage*. This voltage is commonly called AC voltage and is produced by an AC generator. Household voltage is generated by the process described here.

A description of generator theory begins with the familiar rules about a magnetic field crossing a conductor, but there are two differences. The first difference is that the conductor does not pass in one direction only; instead, it *rotates* around a central axis. Second, the conductor is not small; rather, a long conductor, wrapped to form a coil, is used.

Let's begin, however, with a single conductor rotating within a magnetic field, as shown in Figure 20–1. As the conductor moves across the top of the field (point *A*), no current is induced because no lines are crossed. The conductor is moving parallel to the lines. Shortly, however, the conductor begins to cross lines as it rotates clockwise into the field. At point *B*, the conductor is in the most dense section of the field and is moving directly across the lines. This point is a point

FIGURE 20-1
CONDUCTOR ROTATING THROUGH A MAGNETIC FIELD

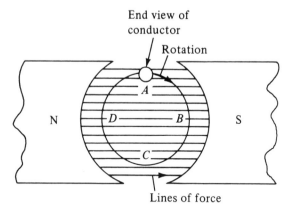

FIGURE 20-2
CURRENT DIRECTION AS A FUNCTION OF CONDUCTOR POSITION

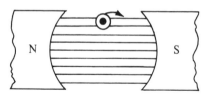

No current

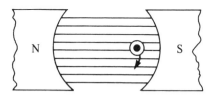

Maximum current toward reader

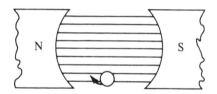

No current

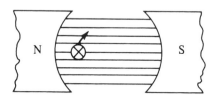

Maximum current away from reader

of maximum induced force since the *strength of force* is proportional to the rate of crossing. Point *C* is similar to point *A* since no current is induced. As the conductor passes through point *D*, maximum current is once again induced.

Now, consider the right-hand rule for induced currents, which was shown in Figure 12–12. For this rule, you hold your right hand so that your thumb, first finger, and second finger point in three directions and at right angles to each other. The rule states that if your first finger points in the direction of the magnetic lines and your thumb points in the direction the magnet is moved, your second finger will point in the direction of the induced current flow. This rule also applies to a generator, as shown in Figure 20–2. In Figure 20–2, assume that a dot in the end of a conductor means that current is coming toward you, while an X means that it is going away. It is as if the current were represented by an approaching or departing arrow.

Figure 20–2 only shows 4 points along the way through the 360° of a complete revolution. Instead of every 90°, though, we will look at every 30°, which gives us 12 points in one revolution, or cycle, as shown in Figure 20–3A. One cycle starts with any particular point and ends at that point. Note that, for our use of the right-hand rule, it is not the magnet that is moving but the conductor. So when the conductor passes down through the 90° point, it is as though the magnet were moving up. Point your thumb up at this part of the analysis. If you apply this rule to the points in Figure 20–3A, you will see that the current approaches you during the first 180°. For convenience, this direction will be called positive on the graph of magnitude versus direction. See Figure 20–3B. The other half of the cycle is negative since the current is moving away.

The graph in Figure 20–3B shows the magnitude and polarity of the induced current. Although only 12 points can be actually plotted for the graph, the graph can be completed by drawing a smooth curve between the points, as shown in the figure.

Assume that the maximum current value at 90° is equal to 1 ampere. The remaining values are equal to the sine of the angle of rotation, leading to the name of the graph, which is called a sine curve or a sine wave. The sine function is a trigonometric function; some of the values of the sine function are given in Table 20–1.

The last consideration in generator theory is the additional turns formed by making a coil. When the turns are wrapped in the manner shown in Figure 20–4A, the induced currents add to each other. Electrons moving toward you in the group of conductors on the right in Figure 20–4B add to those being pushed away in the conductors on the left. As the coil spins around inside the magnetic field, alternating current — electricity — is generated.

The force that rotates the coil in a generator can come from many sources. Automobile generators (called alternators) are turned by the engine with a fan belt. Some homes get electricity through the force of the wind by using a generator with blades mounted on a tower. Hydroelectric power plants use flowing water to turn large generators. Steam-powered generators are the most common electricity-producing facilities. The steam results from heating water by burning coal or oil or

by using a nuclear reactor. High-pressure steam is then applied to fan blades in the turbine, causing shaft rotation. This action, in turn, rotates the generator, which produces electricity. Output levels can be in excess of 100 kilovolts, with a power capability of 1 megawatt.

FIGURE 20-3
MAGNITUDE AND DIRECTION OF INDUCED CURRENT

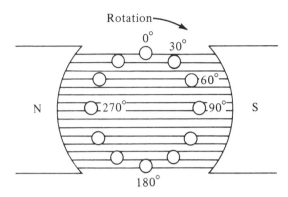

A. One cycle of induced current

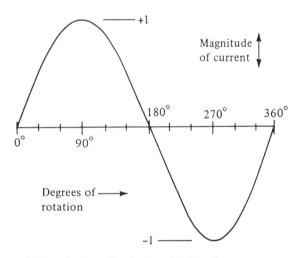

B. Magnitude and polarity of induced current

TABLE 20-1
SINE FUNCTION

Angle (in Degrees)	Magnitude
0	0.0
30	0.5
45	0.707
60	0.866
90	1.0
120	0.866
135	0.707
150	0.5
180	0.0
210	−0.5
225	−0.707
240	−0.866
270	−1
300	−0.866
315	−0.707
330	−0.5
360	0.0

AC VOLTAGE

Figure 20–3B shows that an AC sine wave is continually varying in size and polarity. Because the sine wave is continually changing, it is difficult to give it a value. Therefore, AC voltages have many values. In this section, we will discuss these values and define some terms associated with them.

Values of AC

We begin by looking at a typical voltage sine wave. The sine wave in Figure 20–5 represents a voltage that is continually varying up to +3 volts, down through 0 volts to −3 volts, and back up again. Note that the voltage does not really begin at point *A* but has been going on before that point. And it will continue after passing point *B*. However, it is common practice to look at only one cycle of a voltage, as pictured in Figure 20–5. One **cycle** (abbrevi-

FIGURE 20-4
INDUCED CURRENT IN A ROTATING COIL

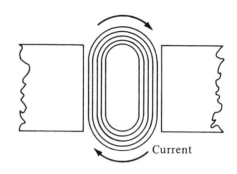

Current

A. Top view

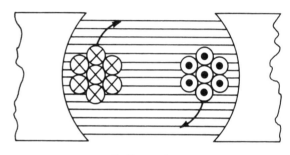

B. Front view

FIGURE 20-5
PERIOD OF A SINE WAVE

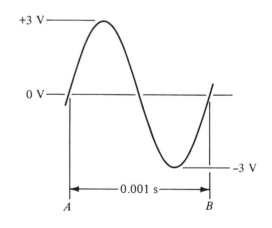

ated c) is generally considered to be the activity that occurs from the time when the voltage passes through 0 in a positive direction (point *A*) to the time that the voltage passes through 0 again in the positive direction (point *B*). The **period** of a voltage is defined as the time it takes to go through one cycle. The wave in Figure 20–5 has a period of 1 millisecond (abbreviated ms).

The **frequency** of a voltage describes the number of cycles the voltage goes through in 1 second. Frequency will be considered in greater detail in the next chapter. The unit of frequency is the **hertz** (Hz). The voltage pictured in Figure 20–5 has a period of 1 millisecond, or 0.001 second, and its frequency (*f*) is calculated as follows:

$$f = \frac{1}{t} = \frac{1}{0.001 \text{ s}} = \frac{1000 \text{ c}}{1 \text{ s}} = \textbf{1000 Hz}$$

where *f* is the symbol used for frequency and *t* is the symbol for time.

The highest level a voltage reaches is called the **peak voltage.** This voltage has a peak value of 3 volts. However, AC voltages alternate, reaching a peak value in both directions. So this voltage peaks at 3 volts one way and then at -3 volts the other. This alternation produces a peak-to-peak voltage of 6 volts. **Peak-to-peak voltage** is the absolute magnitude of the voltage from the positive peak to the negative peak.

As shown in Figure 20–5, the level, or **amplitude,** of an AC voltage varies continuously between the peak values. This voltage goes up and down 1000 times per second. Thus, its amplitude at any specific instant of time can be any value between $+3$ volts and -3 volts. This instantaneous value can only be determined if the specific point on the cycle is indicated. The sine function is used to indicate this specific point. The **instantaneous voltage,** then, is the voltage at a specified point in the cycle, and it can have a magnitude anywhere

between the positive and negative peak voltages.

Instantaneous voltages are represented by small letter *v*. And the Greek letter theta (*θ*) represents the angle of rotation. Thus, the equation for instantaneous voltage is

$$v = V_{\text{peak}} \times \sin \theta$$

Notice that the sine function is written as sin in equations. Table 20–1 can be used to find values for the sine function.

Let's do some calculations for the voltage shown in Figure 20–5. At 0°, the instantaneous voltage is

$$v = V_{\text{peak}} \times \sin \theta = 3 \text{ V} \times 0 = \textbf{0 V}$$

At 45°, the instantaneous voltage has the following value:

$$v = V_{\text{peak}} \times \sin \theta = 3 \text{ V} \times 0.707$$
$$= \textbf{2.121 V}$$

And at 240°, the value is

$$v = V_{\text{peak}} \times \sin \theta = 3 \text{ V} \times -0.866$$
$$= \textbf{-2.598 V}$$

These values are shown in Figure 20–6.

With all these values for just this one voltage sine wave, how do we know what size lamp to use? There are so many numbers we can use to describe an AC voltage that one must be selected for the most common use. That value is called rms (for "root mean square"), or effective, value. The **rms, or effective, voltage** is that value of AC voltage that will do the same amount of work as an equal amount of DC voltage. For example, an incandescent lamp will operate on both AC and DC. It will glow equally bright at 6 volts rms and at 6 volts DC. The rms value of the voltage in Figure 20–6 can be determined this way:

$$V_{\text{rms}} = 0.707 \times V_{\text{peak}} = 0.707 \times 3 \text{ V}$$
$$= \textbf{2.121 V}$$

FIGURE 20-6
INSTANTANEOUS VOLTAGE VALUES OF A
SINE WAVE

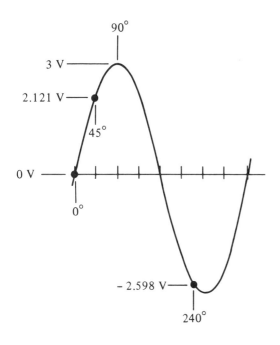

FIGURE 20-6
INSTANTANEOUS VOLTAGE VALUES OF A
SINE WAVE

The rms voltage is so commonly used that when a device is rated at 120 volts AC (120 V AC), it is understood that it means volts rms. While other values are considered for certain situations, you will almost always use the rms value.

The root mean square voltage is derived in the following way: The first step is to square the instantaneous values of the sine wave and draw a sine-squared curve, as shown in Figure 20–7. Some of the new points on this graph are $0^2 = 0$, $+3^2 = +9$, and $-3^2 = +9$. This new curve does not cross zero, and it goes through two cycles, as indicated in the figure.

The next step is to find the mean, or average, of this curve. The mean of a sine-shaped curve is a line through the center. The mean value, then, of this sine-squared curve is

4.5 volts. Note that the average of the original curve was 0 volts.

The third step is to calculate the square root of the mean. The square root of 4.5 volts is 2.121 volts.

So 2.121 volts is the root of the mean of the square of all instantaneous values of a sine wave voltage that is also equal to 3 volts peak and 6 volts peak to peak in amplitude. Most important, this AC voltage with many values has the same work capacity as 2.121 volts of DC voltage.

AC versus DC

DC voltage is easy to produce chemically with batteries, so it is desirable for portable uses. Radios and automobiles are two of the most common uses for a portable voltage-producing device. Even the automobile, though, still needs another voltage source to recharge the battery once the battery is used to start the engine. The automobile alternator produces AC, which is then changed to DC to keep the engine running and to recharge the battery. This particular AC-to-DC process will be described later in Chapter 30.

Two advantages of AC are its general availability almost everywhere at electrical outlets and its ability to be increased and decreased with transformers. (Transformers will be described in the next chapter.)

Some devices only work on AC or on DC, while other devices will work on both. A transistor radio is a DC device that usually operates on internal batteries. When it is connected to a wall socket, the AC is reduced to a lower voltage and changed to DC.

One way to visualize the comparison of DC and AC is to consider a river and a tidal basin. The river is like DC in that there is a constant rate of flow. The water flows, like electrons, in one direction and at the same speed all day long. Thus, a paddle wheel placed in the river can operate some mechan-

FIGURE 20-7
DETERMINING THE ROOT MEAN SQUARE
VALUE OF A SINE WAVE

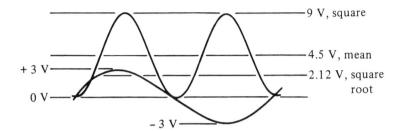

ical device, which would, in turn, perform some task. At one time, grain was processed this way.

The tidal basin is like AC in that the water flows in and out. A rising tide brings the water in, and the falling tide allows the water to leave. Like electron flow in an AC circuit, the water flow is continually changing in speed and direction. The period of the flow is about 12 hours, and the frequency is about 2 cycles per day. Sometimes, of course, the tidal water is not moving, and sometimes, it moves faster than the river. However, over the course of a day, its paddle wheel can do as much work as the one in the river if its *effective* rate of flow is the same as that of the river.

MEASURING AC VOLTAGES

Before you begin to measure an AC voltage, you must know which value you want and which value the instrument measures. Do you want rms or do you want peak-to-peak voltage? Meters measure rms, and oscilloscopes measure peak-to-peak voltage. If you measure one and want the other, you must calculate it. This calculation will be discussed shortly.

Frequency is another concern when you measure an AC signal. Instruments do not have unlimited ranges. If you attempt to measure a voltage beyond the frequency range of the instrument, you will get incorrect results. In addition, you may not know that the results are wrong. How do you resolve this problem? As usual, read the manual. You will be surprised to find the limited frequency range of some otherwise very desirable instruments. Finally, don't forget to inform the instrument that the voltage is AC. If you do not, the instrument may not work.

Using a Multimeter

All the rules for measuring DC with a multimeter apply in measuring AC. However, there are a few additional concerns. The meter should be on an AC range, as should any switch on the test lead probes. Before you attempt to measure current, be sure the meter can do it. Some meters will measure direct current but not alternating current. So, again, read the manual.

Another difference involves the scale used for AC and DC. This difference is most noticeable at low voltages. Compare the voltage ranges on the meter in Figure 20–8. The same scales are used for AC and DC voltage except for the 2.5-volt range. This meter, like most, can only measure DC, so it changes AC to DC

FIGURE 20-8
VOLTMETER SCALE

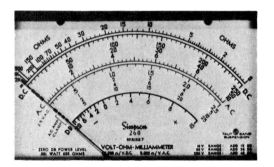

$$V_{peak} = V_{rms} \times 1.414$$
$$= 120\ V_{rms} \times 1.414$$
$$= \textbf{169.7 V peak}$$
$$V_{pp} = V_{peak} \times 2$$
$$= 169.7\ \text{V peak} \times 2$$
$$= \textbf{339.4 V pp}$$

The remaining procedures are quite simple. You just follow the rules you followed while measuring DC voltages.

If, however, you are measuring line voltage, you must be especially concerned about grounds. One side of an electric outlet is connected to ground (the earth). The floor you are standing on is connected to the earth through materials that are not excellent conductors but are not excellent insulators either. You must know how grounds are connected in order to avoid electric shock. Keep in mind that a few milliamperes can cause death.

Suppose you try to measure line voltage with a metal case instrument that has one of its leads connected to that case, a not uncommon situation. You have a 50–50 chance of connecting the case of that instrument to the hot side of the line. In that situation, there are 115 volts or so between the case, which you are certainly going to touch, and the floor you are standing on. The message is this: Don't measure line voltage with a metal case instrument. But if you must, identify the line voltage ground or the common first, and then connect the instrument ground to it.

Grounds are also a concern with chassis that are connected to the line. Some home entertainment products have one side of the line connected to the chassis. While this practice has been discontinued, there are many old units still around. Depending on the direction the cord is plugged into the wall outlet, there is, again, a 50–50 chance that the case will be 115 volts above the ground. Thus, you can be shocked by just touching the chassis. If you

internally when asked to measure it. But the rectifiers used for this change become nonlinear at low ranges. Thus, the meter has a different scale for the 2.5-volt AC range. Remember to check to see whether there is a special range for AC, and then use it. If you do not, you may obtain a wrong answer.

The value given by a meter is usually an rms value. Suppose you want to know peak or peak-to-peak value. The formulas for these calculations are variations of one you already know. Since $V_{rms} = 0.707 \times V_{peak}$, we have the following formula for V_{peak}:

$$V_{peak} = \frac{V_{rms}}{0.707}$$

or

$$V_{peak} = V_{rms} \times 1.414$$

And since the peak-to-peak voltage is twice the peak voltage, we have the following formula, where V_{pp} represents the peak-to-peak voltage:

$$V_{pp} = 2 \times V_{peak}$$

As an example, a meter indication of 120 volts rms is equal to the following peak and peak-to-peak values:

are also touching a well-grounded test instrument, you can be electrocuted. So, in working on these systems, be careful, and use an isolation transformer.

Using an Oscilloscope

Oscilloscopes usually have grounded cases and connect one side of the unknown voltage to the case. So all the precautions just discussed apply here, too. To measure AC voltage, you begin with all the steps given in Chapter 19 for measuring DC voltage. The basic adjustments must be made first, and the oscilloscope must be calibrated. The input switch must be on AC, of course, and the vertical position must be set so that there is a solid trace across the center of the screen. Select a high voltage range and connect the unknown voltage to the vertical input and ground terminals. Three steps must now be taken.

The vertical range control must be switched to provide the tallest image that does not go beyond the grid markings. Do *not* change the gain, because doing so will uncalibrate the oscilloscope. Next, the sweep range and variable controls must be adjusted until 1 or 2 cycles appear. The trigger selector must be on internal + in order to stop the trace. An adjustment of the trigger level control may also be necessary before you can stop the trace. Now, you can determine the voltage value.

One step may add some convenience and reduce error. Adjust the vertical position to place the lower peaks on the bottom line. Now, also adjust the horizontal position to place an upper peak on the line going up through the center. This line often has subdivisions on it. Now, just count the number of centimeters from the trace bottom to top. The trace in Figure 20–9 is 4.5 centimeters high. Since the oscilloscope is on the 10-volts-per-centimeter range, the voltage has a value of 45 volts peak to peak. That is,

FIGURE 20-9
AC VOLTAGE WAVEFORM

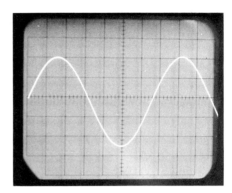

$$10 \text{ V/cm} \times 4.5 \text{ cm pp} = \textbf{45 V pp}$$

The rms and peak values can be found by using the procedures described earlier. The calculations are

$$V_{peak} = \frac{V_{pp}}{2} = \frac{45 \text{ V}}{2} = \textbf{22.5 V peak}$$
$$V_{rms} = 0.707 \times V_{peak} = 0.707 \times 22.5 \text{ V}$$
$$= \textbf{15.9 V rms}$$

As another example, an oscilloscope having a 3-centimeter, peak-to-peak display while on the 50-millivolts-per-division range is indicating a voltage of

$$50 \text{ mV/division} \times 3 \text{ divisions pp}$$
$$= \textbf{150 mV pp}$$

The peak value is

$$V_p = \frac{V_{pp}}{2} = \frac{150 \text{ mV pp}}{2} = \textbf{75 mV peak}$$

The rms value is

$$V_{rms} = 0.707 \times V_p$$
$$= 0.707 \times 75 \text{ mV peak} = \textbf{53 V rms}$$

Often, when you measure an AC voltage with an oscilloscope, you also measure its frequency. The generation and measurement of different frequencies is the topic of the next chapter.

SUMMARY

Alternating current, like direct current, consists of electrons moving from one location to another. And its movement can cause work to be done. But unlike direct current, alternating current does not flow in a constant direction and at a constant rate. Instead, it is continually changing in amplitude and direction. Some devices function equally well on both AC and DC, but other devices will operate on only one. The main advantage of AC is its ease of generation; its main disadvantage is a lack of portability.

When a conductor is moved through a magnetic field, a force to move electrons is created. The strength of this field depends on the density of the field, the number of conductors, and the rate at which the conductors cross the lines. The polarity of this induced force depends on the direction of the motion and the lines. By using the right-hand rule for induced currents, we can determine the direction of induced current.

AC generators produce alternating voltages by rotating a coil through a stationary magnetic field. The force to rotate this coil can come from flowing water, a windmill, or many other sources. One very popular source is steam generated by coal, oil, or nuclear power.

There are many ways to describe the level of an AC voltage. Since the voltage is always changing, there are an infinite number of instantaneous values. These values can be found between the positive peak and the negative peak, an amount that is called the peak-to-peak voltage. The AC value used most often is the rms value, which is 0.707 times the peak value. The rms voltage will do the same amount of work as an equal value of DC voltage.

AC voltages are measured with a multimeter in essentially the same manner as DC voltages. Of course, you must instruct the instrument and its probes that the voltage is AC. And some meters will measure AC voltage but not current, so you must read the manual before you begin. Another difference between the two measurements is the low-voltage scale. Many meters have separate scales for AC and DC. Using the wrong scale can create an error of 10% or more in your results.

Oscilloscopes must also be instructed that the voltage being measured is AC. Sweep and trigger adjustments must be made so that you can view 1 or 2 cycles of the unknown voltage. Once all the adjustments have been made, including calibration, the peak-to-peak value of the voltage can be determined. If you are interested in the rms value, you must calculate it. You can, however, get that value directly by using a multimeter.

CHAPTER 20

REVIEW TERMS

alternating current (AC): current that alternates or continually changes direction

amplitude: level of an AC voltage

cycle: one complete performance of a repeating or periodic process

effective voltage: same as rms voltage

frequency: number of cycles in 1 second

hertz: unit of measure for frequency

instantaneous voltage: voltage at a specified point in the cycle

peak voltage: highest positive or negative level reached by a voltage alternation

peak-to-peak voltage: absolute magnitude of the voltage from the positive peak to the negative peak

period: time for one cycle

root mean square (rms) voltage: that value of AC voltage that will do the same amount of work as an equal amount of DC voltage

REVIEW QUESTIONS

1. Define alternating current.

2. Describe alternating voltage.

3. Describe an AC generator.

4. Define the period of an AC voltage.

5. Define instantaneous value.

6. Define rms, or effective, value.

7. Describe peak-to-peak value.

8. Name two common sources of AC voltage.

9. Describe the basic difference between AC and DC.

10. What precautions must be observed when you are measuring AC line voltage?

11. What is meant by induced current?

12. Why is the average value of AC voltage not a useful number?

13. Name two advantages of AC over DC.

14. Why is rms voltage the most commonly used value for AC?

15. Describe how you can complete a circuit by touching only one side of a line voltage wall outlet.

REVIEW PROBLEMS

1. What value of DC voltage has the same effective value as 220 volts peak-to-peak AC?

2. What value of DC voltage has the same effective value as 48 volts peak-to-peak AC?

3. What is the instantaneous value of a 100-volt peak signal at 30°? At 180°? At 270°?

4. What is the instantaneous value of a 16-volt peak signal at 45°? At 210°? At 330°?

FIGURE 20-10

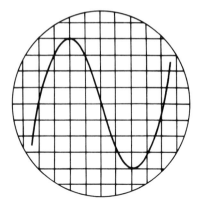

FIGURE 20-11

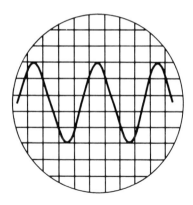

5. What is the frequency of a signal with a period of 1.25 microseconds?

6. What is the frequency of a signal with a period of 0.4 millisecond?

7. What is the period of a signal with a frequency of 60 hertz?

8. What is the period of a signal with a frequency of 680 megahertz?

9. What peak current will occur if 120 volts rms is connected across 34 kilohms?

10. What peak current will occur if 24 volts rms is connected across 680 ohms?

11. In Figure 20–10, what peak-to-peak voltage is indicated if the range is 0.1 volt per division? What is the rms value?

12. What peak-to-peak voltage is indicated in Figure 20–10 if the range is 20 millivolts per division? What is the rms value?

13. In Figure 20–11, what peak-to-peak voltage is indicated if the range is 20 volts per division? What is the rms value?

14. What peak-to-peak voltage is indicated in Figure 20–11 if the range is 50 microvolts per division? What is the rms value?

21

FREQUENCY AND TRANSFORMERS

Define frequency, and describe common frequency ranges.

State the basic functions of an audio frequency oscillator.

Determine the output voltage and frequency of an audio frequency oscillator by using an oscilloscope.

Describe the operation of a transformer, including turns ratio and impedance ratio.

Identify the various types of transformers and state their use.

FIGURE 21-1
FREQUENCY RESPONSE CURVE FOR AN
AUDIO AMPLIFIER

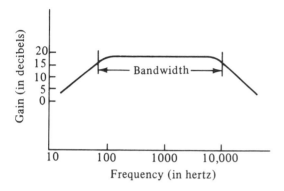

FIGURE 21-1
FREQUENCY RESPONSE CURVE FOR AN
AUDIO AMPLIFIER

FREQUENCY

The objective of this chapter is to introduce the concept of frequency, the sources of various frequencies, and the operation of transformers. We will begin with an overview of frequency.

Frequency is one way of describing an AC voltage. **Frequency** is a number that indicates how many cycles a voltage goes through in 1 second. The unit of frequency is the **hertz** (Hz); 1 hertz is equal to an alternation rate of 1 cycle per second. The line voltage to a home goes through 60 cycles per second, so it has a frequency of 60 hertz. This factor must be considered in electronic systems since some components are very frequency sensitive. While a resistor may not react to a frequency change from 60 to 50 hertz, inductors and capacitors will. We will learn more about such sensitivity in Chapters 26 and 27.

Common Frequencies

A specification found on most stereo systems is the term **frequency response,** which de-scribes the range of frequencies that the system will respond to or reproduce. These frequencies are recorded on a tape and disk and should be reproduced by the speaker. Figure 21–1 shows the frequency response curve of a typical audio amplifier. Amplifier gain is on the vertical axis, and frequency is on the horizontal axis.

The typical frequency response of an audio system is 20 to 20,000 hertz; this range is the normal audio frequency range. Energy at this frequency produces sound that people can generally hear. There is no need for the system to go beyond that range because those are the limits of the average person's hearing range. And as a person ages, the higher limit will not be quite so high.

As the frequency of energy increases, interesting things begin to occur. For example, alternating frequencies around 1 megahertz (MHz) are fed by AM radio stations into their antennas. The large changing current produces a large changing magnetic field that radiates into space. This changing magnetic field travels many miles and crosses the windings in a radio's antenna. It, in turn, produces a changing current that represents the information transmitted from the studio. This radio frequency range can be as low as 50 kilohertz (kHz) or as high as 420 megahertz. American AM broadcast stations operate in the range between 550 and 1600 kilohertz, or 0.5 to 1.6 megahertz. At this frequency range, signals can travel a few thousand miles because they follow the curvature of the earth.

A little higher up on the frequency spectrum are the FM radio stations. These stations are in the 88 to 108 megahertz range. FM radio signals do not follow the curvature of the earth, as AM radio signals do. Instead, they beam in a straight line, the line of sight. For this reason, high towers are used to raise FM antennas to a high level, increasing the line of sight and, therefore, the range. Since AM broadcast signals are not limited by line of sight, high towers are not needed. So with

FM, an antenna is mounted on the top or side of a high tower. With AM, a smaller tower *is* the antenna. (The subject of wave propagation will be discussed in greater detail in your communications course.)

As the frequency increases even further, the light spectrum is soon reached. Whereas changing the audio frequency produces different tones, changing frequency here produces different colors. A further increase in frequency produces invisible light and X rays.

Waveshape

The generation of an AC voltage was described in the preceding chapter. There, we saw that a sine-shaped voltage waveform was the easiest and most natural to produce. However, all waves are not sine waves. Consider a note on a piano and the same note on a trumpet. They do not sound the same. But if they are playing the same note, their frequencies must be the same. What, then, is the difference? The difference is in their waveshapes. That is, the same note from these two instruments produces different waveshapes.

One factor that determines the shape of a wave is the harmonic content. **Harmonics** are multiples of a given frequency. For example, the 2nd harmonic of 60 hertz is 120 hertz, and the 3rd harmonic is 180 hertz. Complex waveforms are produced by adding specific amounts of harmonics to the original signal.

A 60-hertz square wave is produced by starting with a 60-hertz sine wave, as shown in Figure 21–2A. Then, decreasing amounts of odd harmonics are added to it. That is, ⅓ as much voltage is added at the 3rd harmonic, ⅕ as much voltage at the 5th harmonic, and so on. As the figure shows, a square wave is a complex wave.

Some of the other common waveshapes are ramps, triangles, and pulses, shown in Figure 21–2B. As you can see, some waves are symmetrical. That is, like a sine wave, they have the same shape on both sides of zero.

FIGURE 21-2
WAVEFORMS

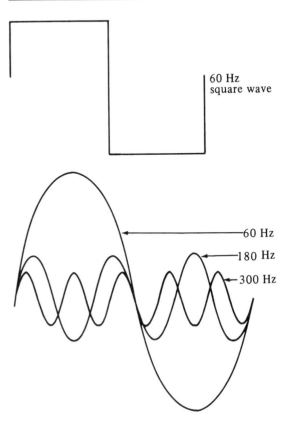

A. Square wave produced from sine wave's odd harmonics

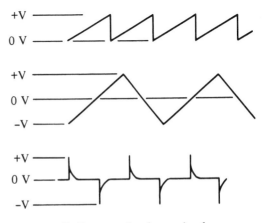

B. Ramps, triangles, and pulses

FIGURE 21-3
AUDIO FREQUENCY OSCILLATOR

Others are nonsymmetrical, and still others do not pass through zero. Devices that produce these voltages are described next.

OSCILLATORS AND GENERATORS

Multimeters and oscilloscopes are used to measure voltages or signals in electronic products and systems. Sometimes, though, with systems like stereos, it is necessary to produce more controlled signals than those produced by a radio or by tapes. Therefore, **oscillators** and **generators** are used to produce test signals. We begin our study with the audio frequency oscillator.

Audio Frequency Oscillator

An **audio frequency oscillator,** shown in Figure 21–3, is an instrument that generates voltages in the audio frequency range. These voltages are used as test signals in the evaluation of electronic products. For example, an audio frequency oscillator can be used to test the frequency response of stereo amplifiers.

Rather than use music with its many complex waveshapes, a technician applies simple sine waves of selected frequencies to the amplifier input while measuring the output voltage across the speaker with an oscilloscope.

The advantage of using an audio frequency oscillator is the ability to get a good-quality sine wave of known frequency and amplitude. Some generators also provide square waves. Since a square wave is difficult to produce, the quality of the wave depends on the price of the oscillator. The accuracy of the frequency is also a cost factor.

The operation of an audio frequency oscillator is quite simple. You simply select the desired frequency and the amplitude. If a square wave is also available, you select the waveshape. Two terminals are usually available for output connection. One may already be grounded, or there may be a third terminal for optional grounding.

The desired frequency is usually selected by using a range switch and a vernier dial. You connect a voltmeter across the output terminals and adjust the amplitude control in order to get the desired output voltage. Most manufacturers offer models that have built-in meters or dials to indicate output voltage level.

Oscillators like the one in Figure 21–3 cover the audible range of the audio frequency spectrum and perhaps a little more. The lower end of the available band is generally 20 hertz, although some go lower than 1 hertz. The upper end is usually about 30 kilohertz, although some go above 100 kilohertz. Output voltages are generally adjustable from a few millivolts to somewhere around 10 volts.

Radio Frequency Generator

Like oscillators, generators produce waveforms over a wide frequency range. However, generators are capable of producing complex waveforms. **Radio frequency generators** produce signals in the megahertz range and above. Their waveforms include both a base of a selected radio frequency signal and an added audio frequency signal. The process of adding an audio signal onto a radio frequency signal is called **modulation.**

Radio frequency generators, like the one shown in Figure 21–4, are used in the alignment of radio and television equipment. Although these procedures will not be discussed here, we can describe the generator. Like an audio oscillator, a radio frequency generator has a range switch and a vernier dial for selecting the desired frequency. An output voltage control is available for adjusting the output, and a connector is available for access. Coaxial cable is used with the output since the output is a radio frequency signal. This cable is much like the coaxial cable used to bring the radio frequency signal from a car antenna to the radio.

Modulation selection and level controls are also available. The modulation process is a little too complicated to explain at this time. What we suggest is that later, when you are looking at the output of an audio frequency oscillator, you also view the output of one of these generators. Look at the quality of the

FIGURE 21-4
RADIO FREQUENCY GENERATOR

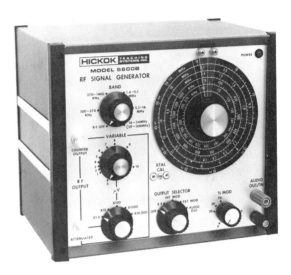

signal and see what effect the modulation control has on the output waveform.

Function Generator

A function generator, shown in Figure 21–5, is often a more expensive device than an oscillator or a radio frequency generator. There are several reasons for the extra cost. The accuracy and precision of the frequency and voltage and the quality of the output waveshape may be better than those of other generators. Furthermore, **function generators** produce a variety of waveshapes, such as sine, square, ramp, and pulse waveforms. These instruments are used to evaluate circuits by simulating different operating conditions with a variety of inputs.

One additional feature of a function generator should be mentioned here. The pulses available do not go up and down around zero as some of the symmetrical waveshapes do.

FIGURE 21-5
FUNCTION GENERATOR

Instead, pulses go from zero up to a selected level and back to zero as positive pulses, or from zero in the other direction as negative pulses. Their rate is called the repetition rate rather than frequency. That term still refers to the number of pulses per second, and it is determined by dividing 1 by the time t that elapses from the beginning of one pulse to the beginning of the next. This time is indicated in Figure 21–6, as is the pulse width, which is the length of time each pulse exists. Pulses such as these play an important role in the operation of digital computers.

FREQUENCY MEASUREMENT

Following voltage, frequency is the next most commonly measured circuit parameter. Two approaches can be used to determine fre-quency. One method is to measure the time that elapses for one cycle and then calculate frequency. The other method is to measure frequency directly. The method selected gen-erally depends on the type of instrument avail-able and the accuracy required. Direct mea-surement will be discussed first.

Using a Counter

By far, the easiest instrument to use in mea-suring frequency is an electronic frequency counter, like the one shown in Figure 21–7. It counts the number of cycles in a given amount of time, such as 1 second. Counters come in a variety of sizes and prices, with most of the difference among models being in the fre-quency range provided. Because of the way they operate, accuracy is usually excellent. For example, an electronic counter can usu-ally measure a frequency of 1 megahertz within 1 hertz, for an accuracy of 1 part per million, or 0.0001%.

Counters can often also measure the time it takes for an event to occur, so you must be certain to let it know that you are measuring frequency. Also, be sure that the voltage level of the unknown frequency does not exceed the input capacity of the counter, which often is less than 1 volt. Once the unknown signal is connected, turn the range switch to the posi-tion that gives the greatest number of digits. The results can be read directly from the dis-play.

Using an Oscilloscope

While counters are the most accurate fre-quency-measuring instruments, they are not always available. And there are times when this degree of accuracy is not needed or when the waveshape must be observed. In these cases, you can use an oscilloscope. Measuring frequency with an oscilloscope begins with all

FIGURE 21-6
PULSE WAVEFORM

FIGURE 21-6
PULSE WAVEFORM

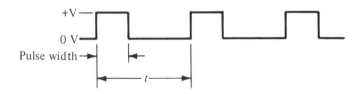

FIGURE 21-7
ELECTRONIC COUNTER

the steps outlined for measuring AC voltage. Then, a few additional steps must be taken.

To measure frequency, you must use an oscilloscope with a calibrated sweep circuit. This feature is indicated by horizontal sweep controls marked in milliseconds per centimeter and so forth. The first step is to calibrate the sweep time circuitry. To do so, place the sweep variable control in the preset or calibrated position. Then, select a sweep range that allows you to view at least one complete cycle. Position the wave vertically so that it is centered between the top and bottom. Then, position it toward the left so that it crosses

zero in the positive direction on the left vertical line.

Now, you are ready to measure the frequency. Begin by counting the number of centimeters, or divisions, for one cycle. Let's assume that there are 4. Then, multiply that number by the range. For example, we'll assume that the range is 50 milliseconds per centimeter (or division). Thus, the time for 1 cycle (c) is 200 milliseconds, or 0.2 second. Since frequency equals 1 divided by the time for 1 cycle, the frequency is

$$f = \frac{1}{t} = \frac{1}{0.2 \text{ s}} = 5 \text{ c/s} = \textbf{5 Hz}$$

Now, let's consider a specific example. Refer to Figure 20–9. As the waveshape in Figure 20–9 rises, it crosses the horizontal centerline at the left end of the screen. It crosses the line on the way up again, 6.6 divisions later. Thus, there are 6.6 divisions per cycle since that is the distance between two occurrences of an event (like passing through 0 volts in the positive direction). This distance is the same as that between the two positive peaks of the waveform, although the latter is more difficult to read.

If the oscilloscope presenting this display is on the 2-milliseconds-per-division range, the waveform period indicated is

$$t = 6.6 \text{ divisions} \times 2 \text{ ms/division}$$

$$= \textbf{13.2 ms}$$

FIGURE 21-8
MEASURING AN UNKNOWN FREQUENCY
WITH LISSAJOUS PATTERNS

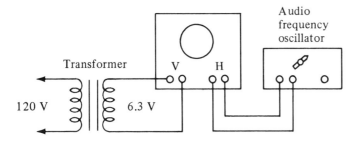

The frequency of this waveform is

$$f = \frac{1}{t} = \frac{1}{13.2 \times 10^{-3} \text{ s}} = \textbf{75.8 Hz}$$

If the horizontal range of the oscilloscope in Figure 20–9 is on the 50-microsecond range, the period and frequency are

$$t = 6.6 \text{ divisions} \times 50 \text{ μs/division}$$
$$= \textbf{330 μs}$$
$$f = \frac{1}{t} = \frac{1}{330 \times 10^{-6} \text{ s}} = \textbf{3.03 kHz}$$

Using Lissajous Patterns

Another way to determine an unknown frequency is by comparing it with a known frequency. A druggist often uses a similar method to measure weights with a balance. Oscillators can be calibrated by comparing their outputs with that of high-quality generators. An oscilloscope is used as a viewing device but is not relied on for accuracy. Just as the druggist's balance relies on external weights for its accuracy, this system relies on a known frequency as its standard.

The process uses **Lissajous patterns,** which are oscilloscope waveforms that show the relationship between two frequencies. They can be produced by connecting a circuit like the one shown in Figure 21–8. The transfomer is used to represent some unknown frequency, while the oscillator with its calibrated dial serves as a known frequency. Switch the horizontal controls from the sweep position to external to prepare the oscilloscope for this experiment. This step removes the internal sweep generator as the source of horizontal beam movement and replaces it with the external audio frequency oscillator. Once everything is turned on, check to see whether the equipment is operating by varying the oscillator amplitude control. It should vary the width of the trace.

The vertical, horizontal, and vertical gain controls should be adjusted for a trace of equal height and width. Note that oscilloscope calibration has no significance in this test. The point of the experiment is to adjust the known frequency of the oscillator until it is equal to the unknown frequency of the transformer. The frequencies are equal when a circle appears on the CRT, as shown in Figure 21–9. Now, since they are equal, the unknown frequency can be read from the oscillator dial. It should indicate 60 hertz.

There are times when an oscillator does not have the range to reach the unknown frequency. For instance, it may only be able to go as low as two or three times the unknown frequency or as high as ½ or ⅓ the unknown frequency. In such cases, a complex pattern will result. The best way to understand what is happening is to change the oscillator frequency from 60 to 30 hertz. As you approach 30 hertz, a 2 : 1 pattern, as shown in Figure 21–10, should appear. This pattern means that one signal is twice the other. The question is, though, Which is twice the other?

There are formulas that can be used to interpret the pattern. But the best way to interpret it is to think about what is happening. Consider these questions: How many times does the beam go to the top, and what makes it go to the top? How many times does it go to one side, and what makes it go to the side? The pattern in Figure 21–10 touches the top two times, and the vertical movement makes it do that. It touches the side one time, and the horizontal movement makes it do that. So we can say that the vertical equals 2 and the horizontal equals 1, which can be expressed another way:

$$\frac{vertical}{horizontal} = \frac{2}{1}$$

Since the oscillator is connected to the horizontal control and is used as the known frequency, its dial reading can be used in the equation. The expression can now be completed as follows:

$$\frac{vertical}{30 \text{ Hz}} = \frac{2}{1}$$

$$vertical = \frac{2 \times 30 \text{ Hz}}{1} = \mathbf{60\ Hz}$$

If you change the oscillator to 180 hertz, you should get a pattern with three loops in one direction, like the pattern shown in Figure 21–11. This time, the top is touched once for

FIGURE 21-9
1 : 1 LISSAJOUS PATTERN

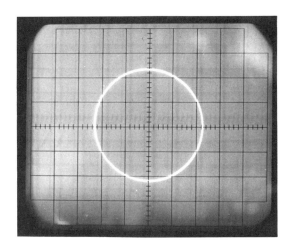

FIGURE 21-10
2 : 1 LISSAJOUS PATTERN

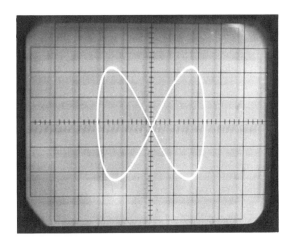

FIGURE 21-11
1 : 3 LISSAJOUS PATTERN

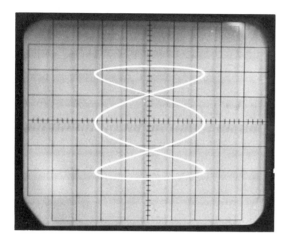

FIGURE 21-12
TRANSFORMER SYMBOLS

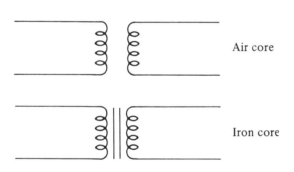

Air core

Iron core

every three times the side is touched. Thus, we can calculate the unknown this way:

$$\frac{\text{vertical}}{\text{horizontal}} = \frac{1}{3}$$

$$\frac{\text{vertical}}{180 \text{ Hz}} = \frac{1}{3}$$

$$\text{vertical} = \frac{180 \text{ Hz}}{3} = \mathbf{60 \text{ Hz}}$$

In all these cases, 60 hertz was used as the unknown so that you could check the accuracy of your results. For more practice, use two oscillators, one as the known and one as the unknown.

TRANSFORMERS

Transformers, like the one used in the previous experiment, consist of two inductors close enough to be within each other's magnetic field. Thus, a **transformer** can be defined as two or more coils that are electrically separated and magnetically connected. The schematic symbols in Figure 21–12 give an impression of that definition.

Transformer Operation

To understand how a transformer works, you must remember that it is an AC device and that its most common use is to change the level of a voltage. While it can be operated in either direction, its input side is called its **primary,** and the output side is the **secondary.** Let's assume that we have the transformer used in the Lissajous pattern experiment. It had an input voltage of about 120 volts and an output of about 6 volts. That is, it had a step-down ratio of about 20 : 1. The 120 volts applied to the primary vary continuously in amplitude and polarity. As shown in Figure 21–13, the top connection is positive and the bottom negative for half of the cycle, with the polarity reversing for the second half. This alternation causes a current in the primary that varies continuously, in amplitude and direction, along with the polarity and amplitude of the applied voltage.

Now, let's add an iron core, a secondary winding, and a 6-volt lamp to serve as a load that will use some of the power, as shown in Figure 21–14A. During the first half of the cycle, electrons leave the lower terminal of

FIGURE 21-13
TRANSFORMER PRIMARY CURRENT

FIGURE 21-14
SECONDARY CURRENT

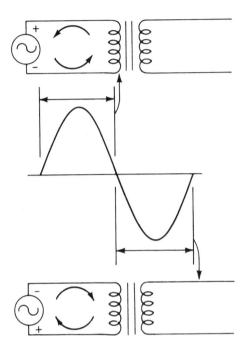

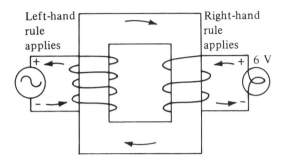

A. First half cycle

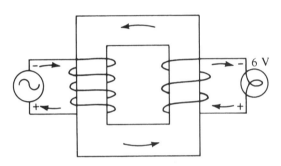

B. Second half cycle

the voltage source and work their way up through the winding. They will not enter the core because the winding is insulated with a thin coat of enamel. And since the insulation is thin, many turns can be wrapped around the core in a small space.

As this current increases, it creates a magnetic field, which, according to the left-hand rule, flows clockwise through the core. However, according to the right-hand rule for induced currents, the increasing and then decreasing magnetic field induces an increasing and decreasing current in the secondary and on through the lamp. Now, the polarity of the voltage source reverses, as shown in Figure 21–14B.

The direction of the primary current changes, too, causing a reversal in the magnetic field. This reversal, in turn, changes the direction of the secondary current. Note

that this process only works because a continually changing current produces a continually changing magnetic field, which, in turn, produces the secondary current. With pure DC from a battery applied, this process would not occur, and no secondary voltage would be induced.

Another consideration for a transformer is the number of turns in each winding. The number of magnetic lines produced depends on the number of turns in the primary winding. Likewise, the current induced in the secondary depends on the number of magnetic lines and the number of secondary turns. If the number of turns were equal and the magnetic linkages perfect, we would expect the primary and secondary voltages to be equal.

Also, we would expect a secondary with half as many turns to have half as much voltage. This relationship is called the **turns ratio** and is expressed as

$$\frac{V_p}{V_s} = \frac{n_p}{n_s}$$

where V_p is the primary voltage, V_s is the secondary voltage, n_p is the number of turns in the primary, and n_s is the number of turns in the secondary. For the transformer we are considering, the turns ratio, then, is

$$\frac{V_p}{V_s} = \frac{n_p}{n_s} = \frac{120 \text{ V}}{6 \text{ V}} = \frac{20}{1}$$

which is a turns ratio of 20 : 1, as we mentioned earlier. Similarly, if a transformer has a primary voltage of 120 volts and 5 secondary turns for each 2 primary turns, its secondary voltage is found as follows:

$$\frac{V_s}{V_p} = \frac{n_s}{n_p}$$

$$V_s = \frac{V_p n_s}{n_p} = \frac{120 \text{ V} \times 5}{2} = 300 \text{ V}$$

This transformer is a step-up transformer with a turns ratio of 1 : 2.5.

The power used in the secondary must be provided by the primary. If the 6-volt lamp described earlier has a current of 0.5 ampere, the secondary power P_s is

$$P_s = V_s I_s = 6 \text{ V} \times 0.5 \text{ A} = 3 \text{ W}$$

Since this power must be provided by the primary and its voltage is already known, the primary current can also be found, as follows:

$$P_p = P_s = 3 \text{ W}$$

$$P_p = V_p I_p$$

$$I_p = \frac{P_p}{V_p} = \frac{3 \text{ W}}{120 \text{ V}} = 0.025 \text{ A}$$

As these calculations show, a 20 : 1 step-down in voltage also appears as a 1 : 20 step-up in current.

The load of the lamp appears as 12 ohms to the secondary because of this Ohm's law evaluation:

$$R_{lamp} = \frac{V_s}{I_s} = \frac{6 \text{ V}}{0.5 \text{ A}} = 12 \ \Omega$$

However, the voltage source sees it differently because it is looking at the load through the transformer. This resistance is what the power source sees:

$$R = \frac{V_p}{I_p} = \frac{120 \text{ V}}{0.025 \text{ A}} = 4800 \ \Omega$$

The resistance that the power source sees can be determined another way, through a relationship called the **impedance ratio (Z)**, which is as follows:

$$\frac{Z_p}{Z_s} = \left(\frac{n_p}{n_s}\right)^2$$

The resistance the power source sees, the impedance of the primary, then, is as follows:

$$Z_p = Z_s \left(\frac{n_p}{n_s}\right)^2 = 12 \ \Omega \times 20^2$$

$$= 12 \ \Omega \times 400 = 4800 \ \Omega$$

The subject of impedance as it relates to transformers and other components will be discussed further in later chapters.

One other consideration for transformers must be discussed. The power transfer from primary to secondary has been described as if it were perfect. It is not, as we will see next.

Cores and Losses

The efficient coupling of the primary lines with the secondary turns is the job of the transformer core. Therefore, the quality of the core is very important. One way to describe its effectiveness is with a **coefficient of coupling** *(k)*. This coefficient is based on the number of magnetic flux lines in the primary that cut across the secondary. If 4000 lines are produced in the primary and 3600 of them cut across the secondary, then the coefficient of coupling is 0.9:

$$k = \frac{\text{secondary lines}}{\text{primary lines}} = \frac{3600}{4000} = \mathbf{0.9}$$

The loss during coupling is not as great with an iron core as it is with an air core.

Energy is also lost because of copper losses, which are caused by the resistances of the copper windings. These resistances are shown in Figure 21–15 as the primary and secondary resistances, R_p and R_s. In transformers with large amounts of current, the power dissipated by these resistances can cause the transformer to get quite warm.

Another loss occurs because of the residual magnetism in the core. As shown in Figure 21–16, the strength of the field (the flux density) increases as its cause (the current times the number of turns, or the ampere-turns) increases. Since many molecular magnets must be turned around twice during each cycle, the field does not begin to reverse the instant the current begins to reverse. This slight delay is the cause of a magnetic loss called **hysteresis.**

Finally, the varying magnetic field in the core tends to induce currents in the core as well as in the secondary. This circulation of electrons in the core is called **eddy current,** and it is undesirable because it is a waste of energy and possibly harmful to the transformer. The eddy current is reduced by using a laminated core made out of thin sheets of metal. The laminations are coated with an electrical insulation that reduces core currents without affecting the magnetic flow.

Figure 21–17 shows an impedance-matching transformer, similar in construction to

FIGURE 21-15
COPPER LOSSES FROM WINDING RESISTANCE

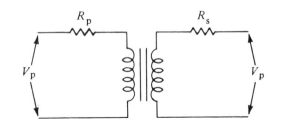

FIGURE 21-16
TRANSFORMER HYSTERESIS CURVE

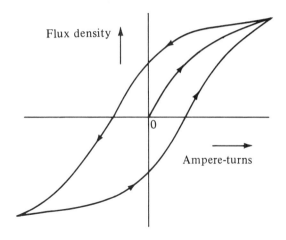

many transformers you will work with. The metal mounting frame covers and holds the laminations, while a heavy paper cover surrounds the coil winding. Solder-type terminals provide for connection with the coil.

Uses of Transformers

Power transformers used in televisions are good examples of a common use of transformers. They change the line voltage to other

FIGURE 21-17
IMPEDANCE-MATCHING TRANSFORMER

FIGURE 21-18
POWER TRANSFORMER

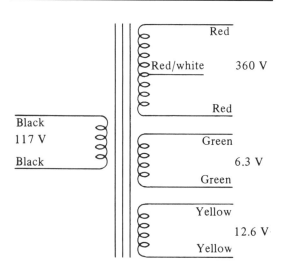

voltage levels required by the circuitry. These transformers are much like the 6-volt transformer described in the previous sections, although they have one major difference. Power transformers have several secondary windings. A television schematic, for instance, might include a transformer like the one indicated by the schematic in Figure 21–18.

While there are some common practices in transformer lead color coding, not all manufacturers use them. There is significance to the colors, though, so you should check the schematic or catalog to find out what they mean for the transformer you are working with. The example in Figure 21–18 uses black as the input or primary. Since both leads have the same color, they are interchangeable. The high-voltage secondary places 360 volts between the red leads. The connection to the center of the winding is called the **center tap** and is usually represented by adding a stripe to the basic winding color. In this case, it is a red lead with a white stripe. Two low-voltage windings are also available. The yellow pair in Figure 21–18 provides 12.6 volts, and the green pair provides 6.3 volts.

Other symbols may also be used. A dotted line around the outside of the transformer means that the case is shielded. Internal connections may be added. The schematic symbol may also indicate that additional components are added for filtering inside the transformer housing.

The autotransformer, whose schematic is shown in Figure 21–19, is simply an adjustable power transformer that can provide an output voltage that is variable from 0 to about 140 volts. It is one more example of transformers used for voltage change.

There is another common use of transformers, **impedance matching.** Audio and radio frequency interstage transformers are used for impedance matching. The terms *interstage* and *impedance matching* mean that the transformer is used between stages or circuits in a device to match one stage to the next one. The stages can be one amplifier to another, an amplifier to a speaker, or an antenna to a television. Figure 21–20 shows the schematic of a typical interstage transformer used between two stages in a radio frequency amplifier. One stage is connected to the primary winding, and the next stage is connected to the secondary winding. Since the signal goes from primary to secondary, these windings are often called the input and output of the transformer. In this particular circuit, an adjustable core in the transformer is used to select those frequencies that will be passed and those that will be blocked.

Although the subject of impedance and frequency rejection will be discussed extensively in Chapters 24 through 27, an introduction is presented here. **Impedance** is the algebraic sum of the oppositions in a circuit, not only those caused by resistors but also those caused by inductors and capacitors.

For one circuit to pass on its information to the next, their impedances should look the same to each other. That is, the output impedance of the first circuit should be the same as the input impedance of the next. If the first looks ahead and sees 12 ohms, then the second should look back and see 12 ohms. When the impedances are the same, maximum power transfer occurs. When they are not matched, there can be significant losses.

The circuit described previously in Figure 21–14 has a turns ratio of 20 : 1, which provides a voltage step-down from 120 to 6 volts. As we saw earlier, squaring the turns ratio of 20 : 1 determines the impedance ratio, which is 400 : 1. That ratio means that the 12-ohm lamp looks like 4800 ohms at the primary side. Impedance matching is the reason a transformer is used to match an 8-ohm speaker with an amplifier that has an output impedance that is much higher.

FIGURE 21-19
AUTOTRANSFORMER

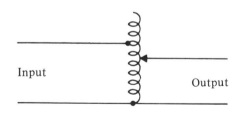

FIGURE 21-20
ADJUSTABLE-CORE, RADIO FREQUENCY INTERSTAGE TRANSFORMER

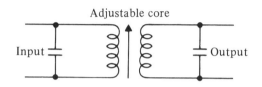

SUMMARY

Frequency is one of the dimensions used to describe an alternating voltage or current. Frequency describes the rate at which the signal alternates or, more specifically, the number of alternations per second. The unit of frequency is the hertz; 1 hertz is equal to 1 cycle per second. Some electronic circuits operate at audio frequencies, which can range from 20 to 20,000 hertz; others operate at radio frequencies higher than 400 megahertz.

Audio frequency oscillators are used by test technicians to evaluate audio frequency circuits. The advantage of this device is that it provides a good-quality test signal of selected frequency and amplitude. By using this test signal, a technician can determine how a circuit responds to a specific frequency — that is, the circuit's frequency response. If an oscilloscope is used as the output-monitoring device and an oscillator is used for a pure sine wave input, distortion can be found. The technician can check the signal as it moves through the circuitry to determine where the sine wave loses its shape.

Radio frequency generators have a similar function, but their test signals are smaller and at a much higher frequency. Another difference is that their signal can be modulated. Function generators provide an assortment of signal shapes or waveshapes, not just sine waves. Square, ramp, triangle, and trapezoidal waves are usually available. Another advantage of the function generator is that the output frequency and voltage are of better quality and more accurately indicated than they are in other frequency generators.

There are several ways to measure the frequency of a signal. The fastest and most accurate way is with a counter. It is fast because the counter is easy to use. It is accurate because of the circuit design and the digital display, which leaves little room for technician error. Frequency can also be measured with an oscilloscope by measuring the time for 1 cycle with a calibrated horizontal sweep circuit. In a third method of frequency measurement, an unknown signal is compared with a known signal from a calibrated instrument such as an oscillator. An oscilloscope is used only as a display device for showing a Lissajous pattern created by the two signals. When both signals have the same frequency, a circle will appear.

Transformers are components in which two or more coils are placed close enough together so that they share a common magnetic field. The conductor in each coil or inductor is electrically insulated from but magnetically connected with the others. The current in one coil, called the primary, produces a changing magnetic field that cuts across and induces a voltage in all the other coils, called secondaries. The level of the secondary voltages depends on the turns ratio of the transformer. If a secondary has three times as many turns as the primary, it will have three times the voltage. This turns ratio of 3 can be squared to produce an impedance ratio of 9. This ratio indicates that when the primary voltage is $\frac{1}{3}$ the secondary voltage, any resistance connected across the secondary appears $\frac{1}{9}$ as large at the primary.

The magnetic properties of the core determine the efficiency with which the primary voltage is transferred to the secondary. Iron is a more efficient core than air, although iron introduces some problems. For example, the rate at which the magnetic field can be reversed is lower in iron than in air, so an energy loss called hysteresis is produced. Electron movements called eddy currents are produced in the iron core because it is an electric conductor. Laminating the core reduces this problem. Finally, the resistance of the copper windings causes copper losses, which tend to make transformers warm.

There are two common applications for transformers in electronic products: voltage change and impedance matching. Power line transformers change line voltage to levels in the tens and hundreds. Color-coded leads can also provide a number of different voltages at the same time from one transformer. The matching of impedances between adjacent circuits is necessary for maximum power transfer. Just as the turns ratio provides desirable voltage change, the impedance ratio provides desirable impedance change. The transformer is the component used to match the impedance of one circuit with the impedance of another circuit.

CHAPTER 21

REVIEW TERMS

audio frequency oscillator: instrument that generates voltages in the audio frequency range

center tap: connection to the center of the winding of a transformer

coefficient of coupling: number indicating the effectiveness of primary-to-secondary transfer in a transformer

eddy current: induced current in the core of a transformer caused by the varying magnetic field

frequency: number indicating how many cycles a voltage goes through in 1 second

frequency response: specification describing the range of frequencies that a system will reproduce or respond to

function generator: generator capable of producing complex waveshapes

generator: oscillator with the added capability of modulation

harmonic: multiples of a given frequency

hertz: unit of measure for frequency

hysteresis: magnetic loss in a transformer caused by residual magnetism in the core

impedance: algebraic sum of the oppositions in a circuit

impedance matching: use of transformers between stages (circuits) in a device to match one stage to the next

impedance ratio: relationship between the impedance of the primary of a transformer and the impedance of the secondary; used to determine resistance

Lissajous pattern: oscilloscope waveform that shows the relationship between two frequencies

modulation: process of adding an audio signal onto a radio frequency signal

oscillator: instrument that produces sinusoidal test signals

primary: input side of a transformer

radio frequency generator: instrument that produces signals in the megahertz range

secondary: output side of a transformer

transformer: two or more coils that are electrically separated and magnetically connected for the purpose of voltage transfer

turns ratio: relationship between the number of turns in the primary of a transformer and the number of turns in the secondary

REVIEW QUESTIONS

1. Give the range of audio frequency.

2. Give the range of radio frequency.

3. What is the relationship between the frequency and the period of a sine wave?

4. What is meant by frequency response as a specification of an audio system?

5. What is the difference between an oscillator and a generator?

6. Describe the output and applications of a function generator.

7. Explain what is meant by line-of-sight wave travel.

8. Describe transformer operation.

9. Define turns ratio.

10. Discuss transformer losses.

11. Explain why the left-hand rule applies to the primary side of a transformer while the right-hand rules applies to the secondary side.

12. What is the distinction between a transformer's voltage ratio and its impedance ratio?

13. Explain the significance of the various colors used on power transformer leads.

14. Name two common applications of transformers.

15. What is meant by Lissajous pattern?

16. Discuss the difference between AM and FM radio signals.

REVIEW PROBLEMS

1. Sketch a square wave with a repetition rate of 1000 hertz and a pulse width of 0.0002 second.

2. Sketch a square wave with a repetition rate of 2500 hertz and a pulse width of 0.00025 second.

3. Calculate the secondary voltage of an 8 : 1 step-down transformer with a 120-volt primary.

4. Calculate the secondary voltage of a 1 : 3.5 step-up transformer with a 120-volt primary.

5. What is the turns ratio of a transformer with a primary voltage of 120 volts and a secondary voltage of 24 volts?

6. What is the turns ratio of a transformer with a primary voltage of 120 volts and a secondary voltage of 360 volts?

7. What impedance will a secondary load of 1000 ohms appear as at the primary of the transformer in Problem 5?

8. What impedance will a secondary load of 18 kilohms appear as at the primary of the transformer in Problem 6?

9. What is the third harmonic of 60 Hertz? What is the fifth harmonic?

10. Determine the frequency indicated in Figure 20–10 if the range is 100 milliseconds per division.

11. Determine the frequency indicated in Figure 20–11 if the range is 40 microseconds per division.

12. Determine the vertical frequency for the pattern shown in Figure 21–21.

13. Determine the vertical frequency for the pattern shown in Figure 21–22.

FIGURE 21-21

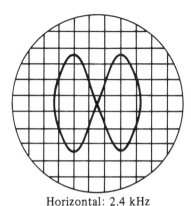

Horizontal: 2.4 kHz

FIGURE 21-22

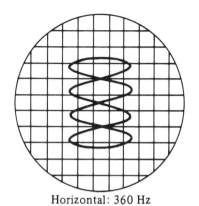

Horizontal: 360 Hz

14. What pattern will a 60-hertz horizontal signal and a 40-hertz vertical signal produce on an oscilloscope?

15. Calculate the primary current and power for the circuit shown in Figure 21–23 if the primary voltage is 24 volts, the load is 30 ohms, and the transformer has a 1 : 2.5 step-up ratio.

16. Calculate the primary current and power for the circuit shown in Figure 21–23 if the primary voltage is 120 volts, the load is 600 ohms, and the transformer has a 4 : 1 step-down ratio.

FIGURE 21-23

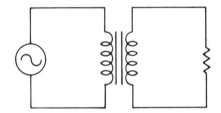

MULTIMETER THEORY AND APPLICATIONS

Describe the operation of a D'Arsonval movement.

Describe the functions and characteristics of an ammeter, a voltmeter, and an ohmmeter.

Determine the values for multipliers and shunts in an analog meter.

Describe the basic operation of a digital multimeter.

Compare the characteristics of analog, digital, and FET multimeters.

ANALOG METERS

The objective of this chapter is to introduce the theory of operation and characteristics of analog and digital multimeters. Analog meters will be described first.

D'Arsonval Movement

The heart of most analog meters is an electromagentic device called a **D'Arsonval meter movement.** It operates on the principle that electric current produces magnetism that produces a force to move a pointer. The amount of pointer movement depends on the amount of current, which makes pointer deflection an analog of current flow. Figure 22–1 shows the basic features of the D'Arsonval movement.

As shown in Figure 22–1, a coil of wire is wound around a lightweight metal core and connected to external terminals on the meter case. As current flows through this coil, an electromagnetic field is temporarily established. According to the left-hand rule, this field has a north polarity at the left end and a south polarity at the right. Since like poles repel, the temporary magnetic field opposes the one already established by the permanent magnet. The result is that the coil rotates in a clockwise direction. This action, in turn, moves the pointer up the scale a distance proportional to the amount of current flowing. Since pointer deflection is proportional to current flow, the meter is called an **analog meter.** Figure 22–2 shows a typical analog multimeter.

If the polarity of the applied signal were reversed, the pointer would attempt to move in the opposite direction. Stops are installed to prevent this motion.

In some meters, two spiral springs are used to keep the pointer at its normal zero location when no voltage is applied, as shown in

FIGURE 22-1
BASIC PARTS OF D'ARSONVAL METER
MOVEMENT

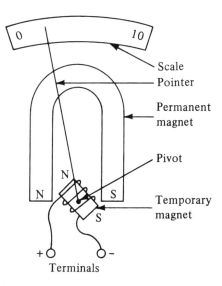

Figure 22–3. They also serve as current carriers between the moving coil and the remainder of the fixed assembly. Precision jeweled bearings ensure that the pointer axle moves easily with slight current changes.

Other meters use a twisted ribbon in place of the pivot, springs, and bearings to hold the coil and pointer. Called a taut-band suspension, this system has less friction and looseness from wear.

Voltmeter

Since the D'Arsonval movement responds to current flow, we can use it to measure voltage. To do so, we need only apply Ohm's law and series circuit rules. Assume that this movement has a coil resistance of 1000 ohms and that 1 milliampere of current will cause it to deflect full scale. Thus, 1 volt connected across it will also cause it to move full scale.

FIGURE 22-2
ANALOG MULTIMETER

FIGURE 22-3
ADDITIONAL DETAIL OF THE METER
MOVEMENT

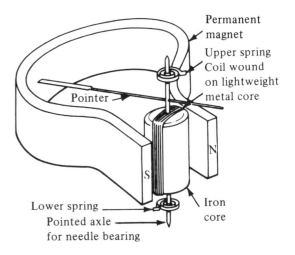

current flow. For a 10-volt range, the resistance of the multiplier must be 9000 ohms, as the following calculations show:

$$R_T = \frac{V_T}{I_T} = \frac{10 \text{ V}}{0.001 \text{ A}} = 10{,}000 \ \Omega$$

$$R_T = R_{meter} + R_{multiplier}$$

$$R_{multiplier} = R_T - R_{meter}$$

$$= 10{,}000 \ \Omega - 1000 \ \Omega = \textbf{9000} \ \boldsymbol{\Omega}$$

The following calculations show that a 99,000-ohm multiplier is required for a 100-volt range and a 249,000-ohm multiplier for a 250-volt range. For the 100-volt range, we have

$$R_T = \frac{V_T}{I_T} = \frac{100 \text{ V}}{0.001 \text{ A}} = 100{,}000 \ \Omega$$

$$R_{multiplier} = R_T - R_{meter}$$

$$= 100{,}000 \ \Omega - 1000 \ \Omega$$

$$= \textbf{99{,}000} \ \boldsymbol{\Omega}$$

It is therefore a 1-milliampere, or 1-volt, meter, as shown in Figure 22–4.

Now, suppose we want to extend the range of this meter to make it a 10-volt, full-scale meter. That is, when 10 volts are connected across the terminals, we want 1 milliampere to flow through the movement and cause it to deflect full scale; see Figure 22–4B. For this situation, a resistor called a **multiplier** is connected in series with the meter movement to extend, or multiply, the voltage range. It simply adds to the movement resistance to produce a resistance large enough to limit the

FIGURE 22-4
CIRCUIT OF A BASIC VOLTMETER

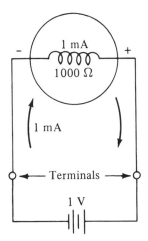

A. 1-milliampere, or 1-volt, range

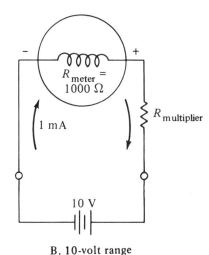

B. 10-volt range

For the 250-volt range, we have

$$R_T = \frac{V_T}{I_T} = \frac{250 \text{ V}}{0.001 \text{ A}} = 250,000 \ \Omega$$

$$R_{multipler} = R_T - R_{meter}$$

$$= 250,000 \ \Omega - 1000 \ \Omega$$

$$= \mathbf{249,000 \ \Omega}$$

Another important point to note is that this meter is a DC instrument. It works by averaging the applied signal. Since the average value of one cycle of AC is zero, this instrument will not measure AC directly.

If a D'Arsonval meter is to measure the value of an AC signal, the signal must first be converted to DC. This conversion is done internally through a process called *rectification,* to be described in Chapter 30.

The **sensitivity** of a voltmeter was briefly mentioned in Chapter 7. It is an indication of the demands (the load) a meter places on the circuit it is testing. A very sensitive meter responds easily with little demand; a less sensitive meter demands more. The sensitivity of a meter is expressed in ohms per volt and depends on the current and resistance rating of the movement.

Meter sensitivity is equal to its total resistance on a given range divided by the full-scale voltage of that range. Consider the meter just described. Its sensitivity is calculated as follows:

$$\text{sensitivity} = \frac{\text{ohms}}{\text{volt}} = \frac{1000 \ \Omega}{1 \text{ V}}$$

$$= \frac{10,000 \ \Omega}{10 \text{ V}} = \frac{250,000 \ \Omega}{250 \text{ V}}$$

$$= \mathbf{1000 \ \Omega/V}$$

As you can see, a meter has the same sensitivity regardless of range. The meter just described is not a very sensitive meter. A more sensitive meter will have a sensitivity of, say, 20,000 ohms per volt.

A 20,000-ohms-per-volt meter is considered to be more sensitive because it demands less current for full-scale deflection:

$$I = \frac{V}{R} = \frac{1 \text{ V}}{20,000 \ \Omega} = 0.00005 \text{ A}$$

$$= \mathbf{0.05 \text{ mA}}$$

FIGURE 22-5
CIRCUIT OF A BASIC AMMETER

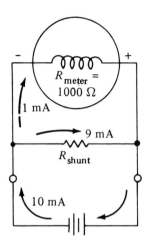

On the 100-volt range, this 20,000-ohms-per-volt meter appears as a resistance of 2 meg-ohms, since

$$100 \text{ V} \times \frac{20{,}000 \text{ }\Omega}{1 \text{ V}} = \textbf{2 M}\mathbf{\Omega}$$

Remember that the 1000-ohms-per-volt meter had a total resistance of 100 kilohms on the 100-volt range. When you measure voltage with a meter, the meter appears as a resistance in parallel with the circuit under test. So this 2-megohm meter will have far less impact than the 100-kilohm meter when placed across a circuit. With its higher resistance, it will draw less current.

Here is the significant point about meter sensitivity.

Key point: The less current a voltmeter demands from a circuit under test, the less disruptive the test is likely to be.

You should always be aware of the characteristics of your test circuit and meter when making any measurements.

Ammeter

The meter we have been considering can also be used directly, as is, as a 1-milliampere current meter. To extend its range, we use a parallel resistance called a **shunt.** That is, we want to measure an amount of current with the meter that is greater than the capacity of the movement. To do so, we connect a smaller resistance shunt in parallel with the movement, as shown in Figure 22–5.

For full-scale deflection, the meter (and shunt, since they are in parallel) will always have a 1-volt drop and 1 milliampere of current, since

$$V_{\text{meter}} = I_{\text{meter}} \times R_{\text{meter}}$$
$$= 0.001 \text{ A} \times 1000 = \textbf{1 V}$$

Here is how the meter can be converted to have a full-scale range of 10 milliamperes. For a full-scale deflection when 10 milliamperes flows, the difference between this 10 milliamperes and the 1-milliampere, full-scale current of the meter must be bypassed through a shunt. That is,

$$I_{\text{T}} = I_{\text{meter}} + I_{\text{shunt}}$$
$$I_{\text{shunt}} = I_{\text{T}} - I_{\text{meter}} = 10 \text{ mA} - 1 \text{ mA}$$
$$= \textbf{9 mA}$$

For a shunt current of 9 milliamperes, the shunt resistance must be

$$R_{\text{shunt}} = \frac{V_{\text{shunt}}}{I_{\text{shunt}}} = \frac{1 \text{ V}}{0.009 \text{ A}} = \textbf{111 }\mathbf{\Omega}$$

The following calculations show that a 100-milliampere meter will use a 10.1-ohm shunt and a 1-ampere meter will require a 1-ohm shunt. For a 100-milliampere range, we have

$$I_{\text{shunt}} = I_{\text{T}} - I_{\text{meter}} = 100 \text{ mA} - 1 \text{ mA}$$
$$= \textbf{99 mA}$$

$$R_{shunt} = \frac{V_{shunt}}{I_{shunt}} = \frac{1\text{ V}}{0.099\text{ A}} = \mathbf{10.1\ \Omega}$$

For a 1-ampere range, we have

$$I_{shunt} = I_T - I_{meter} = 1.0\text{ A} - 0.001\text{ A}$$

$$= \mathbf{0.999\ A}$$

$$R_{shunt} = \frac{V_{shunt}}{I_{shunt}} = \frac{1\text{ V}}{0.999\text{ A}} = \mathbf{1\ \Omega}$$

One problem arises as the range of the meter gets extended even further. Shunt resistor values become much smaller, and thus, the resistance of their connection can become a significant factor in the circuit.

Ohmmeter

The most common method for measuring resistance with an ohmmeter is to connect a battery of known voltage across the unknown resistance to see how much current will flow. For that reason, 0 ohms is often indicated by full-scale deflection, while an open circuit is indicated by no current — that is, no deflection. When we zero an ohmmeter, we are just showing the meter what zero resistance looks like. This process allows us to adjust the circuit as the battery ages.

Figure 22–6 shows the circuit of a basic series ohmmeter. A voltage divider is formed with the meter coil, the range resistor, the resistor being measured (unknown), and the battery. Voltage dropped across the zero-adjust resistor and meter coil causes the deflection of the meter. Since this voltage is a function of resistance ratios involving all of these resistors, and since the unknown resistance can vary over a wide range, the range resistor values must also be changeable. A range switch is therefore used to select the range resistor value. The selected value should produce a meter voltage drop so that deflection is near the center of the scale.

Battery size depends on resistance range

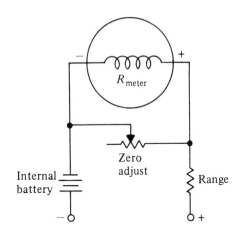

FIGURE 22-6
CIRCUIT OF A BASIC OHMMETER

and brand but tends to range from 1½ to 30 volts. When the ohmmeter terminals are shorted together, the voltage divider should produce just enough voltage to cause full-scale deflection. Because of battery aging, a **zero-adjust potentiometer** is required. This device varies the current caused by the divider and limits it to an amount equal to full-scale deflection.

One characteristic of ohmmeters is very important to remember. As we mentioned earlier, resistance measurement includes current flowing through the unknown resistance. This procedure leads to some problems. For instance, if you are checking a fractional-amperage fuse, you may burn it out with the current used by an insensitive ohmmeter. Or in the test of a transistor, the voltage used by an ohmmeter may be close to the threshold level and cause erratic indications. Once again, you should know your equipment and read the manual.

Notice the terminal polarity markings in Figure 22–6. The minus terminal of the meter is connected to the positive side of the battery.

FIGURE 22-7
DIGITAL MULTIMETER

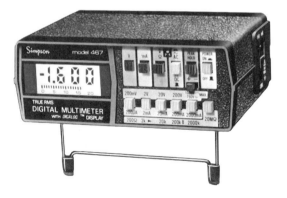

This connection does not cause a problem for the meter since the minus side of the movement is connected to the negative side of the battery. The pointer will move the correct way. However, testing semiconductors with this meter requires special attention, as we will see in Chapter 28.

FET MULTIMETER

An **FET multimeter** is basically a D'Arsonval meter circuit with a field effect transistor (FET) amplifier preceding the measuring circuitry. One significant change, an obvious advantage, results from this addition. The amplifier isolates the loading effect of the movement from the circuit and makes the meter more sensitive. That is, in this meter, the voltage being measured appears as a reference for the amplifiers that drive the pointer upscale. So this meter is less demanding on a circuit than the multimeters described earlier. They, instead, draw power from the circuit under test in order to move the pointer.

Most multimeters, like the ones we considered earlier, have a sensitivity between 10,000 and 20,000 ohms per volt. Thus, they may appear as a load of between 1000 ohms and 1 megohm or more, depending on the meter and range selected. The typical FET multimeter has a constant input impedance that is typically 10 megohms or more. With these meters, circuit loading is not too much of a concern.

DIGITAL METERS

The most obvious difference between digital and analog meters is the display system. Reading error is almost totally eliminated by the clear and direct indication of parameter values, as shown in Figure 22–7. There is another significant difference between these two types of meter systems: their internal workings. While the analog meter is electromagnetic and electromechanical, the **digital meter** is solid state and static. Other than switches, there are no moving parts.

Comparator Circuit

A variety of circuits can be found in digital multimeters (DMMs) used in the industry today. One of the earlier and more common designs, the ramp generator, will be described here.

A basic element of a DMM is the **oscillator-divider,** a circuit that produces a sine wave of some known and stable frequency. For this circuit, we will assume a frequency of 100 kilohertz (kHz), or 10^5 hertz. Through a series of frequency dividers, this frequency is reduced to 1 kilohertz. It is also reshaped into a square wave in order to provide sharply defined pulses.

Another section of DMM circuitry is a

voltage or current source. This stable and accurate signal is applied to an *RC* circuit called a **ramp generator.** This circuit, in turn, causes the capacitor to begin to charge. As described in Chapter 13, the voltage waveshape of a charging capacitor is nonlinear while it levels off gradually, as illustrated previously in Figure 13–11. However, the first 10% of this curve is quite linear; that is, it is a straight line, as shown in Figure 22–8. It represents a voltage that rises from zero at a constant rate of time. Therefore, if we know the amount of elapsed time, we can determine the amount or level of voltage from the curve.

Other parts of a DMM are the **comparator circuits.** They have two comparisons to make. First, they detect when the capacitor voltage equals zero on the way up. In order for this comparison to be made, the circuit is designed so that the charging curve begins a bit on the negative side of zero, as shown in Figure 22–9. The second point of comparison occurs when the capacitor voltage becomes equal to the unknown voltage.

Another part of a DMM system is the **counter.** This section counts pulses when told to do so by the **gate.** The gate gets its commands from the two comparators. Counting begins when the first comparison occurs, and it stops when the second comparison occurs. The number of pulses counted is then indicated on the digital display. This process is repeated a few times per second.

Here is an example of the sequence of events in voltage measurement. The counter is counting the pulses, 1000 per second. Let's assume that the *RC* circuit has been adjusted so that it charges to exactly 1 volt in 1 second. Thus, if we count the pulses from the time the capacitor voltage goes from 0 to 1 volt, we will have a display of exactly 1000. In fact, we *calibrate* our DMM so that this value is what the count will be. Remember, the DMM has precise voltage and frequency standards inside.

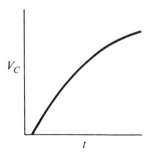

FIGURE 22-8
INITIAL STAGE OF CAPACITOR CHARGING

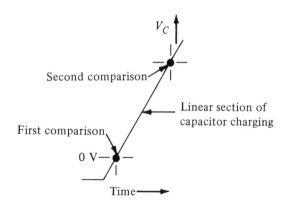

FIGURE 22-9
CAPACITOR-CHARGING VOLTAGE

Overall, its block diagram resembles the one shown in Figure 22–10.

The measurement cycle starts when the capacitor begins to charge, as shown in Figure 22–11. The counter begins to count when the capacitor voltage crosses zero. When the ramp voltage equals the unknown voltage, the comparator detects equality and stops the counter wherever it is. The count is displayed, and the process is repeated. Since a count of 1000 equals 1 volt, a count of 575 represents

┌─ **FIGURE 22-10**
│   **BLOCK DIAGRAM OF A DIGITAL**
│   **MULTIMETER**
└─────────────────────────

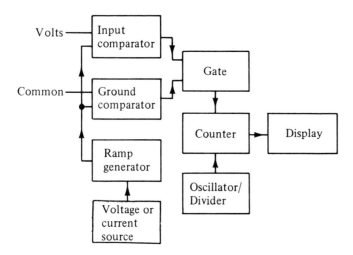

0.575 volt. If the unknown changes, the displayed count also changes.

Ranges

The preceding description assumes that the unknown voltage is probably more than 0.1 volt and certainly less than 1.0 volt. Obviously, that is not always the case. When the unknown voltage is greater than 1 volt, higher ranges are used. The easiest way to obtain higher ranges is to use multipliers (attenuators), as described earlier in the chapter. When the unknown voltage is quite small, voltage amplifiers are used to increase the signal to a measurable size. In either case, amplifiers and attenuators are usually available in steps of ten.

When the unknown voltage is AC, it is changed to DC through some form of AC-to-DC conversion circuit. Then, it is measured by one of a variety of methods. For example, it may be rectified and averaged, or its peak value may be measured. Whichever process is

used, two limitations generally result. Some of the accuracy available for DC is lost, and the frequency response is somewhat lower than that of a D'Arsonval meter.

Current is measured by passing the unknown current through one or more precision resistors and measuring the voltage drop that results. For example, a 1-ohm shunt resistor will have a 1-volt drop when 1 ampere flows through it. A current of 0.53 ampere will produce a voltage of 0.53 volt, and so on. Resistance is measured by comparing the voltage drop across the unknown resistance with that across a standard resistance. The two resistances are connected in series with the built-in voltage source, causing the same current to pass through them. Since the standard resistance has a known value, the unknown resistance can be determined by comparing their voltage drops.

There are a few more range characteristics of DMMs. Many of these instruments have an overload protection that prevents damage from an input voltage in excess of the selected range. Most have a voltage polarity indication that shows the operator whether the unknown voltage is positive or negative with respect to common. Some have automatic range selection, which places the meter on the best range for each measurement.

Displays

Most digital meters use liquid crystal displays (LCDs) as indicators. The most common are called 3- or 3½-digit displays. A 3-digit display indicates any magnitude from 000 up to 999. For more decimal places, a ½ digit is often added at the left, and its range is 0 or 1. This digit extends the highest indication to 1999. Decimal points are located automatically by the circuitry.

As mentioned earlier, the display also indicates polarity. In addition, displays provide other information to the user. For example, the display may indicate when the battery should be replaced or when an overload has occurred. The parameter being measured and its units are also often indicated. Finally, the display is used in the meter calibration process.

METER SPECIFICATIONS

Meter specifications were discussed briefly in Chapter 7, when we first introduced multimeters. Accuracy and sensitivity were also discussed at that time. A voltage accuracy of 1% to 3% for DC and 2% to 4% for AC is typical for analog meters today. Digital meters typically offer an accuracy of 0.2% for DC voltage and 0.5% for AC. The digital meter often adds an error of ±1 digit as the instrument rounds off.

Current and resistance specifications are similar to those for voltage. An analog meter with 3% voltage accuracy probably has a cur-

FIGURE 22-11
POINTS OF COMPARISON IN DIGITAL
MULTIMETER OPERATION

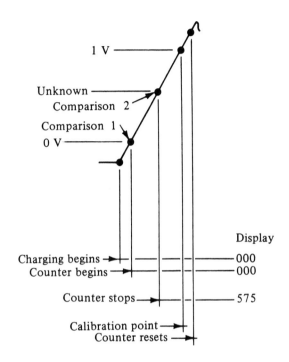

rent and resistance accuracy equal to an equivalent amount of arc on the scale. Digital meters with 0.2% voltage accuracy usually have a resistance and current accuracy that is not much different from the voltage accuracy.

A noticeable difference occurs when the frequency of the AC voltage being measured begins to rise. The specification of concern here is frequency response. **Frequency response** describes the changes in operating characteristics that result as the applied frequency changes. Analog multimeter specifications state that the instrument accurately measures voltages up to a particular frequency, usually about 20 kilohertz. FET multimeters also operate within this range, although 150 kilohertz is not an uncommon upper limit. The frequency response of DMMs is often not that broad.

Before using a meter for a DC voltage measurement, check the frequency response of your test instrument. While some digital meters maintain their accuracy up to 10 kilohertz, others begin to become less accurate above 400 hertz. For this reason, you should know the frequency response of your instrument as well as its sensitivity.

Other specifications of interest are the number of ranges, overload protection, cost, and durability. For instance, some meters can withstand a 4-foot drop; others will be totally destroyed by a 1-foot drop. Some meters have fuses, circuit breakers, or diode limiting for overload protection; others are ruined by a 50% overload. Additional features such as a diode check and audible tones are quite common. The tones are proportional and allow a technician to detect voltage change without continually viewing the display.

Before selecting a meter, read and compare specifications. Notice not only what is mentioned but also what is not mentioned. If a specification is left out by one company, you can assume that that feature is a weak one. Then, decide which characteristics are most important to you and which are least important, and select the multimeter suited to your task.

SUMMARY

Analog multimeters are designed around an electromagnetic and electromechanical system called a D'Arsonval meter movement. Basically, it consists of a coil of wire suspended in the field of a permanent magnet. As the current to be measured flows through the coil, a temporary magnetic field is also established. Its polarity is in opposition to that of the already existing field, which causes a force

to develop. The magnitude of this force is proportional to the current, and the force causes the coil to turn clockwise. As the coil turns, it moves a pointer across a calibrated scale.

The amount of voltage required to move the pointer can be calculated if the coil resistance and full-scale current are known. Thus, the meter movement can be used as a voltmeter. If higher voltage ranges are desired, resistors called multipliers are added in series with the coil. Ohm's law and series circuit rules are all that are required to determine multiplier values.

To use the meter movement for current measurement, you need know only its full-scale current rating. If a higher current range is desired, resistors called shunts are added in parallel. Ohm's law and parallel circuit theory apply here.

An ohmmeter uses the same type of movement as that of the voltmeter but adds a battery to the circuit. Resistance is measured by placing a known voltage across an unknown resistance. The resulting current moves the pointer. Range resistors are placed in series or in parallel with the unknown. A zero-adjust potentiometer is added to allow for adjustment as the battery ages.

Digital multimeters use a variety of electronic circuits. One of the original and more common designs is the ramp circuit. In this system, an oscillator develops a 100,000-hertz sine wave that is reduced to a 1000-hertz square wave through additional circuitry. An electronic counter is used to count the pulses and present the count on an LCD display. This count represents the magnitude of the parameter being measured.

Another circuit in the DMM charges an integrated *RC* circuit. The charge time of this circuit is adjusted so that it rises from 0 to 1 volt in the same time that it takes the counter to count from 0 to 1000. With the aid of a comparator circuit, the counter begins counting when the capacitor voltage is zero and stops counting when the capacitor voltage equals the unknown. Since the counting and charging are linear functions, the count represents the unknown voltage.

Higher voltages can be measured with the addition of attenuators; AC voltage is measured by conversion to DC. Current is measured by determining the voltage drop across a standard resistance. Resistance is determined by measuring the voltage drop across the unknown and comparing it with the drop across a standard resistance in series. Most DMMs have overload protection on all ranges, and many have automatic range selection.

LCDs are used to indicate values in DMMs. The lower-power LCD allows a battery life of more than 1000 hours. The 3-digit display can indicate values from 000 to 999, while the 3½-digit display ranges up to 1999. Polarity, overload, parameter measured, and weak batteries are also indicated by the displays. The most significant characteristics of digital displays are their ease of reading and their durability.

Field effect transistor (FET) analog mul-

timeters include additional electronics between the range selection circuitry and the meter movement. These added components provide the energy to move the pointer, in contrast to other multimeters where the energy is taken from the circuit under test. Because of this feature, FETs increase the meter's sensitivity to 10 megohms per volt or more, whereas a typical analog multimeter has a sensitivity ranging from 10,000 to 20,000 ohms per volt. Consequently, circuit loading is not a concern with FET multimeters.

Each type of meter described in this chapter has good and bad qualities. For instance, some meters are more portable than others, and some are more durable. Some do not measure current. Accuracies can range from 0.1% or better up to 4%. Bandwidth can extend only to 400 hertz or into megahertz values. Sensitivity can be as low as 1000 ohms per volt or greater than 10 megohms. Therefore, before selecting an instrument for a measurement, determine the characteristics that are important for the situation. Then, select a multimeter that can do the job.

CHAPTER 22

REVIEW TERMS

analog meter: meter in which pointer deflection is proportional to current flow

comparator circuits: circuits in a DMM that make comparisons of voltages

counter: electronic circuit in a DMM that counts the number of pulses or cycles in a given period of time

D'Arsonval movement:
electromagnetic system
used in a meter that
deflects a pointer as a
function of current

digital meter (DMM): meter
having a digital display
system and solid-state
operating characteristics

FET multimeter: multimeter
with a high input
resistance that results
from the use of field
effect transistors (FETs)

frequency response:
specification of a meter
that describes the
changes in operating
characteristics that result
as the applied frequency
changes

gate: device in a DMM that
controls the counter and
is controlled by the
comparator circuits

multiplier: resistor placed in
series with a voltmeter to
extend the range of the
meter

oscillator-divider: circuit for a
DMM that produces a
sine wave of some
known and stable
frequency

ramp generator: circuit in a
DMM that causes the
capacitor to charge

sensitivity: indication of the
demand a meter places
on the circuit under test

shunt: resistor placed in
parallel with an ammeter
to extend the range of
the meter

zero-adjust potentiometer:
ohmmeter control used
to counteract the effects
of battery aging on the
meter

REVIEW QUESTIONS

1. Describe a multiplier.

2. Describe a shunt.

3. Describe an analog multimeter.

4. Describe a digital multimeter.

5. Describe voltmeter sensitivity.

6. Describe the construction of a D'Arsonval movement. Make a sketch of the movement.

7. Describe the operation of a D'Arsonval movement.

8. Determine the resistance of a 1000-ohm-per-volt meter on the 10-volt range. On the 25-volt range. On the 150-volt range.

9. Determine the resistance of a 20,000-ohm-per-volt meter on the 10-volt range. On the 25-volt range. On the 150-volt range.

10. Explain the difference between a 1000-ohm and a 20,000-ohm-per-volt meter in the way they affect a circuit.

11. Which meter in Question 10 causes the most circuit loading?

12. Describe the operation of a basic digital multimeter. Sketch a block diagram.

13. Compare the advantages and disadvantages of analog and digital multimeters.

14. Describe the difference between electric and electronic multimeters.

REVIEW PROBLEMS

1. Draw the circuit of a 2000-ohm, 1-milliampere meter modified to measure 10 volts. Determine the multiplier value.

2. Draw the circuit of a 500-ohm, 200-microampere meter modified to measure 10 volts. Determine the multiplier value.

3. Draw the circuit of a 150-ohm, 100-microampere meter modified to have 1-, 5-, 10-, and 50-volt ranges. Determine the multiplier values.

4. Draw the circuit of a 2000-ohm, 1-milliampere meter modified to measure 10 milliamperes. Determine the shunt value.

5. Draw the circuit of a 500-ohm, 200-microampere meter modified to measure 10 milliamperes. Determine the shunt value.

6. Draw the circuit of a 150-ohm, 100-microampere meter modified to have 1-, 10-, 20-, and 100-milliampere ranges. Determine the shunt values.

7. Calculate the sensitivity of the voltmeter described in Problem 1.

8. Calculate the sensitivity of the voltmeter described in Problem 2.

9. Calculate the sensitivity of the voltmeter described in Problem 3.

10. Consider the meter described in Problem 1. Suppose it is connected across one of two 10-kilohm resistors that are connected in series with 20 volts. What will the meter indicate?

11. What is the true error for the meter of Problem 10?

12. Repeat the calculations of Problem 10 for the meter described in Problem 2.

13. What is the true error for the meter of Problem 12?

14. Repeat the calculations of Problems 10 and 11 for a 10-megohm multimeter.

LAB REPORTS

State three purposes of lab reports.

Name the sections of a lab report, and describe the contents of each.

Describe the characteristics of a good lab report.

PURPOSES OF LAB REPORTS

The objective of this chapter is to introduce the methods used in preparing reports of lab experiments. A **lab report** is a written statement about an experiment or study that has been conducted. It is usually prepared by the person who did the experiment, and it describes what was done, why it was done, and the results. Words, graphs, schematics, and calculations are used to ensure that the reader will have no questions unanswered. The ability to prepare a good report is as important to a technician as the ability to measure voltage.

Primary Purpose

As a technician, you will often be assigned the task of conducting an experiment or testing some new circuit. Decisions will be based on your results as you describe them to your supervisor. These decisions often involve considerable time and money for your employer, so your results are very important. Furthermore, the accuracy of your results and the clear expression of them are also important. Once again, the quality of your work is a significant factor in an electronics engineering team. The procedures, results, and conclusions that you report will become the basis for other peoples' decisions. The primary purpose of the lab report, therefore, is to provide others with the information they need to make decisions.

Organizing the Experiment

Another reason for preparing a lab report is to get organized before you begin the experiment. Keeping notes about what you are going to do, how you are going to proceed, and what equipment you will use will make you think carefully about each step of the experiment. As you record each step, you will be forced to think about the selection of equipment and materials. For example, you can consider whether the meter selected is the best one for the job, whether a resistor's tolerance is low enough, and whether a transistor is the correct one.

Furthermore, as you plan your experimental procedure, you will be formulating the objective of the experiment. You will decide what you are trying to determine and how you are going to accomplish it. These decisions will lead you to the procedures that you will follow. You will realize that you will need a schematic of the circuit you are going to assemble and tables for the data you will be gathering. But to arrange these tables, you will see that you must first decide on the type of data you want. This decision, in turn, leads you to consider how the test equipment will be used.

Thus, by thinking about the report before you begin, you organize your experiment. Results will come by design rather than by chance. You will know just what data to get and how to get it. Then, when it is time to write your report, you will have all the data you need. You will not have to reconstruct the experiment in order to get additional data.

Explaining Procedures

A good lab report explains just what you did in such a complete manner that another person reading it will have no questions. If your report does not answer all the reader's questions, it is inadequate. For example, the reader must know what circuit you used and the quality of the equipment. Your data must be available so that your reader can reach the same conclusions you reached. The procedure used in collecting the data must be presented so that it can be verified and repeated, if necessary. The objectives of the experiment must be stated so that the reader can see that your conclusions satisfy the objectives. All this documentation gives the reader confidence in the quality of the experiment and, more importantly, gives the reader all the information he or she needs.

Making a Historical Record

Lab experiments are often conducted to satisfy some immediate need. For instance, your company may have a customer who needs a special model of one of its products. An engineer designs it, and you build and test it. Your lab report indicates that it works, and so it is shipped to the customer. Now, your work on that special project is completed and you begin another. But someday, a few years later, the first special product may begin to malfunction. Since you were the technician who worked on this item, you are asked about it. Did this problem occur before? How did you solve it? It has been so long since you worked on it that you do not remember. Thus, you have to look at your old lab reports and refresh your memory about the product and how it acted. You may find that you had faced and resolved the same problem before. Even if that is not the case, your old lab reports should give you enough information about the device that you can troubleshoot it quickly.

Sometimes, another customer will want the same custom design months or years after the first one was produced. A glance at your old lab reports will quickly tell you how to test it and how it should work. In fact, your notes may indicate some changes that will make the new device more reliable. There may even be data that you decide you would like to have before this product is shipped in order to answer questions that might come up later.

If you go to work for a manufacturer, you will find that your employer patents product designs. This process is time-consuming, com-

plex, and expensive, but it is valuable. The procedures followed by technicians and engineers usually are based on the assumption that a patent may be applied for. Lab reports and notebooks serve a very important purpose in this process: They provide evidence that your company developed concepts and conducted experiments in some certain area.

So lab reports serve many purposes as a historical record. They may be for your own records or for those of your immediate supervisor. They may be used for troubleshooting a device in the future. They may be a requirement of your company and remain as their property. Or they may be used as evidence in patent applications.

SECTIONS OF A LAB REPORT

The style and length of a lab report may be specified by the person requesting it. If not, they may depend on the nature of the experiment and what is required to describe it. In any case, there are several essential parts to all lab reports. The outline provided in this section should satisfy the needs of most situations you will encounter.

Intentions and Equipment

The first part of a lab report is the **objective** — that is, a statement about what you want to determine. It is usually one sentence long, and generally it does not include the procedure. Here are some examples of objectives:

Determine the current and voltage in all parts of a series-parallel circuit.

Calibrate the 20- to 200-hertz band of an audio generator by using line frequency and Lissajous patterns.

Compare the operation of a low-voltage power supply with its specifications.

The next section of a report is called **equipment and materials,** and it identifies all equipment and components used. Test equipment is described by type, brand, model, and serial number. This section confirms the quality of your work. In addition, it allows a specific instrument to be checked if the results of the experiment are not as expected. Brand names are generally not used when listing components unless they are a necessary part of the identification. A sample of this section may look like the following:

Hewlett-Packard model 1220A oscilloscope
Lambda model LPD DC power supply
M3400 IC
Assorted test leads
SPST toggle switches

Procedures

The first two sections of a report describe what you want to determine and what you use in an experiment. The third section is called **procedure,** and it describes just what you did. Schematics and block diagrams can be used with the statements to provide a procedure section that is clear and simple but complete. Here are some samples:

The power supply circuit in Figure 1 was connected, using the specified components, and 115 volts AC were applied to the input. The load resistance R_L was varied in 1500-ohm steps from 6000 to 12,000 ohms, and the output voltage was measured and is recorded in Table 1.

The color-coded value of each of the 10 resistors was determined and recorded in Table 1. Tolerance and range values were then calculated and also entered in Table 1. Each resistor was then measured with a bridge, and its measured value was entered in Table 1. This table provided an easy way to see whether each resistor was within tolerance.

Schematics and block diagrams come next in your report and provide the best explanation of the circuit tested. The type of drawing you use depends on the circuit or item tested. If specific components must be discussed, a schematic is necessary. For example, a schematic should be used if you are measuring all parameters in a series-parallel circuit. A block diagram can be used when you do not need to give details about components. For instance, a block diagram is sufficient if you are testing some characteristics of a stereo system.

Tables are used as a clear and concise way of presenting the data you gather. The results of all measurements and calculations should be entered in a table. You should base your table design on the objective and the data. For example, in the evaluation of resistors, a table having the following column heads will be useful: resistor number, colors, coded value, minimum limit, maximum limit, and measured value. Such a table will provide information for comparing measured and calculated resistance values. The objective of the experiment in this case may have been to determine whether the resistors were within tolerance.

Although every calculation you make does not have to be in your report, samples of each are necessary. Sample calculations, the next section of a report, provide the reader with an example of each formula used and how you used it. The sample calculations should show the formula, the values for each parameter, the result, and the appropriate units. The final answer should be written on a separate line, as shown in the sample calculation that follows:

$$t = \frac{L}{R} = \frac{50 \text{ mH}}{2 \text{ k}\Omega} = \frac{0.050 \text{ H}}{2000 \text{ }\Omega}$$
$$= 0.000025 \text{ s} = 25 \text{ }\mu s$$

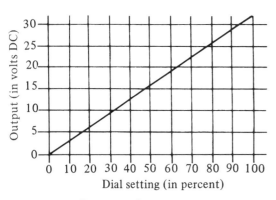

Power supply output

Results and Conclusions

Two sections are left in the preparation of a report. The first is a presentation of actual data and is called **results.** It includes only facts that relate to the objective of your experiment. The second section, called **conclusions,** gives your opinion about the meaning or significance of the results. How you prepare these sections will depend on the type of experiment and data.

One common method for presenting final data is a table, as mentioned earlier in this chapter. Graphs are also commonly used since they provide a very visual way to present information. Although graph design varies with the type of data presented, there are some common practices you can follow. Figure 23–1 gives an example of a graph, which you can refer to as you consider the following points:

1. The independent variable is indicated on the horizontal *(X)* axis. This variable is the one that you control or adjust. In the figure, the independent variable is a dial setting.

2. The dependent variable is the one that changes because of the independent variable. It is indicated on the vertical *(Y)* axis. In the figure, the dependent variable is the output voltage of the power supply.

3. Each axis must be described by name, value, and units. The horizontal axis in the figure represents the dial setting, and it ranges from 0% to 100%, as shown. The vertical axis represents the output voltage that results, and it ranges from 0 to 32 volts DC.

4. At least ten measurements are necessary to ensure that the data truly represent the situation. Points are plotted on the graph exactly as measured, and a smooth line is drawn as a best possible fit. Do not connect the dots unless they fit a smooth line. Place a small circle around each dot to show the original data. See Figure 23–2.

5. Title the graph, add your name, and date it, so that it can stand alone without further explanation. For example, the graph in Figure 23–1 will show the reader how to set the output dial to get any desired voltage between 0 and 32 volts.

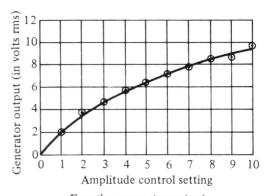

Function generator output

The conclusions section of your report, as mentioned earlier, is a paragraph stating your opinion of the meaning of the results. That is, you had an objective in mind when you conducted the experiment and gathered the data. Now, you must say something about it. Be more specific than "Everything was just fine" or "It did not work properly." Here is a sample of an acceptable conclusion:

Ten resistors were tested with a model 64 multimeter to determine whether they were within tolerance. Their measured values were compared with their color-coded values. All but one were found to be within the indicated range.

PREPARING A LAB REPORT

Now that you know about all the sections of a lab report, you should be ready to write one. There are, however, a few more things to consider before you begin.

Considering the Reader

Who will read your report? You need to know your audience in order to be sure that you include enough, but not too many, facts. Do the intended readers understand the circuit you are working on, or must an explanation of it be included? Will the items, expressions, or procedures be understood by the readers, or should you explain them? As the writer of the report, you must consider who will read it and make sure that those readers will understand it.

Being Complete

Ensuring understanding is not your only responsibility. You must also be complete without being excessive. Stated simply, make sure you include everything the reader must know and perhaps a little that is convenient to know; but include nothing that the reader doesn't need to know. Your report should explain why, how, and what happened for an experiment you conducted. Therefore, as you write your report, ask yourself this question: What does the reader need to know in order to understand the objective, the methods, and the results of this experiment?

Achieving Quality

Quality and clarity ensure a good report. One way to achieve quality is to be *brief*. Schematics, block diagrams, tables, and graphs eliminate words and often express information more clearly than words can. Extra words cause the reader to spend more time getting to the point. Thus, be brief but complete.

Accuracy is another important characteristic of a good report. Be as accurate in presenting your information as you were in obtaining it. *Neatness* is also a quality factor since it gives the reader an indication of your work attitude. A sloppy or messy report will cause a reader to suspect that your work habits — and, therefore, your results — are careless and questionable. In addition, a carelessly prepared report may be difficult for a reader to follow and understand.

The physical aspects of the report should not be overlooked. Prepare your reports in ink. Begin with a title page that gives the experiment name, your name, and the date the experiment was conducted. Follow this page with the remainder of your report, beginning with the objective and ending with your conclusions. Use a template for schematics, and check your spelling. Because of the possibility of patent application, sign and date each page of your report.

SUMMARY

Lab reports communicate the objective, the methods, and the results of an experiment to other persons. The report will serve as a historical record for use at a later date when you or someone else needs to know just what you did.

An advantage of lab reports is that they make you plan your steps carefully, before you begin. Writing down the objective and the materials to be used may lead you to the procedures to be followed. Recording data as you take it may reduce errors caused by haste. A lab report contains the following parts:

— Title page, which contains the title, your name, and date;

— Objective, which states what you wanted to determine;

— Equipment and materials list, which itemizes what you used;

— Procedure, which explains briefly what you did;

— Schematics and block diagrams (if appropriate), which illustrate the circuit or item tested;

— Tables, which present measured or calculated data;

— Sample calculations, which show formulas and results;

— Graphs or tables, which illustrate important results;

— Conclusions, which state your opinion of what the results say about the objective.

As you prepare your report, keep your reader and what the reader needs to know in mind. Your report should answer any questions the reader might have. Use schematics and block diagrams to explain connections. Use tables and graphs to present and compare data. Make sure that the objective is completely and clearly achieved. Finally, be brief but complete, clear, accurate, and neat.

CHAPTER 23

REVIEW TERMS

conclusions: final section of a lab report; presents a subjective summary, comparing results with objectives

equipment and materials: second section of a lab report; identifies all equipment and components used

lab report: written statement about an experiment or study that has been conducted

objective: first section of a lab report; describes what is to be determined in an experiment or study

procedure: third section of a lab report; briefly describes what was done

results: next-to-last section of a lab report; gives specific data gathered from an experiment

REVIEW QUESTIONS

1. List three reasons for writing a lab report.

2. Name the parts of a lab report.

3. Explain how to achieve quality in a lab report.

4. Explain how preparing a lab report helps you organize an experiment.

5. Describe the procedure section of a lab report.

6. Why are equipment brand, model, and serial numbers important in a lab report?

7. Describe the purpose of stating the objective of an experiment.

8. Explain why tables are valuable in a lab report.

9. Why are sample calculations included in a lab report?

10. Which sections in a lab report relate facts and which sections relate opinion?

11. Describe the characteristics of a good graph.

12. Why should the final result of your calculations be written on a separate line?

13. Describe the characteristics of well-written conclusions.

14. Summarize the requirements for a good lab report.

24

INDUCTIVE REACTANCE

Define inductive reactance.

State the relationship among inductance, frequency, and inductive reactance.

Calculate the inductive reactance and impedance of series and parallel *RL* circuits.

Perform a lab experiment to determine the impedance of an *RL* circuit.

INTRODUCTION

The objective of this chapter is to introduce the theory and measurement of inductive reactance. Both vector analysis and mathematical analysis are used in evaluating series and parallel AC *RL* circuits.

Review of Inductance

Inductance is defined as the property of a circuit that opposes change in the current of the circuit. The most significant word in the definition is *change,* because an inductor has its greatest impact when the current changes. In fact, the more frequently the current attempts to change, the stronger the inductor's opposition becomes. We have seen that, if there is an inductor in a DC circuit, the current in the circuit does not rise or fall the instant the switch is opened or closed. A counter emf is produced through self-inductance, and it momentarily opposes the current change. In an AC circuit, there is continual current change, and therefore, there is continual opposition to it. This opposition is called inductive reactance.

A brief review of inductor theory is a good place to begin this chapter. Recall that a magnetic field surrounds all current-carrying conductors. The direction of the field is determined by the current direction, and its strength is determined by the amount of current. The field expands outward as the current rises, and it collapses inward as the current falls. When a long wire is wound to form a coil, the magnetic lines in one part of the conductor interact with those in other sections of the same conductor. Then, a current increase causes the expanding lines to cross other sections of the same conductor. This action, in turn, produces a counter emf that opposes the current increase. When a current decrease begins, the counter emf trys to keep the current going. This entire effect is caused by self-inductance.

Two factors affect the inductance of an inductor: the characteristics of the coil and the material of the core. Inductance increases when we increase the number of turns, the number of layers of wire, the coil diameter, and the ratio of coil diameter to coil length. It also increases when the space between the windings is decreased. Increasing the core permeability will also increase inductance.

Another way to look at inductor action is through the *RL* time constant, *t*. This time constant is defined as the time needed for the current in an inductive circuit to rise to 63% of its maximum value or to fall to 37% of its maximum. It is equal to the circuit inductance, in henrys (H), divided by the resistance, in ohms, or $t = L/R$. The units of the time constant are seconds. An *RL* circuit reaches 99% of its final value after 5 time constants, and this phase is called the steady state. The concept of time constant will come up again when you study advanced circuits.

Inductances, like resistances, add when connected in series. But they have a feature that resistances do not have. This characteristic is called **mutual inductance** *(M)*, and it results when the field of one inductor influences the field of an adjacent inductor. The amount of influence depends on the strength of each field and how close the inductors are to each other. When these adjacent fields aid each other, the mutual inductance increases the total inductance. When the fields oppose each other, the total inductance decreases. The formula for determining total series inductance is

$$L_\mathrm{T} = L_1 + L_2 \pm 2M$$

where L_T is the total inductance, L_1 and L_2 are the adjacent inductances, and *M* is the mutual inductance.

FIGURE 24-1
INDUCTIVE REACTANCE VERSUS
FREQUENCY

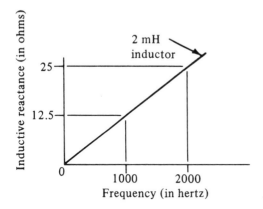

FIGURE 24-1
INDUCTIVE REACTANCE VERSUS
FREQUENCY

FIGURE 24-2
AC RESISTIVE CIRCUIT

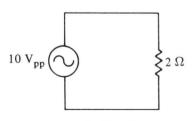

A. Circuit

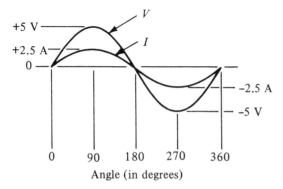

B. Phase in circuit

Inductive Reactance

Inductive reactance is defined as the opposition from an inductor to the flow of alternating current. Its symbol is X_L, and it is measured in ohms. Since AC is always changing, and since inductors oppose change, an increase in frequency will cause an increase in opposition. The opposition, or reactance, also increases with an increase in inductance.

The formula for reactance is

$$X_L = 2\pi f L$$

where f is the frequency (in hertz) and L is the inductance (in henrys). Thus, the amount of reactance caused by a 4-henry inductor at a frequency of 60 hertz is calculated in the following way:

$$X_L = 2\pi f L = 2 \times 3.14 \times 60 \text{ Hz} \times 4 \text{ H}$$
$$= 1507 \ \Omega$$

As another example, the inductive reactance of a 2-millihenry inductor at 1 kilohertz is

$$X_L = 2\pi f L$$
$$= 2 \times 3.14 \times 1 \times 10^3 \text{ Hz} \times 2$$
$$\times 10^{-3} \text{ H}$$
$$= 12.56 \ \Omega$$

Doubling the frequency to 2 kilohertz also doubles the inductive reactance:

$$X_L = 2\pi f L$$
$$= 2 \times 3.14 \times 2 \times 10^3 \text{ Hz} \times 2$$
$$\times 10^{-3} \text{ H}$$
$$= 25.12 \ \Omega$$

Thus, as Figure 24–1 shows, inductive reactance is directly proportional to frequency.

In a pure DC circuit, where the frequency is zero, the inductive reactance is also zero. In such a circuit, the only opposition an inductor would have would be that from the resistance of the inductor itself, and this opposition is usually quite small. In summary, inductive reactance increases with an increase of inductance or an increase in frequency.

Most circuits do not have just DC or just AC but have, instead, a combination of both. Inductors are useful in these circuits because they respond differently to the two types of signals. Basically, they offer little opposition to the flow of DC but offer a lot of opposition to the flow of AC. For this reason, they are used as **filters,** or **chokes,** to reduce the flow of AC in DC circuits. Once again, remember that inductors do have some opposition to DC because they do have resistance. However, it is reasonable to think that an inductor is used because of its reactance. That is, it is used to oppose the passage of some higher frequency while not opposing a lower frequency. In the other common use for inductors, they are connected in series or parallel with capacitors. Fascinating things occur when these components are connected together, as we will see in the next few chapters.

Phase in an Inductive Circuit

In Chapter 8, we saw that the current in a circuit depends on the voltage and the opposition (resistance) in the circuit. In Chapter 20, we saw the way that an alternating voltage is produced and its effect on a circuit. There it was explained that we generally use rms values; but in this chapter, we must once again look at instantaneous values. The combination of Ohm's law, instantaneous voltages, and resistive circuits is a good place to start in our study of inductive circuits.

Consider the circuit in Figure 24–2A. The current depends on the voltage, and the voltage is continually changing. When the voltage reaches a peak value of 5 volts, the current reaches a peak value of 2.5 amperes, as shown

in Figure 24–2B. Ohm's law tells us that. When the voltage is at its negative peak, or -5 volts, the current is at -2.5 amperes because it is flowing in the opposite direction. At $0°$ and $180°$, no current flows because the voltage is also at 0. Figure 24–2B shows all the instantaneous values of voltage and current and how they rise and fall together. They are said to be *in phase* with each other in a purely resistive circuit.

Phase is defined as the relationship between two or more signals, and it is described in degrees. The two signals (voltage and current) in Figure 24–2B are in phase, or have a $0°$ phase difference, because they cross the $0°$ point in the same direction at the same time.

The situation is different with inductors, because they store energy in their magnetic field. Remember: When the applied voltage source begins to get smaller, the collapsing magnetic field keeps the current going a little longer. Thus, we have the following result.

Rule: In an inductor, the current always follows, or lags, the voltage by 90°.

Consider the circuit in Figure 24–3A. The reactance for the circuit is

$$X_L = 2\pi f L = 6.28 \times 32 \text{ Hz} \times 0.010 \text{ H}$$
$$= 2 \ \Omega$$

The voltage is shown in Figure 24–3B, as it was in Figure 24–2B. There is no difference in the voltage of both circuits.

The amount of current is determined by Ohm's law, since it is simply voltage divided by opposition. As the two figures show, the current is also the same for both circuits. That is, the magnitudes (peak to peak) are the same:

$$I = \frac{V}{X_L} = \frac{10 \text{ V}_{pp}}{2 \ \Omega} = 5 \text{ A}_{pp}$$

FIGURE 24-3
AC INDUCTIVE CIRCUIT

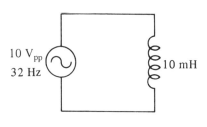

A. Circuit

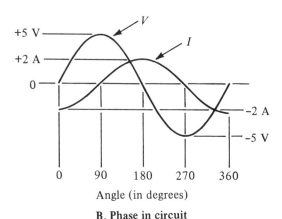

B. Phase in circuit

FIGURE 24-4
VECTOR SUM OF THE DISTANCE TRAVELED

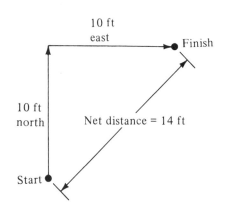

While the magnitudes are the same, the phases are not. In Figure 24–3B, the current crosses 0 volts 90° after the voltage does, and it reaches each peak 90° after the voltage. Energy stored in the magnetic field of the inductor causes the current in the inductor to be 90° behind the voltage across it. That is, the current in an inductor lags the voltage by as much as 90°.

SERIES *RL* CIRCUIT CALCULATIONS

We know that a pair of 2-ohm resistors connected in series will produce a total resistance of 4 ohms. A pair of 10-millihenry inductors in series has a combined inductance of 20 millihenrys. At 32 hertz, these inductors have a reactance of 2 ohms each, or a combined reactance of 4 ohms. What happens when we connect one of the 2-ohm resistors in series with a 2-ohm inductor? Their oppositions add, but not directly, because the components oppose current flow in slightly different ways. We will see how they add in this section.

Impedance Magnitude and Phase

Suppose you walk 10 feet toward the north, turn, and walk 10 feet toward the east, as pictured in Figure 24–4. You will have traveled a distance of 20 feet, but you will only be about 14 feet from the point where you began. Your *effective distance* is 14 feet because the two distances of 10 feet also have direction, and the directions are 90° apart. Since the distances have both *magnitude and direction*, they are added in a special way.

Two of the ways of adding resistances and reactances are considered in this chapter. Both methods use **vectors,** which are quantities that describe a signal in terms of magnitude and direction. One method is much like that shown in Figure 24–4, where we actually draw the magnitudes and directions on a graph and measure their sum and angle. The second method involves calculations. But in both methods, we will be determining the magnitude and direction, or **vector sum,** of the oppositions. The vector sum of all oppositions in an AC circuit is called **impedance.** It has a magnitude in ohms, a phase angle in degrees, and is represented by the letter Z.

There are two important points to consider before we proceed. Here is the first point.

> **Key point:** The magnitude of impedance will be *more* than the total resistance or the total reactance but *less* than the sum of all of them.

Thus, a series circuit with 2 ohms of resistance and 3 ohms of reactance will have an impedance greater than 3 ohms but less than 5 ohms. The second point involves phase.

> **Key point:** Resistance causes no phase shift, but inductive reactance causes the current to lag the voltage by 90°.

Therefore, the phase in a series RL circuit will be such that the current lags the voltage by more than 0° and less than 90°.

Vector Analysis

The vector analysis method described in this section is not as precise as others, but it may help you visualize what is happening in the circuit. We will go through a process of determining total opposition, current, voltage, and power. These are the factors that we determined before with DC circuits. The fact that we must now also consider phase angle makes these calculations a bit more complicated.

FIGURE 24-5
AC *RL* CIRCUIT

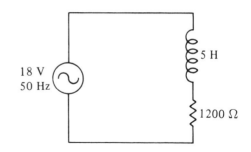

Let's begin with the circuit in Figure 24–5. The first step is to determine the inductive reactance:

$$X_L = 2\pi fL = 6.28 \times 50 \text{ Hz} \times 5 \text{ H}$$

$$= 1570 \ \Omega$$

The second step is to draw a graph showing the 1200-ohm resistance and the 1570-ohm reactance, as is done in Figure 24–6. Both values must be drawn carefully to scale since we are going to measure our answers directly from the graph. Notice that the resistance is drawn directly to the right in the 0° direction, while the inductive reactance is drawn straight up in the +90° direction. Each line on the graph represents 200 ohms.

Once these two oppositions have been drawn on the graph, we can determine their vector sum. We extend a dashed line to the right from the end of the reactance line and another up from the end of the resistance line to a point where they cross, or intersect. Now, we draw a solid line from the origin of the opposition lines to the point of intersection, as shown in Figure 24–7. This solid line represents the magnitude and phase of the circuit impedance. We can measure the length of this line with a compass or scale and then measure the angle with a protractor. Doing so, we find

FIGURE 24-6
RESISTANCE AND INDUCTIVE REACTANCE
IN AN AC CIRCUIT

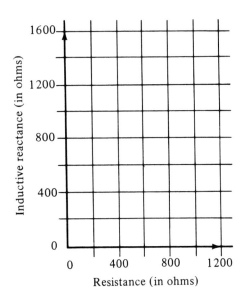

FIGURE 24-7
MAGNITUDE AND PHASE ANGLE OF
IMPEDANCE

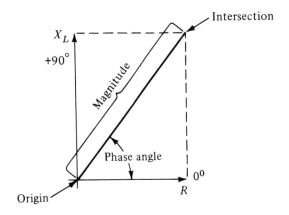

that the circuit has an impedance of 1976 ohms and a phase angle of about 53°. The impedance is written as 1976 $\angle 53°$ ohms, where 1976 gives the magnitude and $\angle 53°$ gives the phase angle.

The next step in the evaluation of a series circuit is to determine the total current. We determine current from Ohm's law by dividing total voltage by total opposition. Remember, also, that these calculations have magnitude and phase. Since the voltage is our reference, or starting, point, its phase is 0°. The calculation procedure has two steps. The magnitude of the current is determined by dividing the numerator (top) magnitude by the denominator (bottom) magnitude. Then, the phase is determined by subtracting the denominator angle from the numerator angle. The calculation is as follows:

$$I = \frac{V}{Z} = \frac{18 \angle 0° \text{ V}}{1976 \angle 53° \text{ } \Omega}$$
$$= \mathbf{0.0091 \angle -53° \text{ A}}$$

Thus, the current lags the voltage in this circuit by 53°.

Now, we can determine the voltage drops V_R (across the resistor) and V_L (across the inductor). Once again, the effect that each component has on phase must be included with its magnitude. These answers are determined by multiplying the magnitudes and adding the angles, as shown next:

$$V_R = IR$$
$$= 0.0091 \angle -53° \text{ A} \times 1200 \angle 0° \text{ } \Omega$$
$$= \mathbf{10.92 \angle -53° \text{ V}}$$
$$V_L = IX_L$$
$$= 0.0091 \angle -53° \text{ A} \times 1570 \angle 90° \text{ } \Omega$$
$$= \mathbf{14.29 \angle 37° \text{ V}}$$

If we were to add these two voltages directly, our answer would exceed the applied voltage of 18 volts. However, we cannot add

FIGURE 24-8
VOLTAGES IN A SERIES AC *RL* CIRCUIT

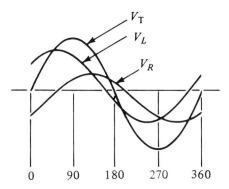

FIGURE 24-9
VECTOR SUM OF CIRCUIT VOLTAGES

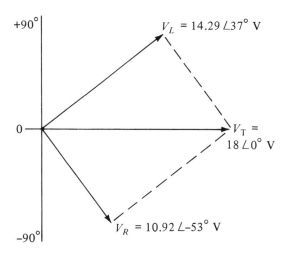

them directly since they are out of phase with each other. Their peaks occur at different times, as shown in Figure 24–8. To add these voltages, we use a graph, as shown in Figure 24–9.

In Figure 24–9, inductor voltage is drawn to scale at 37°, while resistor voltage is at −53°. Once again, a dashed line must be drawn from the end of the resistance line and parallel to the reactance line. Another dashed line is then drawn from the end of the reactance line, parallel to the resistance. A line from the origin to the point where the dashed lines intersect indicates the total voltage. And, as we already knew, the total voltage is 18 ∠0° volts. Also, there is a 90° difference between resistor and inductor voltages.

Mathematical Analysis

The mathematical analysis procedure is faster and more accurate than the vector analysis method, once you become familiar with it. It uses three basic formulas called the sine, cosine, and tangent functions. These formulas simply describe the relationships among R, X_L, Z, and phase, just as Ohm's law describes

V, I, and R. If we know some of the values, the formulas will help us determine the others.

We begin by considering the circuit we just evaluated. The vectors for R, X_L, and Z form three sides of a triangle with the phase angle at the origin, as shown in Figure 24–10. In trigonometry, these sides are called the adjacent (A), the opposite (O), and the hypotenuse (H). For now, we will just call them R, X_L, and Z. The three following formulas describe the relationships we will use, where θ (Greek letter theta) in the formula is the angle between the two sides:

$$\text{sine } \theta = \frac{O}{H} = \frac{X_L}{Z}$$

$$\text{cosine } \theta = \frac{A}{H} = \frac{R}{Z}$$

$$\text{tangent } \theta = \frac{O}{A} = \frac{X_L}{R}$$

Suppose we do not know the phase angle in Figure 24–10, but we do know R, X_L, and

FIGURE 24-10
VECTOR SUM OF CIRCUIT OPPOSITIONS

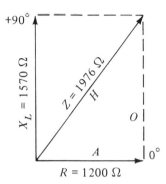

Z. We can use the sine function in the following way:

$$\text{sine } \theta = \frac{X_L}{Z} = \frac{1570\ \Omega}{1976\ \Omega} = 0.7945$$

This sine value, 0.7945, is a unique value for one angle only. Our next step is to find this value or the one nearest to it, in the sine table given in the Appendix. (We can also use the sine function on a calculator.) When we look up 0.7945 in the sine table, we find that 53° is the nearest whole degree answer. Thus, the phase angle θ is 53°.

There are times when we may not know X_L but we do know Z and the phase angle. As we did with Ohm's law, we change the formula around to determine X_L. If we do not know Z or the angle but we do know R and X_L, we use another formula. And sometimes, we may need two formulas.

Let's return again to the original circuit in Figure 24–5 and determine the values with a mathematical approach. Since R and X_L are known, we must begin with the tangent function:

$$\text{tangent } \theta = \frac{X_L}{R} = \frac{1570\ \Omega}{1200\ \Omega} = 1.308$$

From a calculator, we find that $\theta = 52.6°$.

Now that we have R, X_L, and the phase angle, we can use the sine or cosine function to determine Z. We will use the cosine function:

$$\text{cosine } \theta = \frac{R}{Z}$$

or

$$Z = \frac{R}{\text{cosine } \theta} = \frac{1200\ \Omega}{\text{cosine } 52.6°}$$

$$= \frac{1200}{0.6074}\ \Omega = \mathbf{1976\ \Omega}$$

There is one more way that we can mathematically determine impedance when R and X are known. This method uses the Pythagorean theorem and can be demonstrated with the same problem. The formula is

$$Z = \sqrt{R^2 + X_L^2} \quad \text{or} \quad H = \sqrt{A^2 + O^2}$$

Therefore, we have the following calculation for Z:

$$Z = \sqrt{R^2 + X_L^2} = \sqrt{1200^2 + 1570^2}\ \Omega$$

$$= \sqrt{1,440,000 + 2,464,900}\ \Omega$$

$$= \sqrt{3,904,900}\ \Omega = \mathbf{1976\ \Omega}$$

Once impedance and phase are known, we can continue by determining current and voltage, as we did earlier in this chapter.

As another example, consider a circuit with the following values:

$$V_T = 22\ \text{V} \qquad R = 220\ \Omega$$
$$f = 300\ \text{Hz} \qquad L = 250\ \text{mH}$$

First, determine the inductive reactance:

$$X_L = 2\pi f L$$

$$= 2 \times 3.14 \times 3 \times 10^2\ \text{Hz} \times 250$$

$$\times 10^{-3}\ \text{H}$$

$$= \mathbf{471\ \Omega}$$

Next, calculate the impedance:

$$Z = \sqrt{R^2 + X_L^2}$$
$$= \sqrt{(2.2 \times 10^2)^2 + (4.71 \times 10^2)^2} \ \Omega$$
$$= \sqrt{(4.84 \times 10^4) + (22.18 \times 10^4)} \ \Omega$$
$$= \sqrt{27.02 \times 10^4} \ \Omega = \textbf{519.8 } \boldsymbol{\Omega}$$

Now, calculate the current:

$$I = \frac{V_T}{Z} = \frac{22 \text{ V}}{519.8} \ \Omega = \textbf{42.3 mA}$$

Next, determine the voltage drops:

$$V_R = I \times R = 0.0423 \text{ A} \times 220 \ \Omega$$
$$= \textbf{9.3 V}$$
$$V_L = I \times X_L = 0.0423 \text{ A} \times 471 \ \Omega$$
$$= \textbf{19.9 V}$$

Finally, determine the phase angle:

$$\text{cosine } \theta = \frac{R}{Z} = \frac{220 \ \Omega}{519.8 \ \Omega} = 0.4232$$

$$\theta = \textbf{64.96}$$

Since this circuit is inductive, the current lags the voltage by 64.96°.

PARALLEL *RL* CIRCUIT CALCULATIONS

The evaluation of parallel AC circuits is based on some concepts we just discussed for series circuits and others we will introduce now. Some old ideas about parallel resistive circuits will also apply, but others will not. In other words, the subject is getting a bit more complicated. In fact, some people consider this topic of parallel AC *RL* circuits to be the most complicated one encountered in the study of DC and AC circuits. However, a step-by-step approach reduces the complications.

Vector Analysis

Recall that current is constant in a series circuit since it is the same in all parts. However, there are various voltages. In contrast, parallel circuits have only one value of voltage but various values of current. Therefore, the approach to evaluation here is going to be a bit different from the approach in the previous section.

Another difference results from the way we must determine impedance in a parallel circuit. Imagine using the parallel resistor equation with the addition of phase angles. It gets very complicated. For that reason, we will determine impedance by the **total current method.** We will add the branch currents to get a total current; then, we will divide total voltage by total current to determine total impedance.

To begin, let's consider the circuit in Figure 24–11. In the following calculations, we determine the reactance and the two branch currents, I_{XL} and I_R:

$$X_L = 2\pi f L = 6.28 \times 120 \text{ Hz} \times 3 \text{ H}$$
$$= \textbf{2261 } \boldsymbol{\Omega}$$
$$I_{XL} = \frac{V}{X_L} = \frac{24 \text{ V}}{2261 \ \Omega} = \textbf{0.0106 A}$$
$$I_R = \frac{V}{R} = \frac{24 \text{ V}}{1800 \ \Omega} = \textbf{0.0133 A}$$

Our next step is to draw a vector graph of branch currents in order to determine total current. Current I_{XL} is the opposite side, I_R is the adjacent side, and I_T is the hypotenuse, as illustrated in Figure 24–12. The phase angle is between I_R and I_T. Since inductor current lags the total voltage, it is drawn downward. Since the circuit is inductive, the phase angle will be negative.

Now, we measure the length of the hypotenuse. It is equal to 0.017 ampere. Measuring with a protractor gives the phase angle of

FIGURE 24-11
PARALLEL AC *RL* CIRCUIT

FIGURE 24-11
PARALLEL AC *RL* CIRCUIT

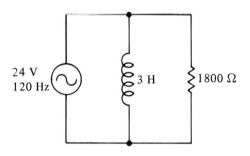

FIGURE 24-12
VECTOR SUM OF CIRCUIT CURRENTS

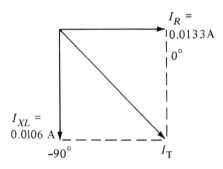

$-38.6°$. Therefore, $I_T = 0.017 \angle -38.6°$ ampere.

Impedance can be determined by dividing total voltage by total current:

$$Z = \frac{V_T}{I_T} = \frac{24 \angle 0° \text{ V}}{0.017 \angle -38.6° \text{ A}}$$

$$= 1412 \angle 38.6° \ \Omega$$

To summarize, impedance in a parallel AC circuit is like resistance in a parallel DC circuit. That is, it is less than the smallest branch opposition. Current in a parallel AC circuit is partly like current in a parallel DC circuit. That is, it is greater than the largest branch current. The difference is that parallel AC cir-

cuit current is the vector sum of the branch currents.

Mathematical Analysis

The same circuit can be evaluated by using the three trigonometric formulas. We first use the tangent function to determine phase angle:

$$\text{tangent } \theta = \frac{O}{A} = \frac{I_{XL}}{I_R} = \frac{0.0106 \text{ A}}{0.0133 \text{ A}}$$

$$= 0.7970$$

Therefore, from a calculator, we find $\theta = -38.6°$.

Once the phase angle is known, the sine function can be used for determining total current, as follows:

$$\text{sine } \theta = \frac{O}{H} = \frac{I_{XL}}{I_T}$$

or

$$I_T = \frac{I_{XL}}{\text{sine } \theta} = \frac{0.0106 \text{ A}}{0.6238}$$

$$= 0.017 \angle -38.6° \text{ A}$$

The magnitude of total current can also be determined by using the Pythagorean theorem, as follows:

$$H = \sqrt{A^2 + O^2} \quad \text{or} \quad I_T = \sqrt{I_R^2 + I_{XL}^2}$$

Therefore, the magnitude of I_T is

$$I_T = \sqrt{I_R^2 + I_{XL}^2}$$

$$= \sqrt{0.0133^2 + 0.0106^2} \text{ A}$$

$$= \sqrt{0.0001768 + 0.0001123} \text{ A}$$

$$= \sqrt{0.0002891} \text{ A} = 0.017 \text{ A}$$

As another example, consider the following values, which we used earlier for a series connection:

$$V_T = 22 \text{ V} \qquad R = 220 \ \Omega$$
$$f = 300 \text{ Hz} \qquad L = 250 \text{ mH}$$

TABLE 24-1
COMPARISON OF SERIES AND PARALLEL
CONNECTIONS

Connection	I_T	Z	θ
Series	42.3 mA	519.8 Ω	65°
Parallel	110.4 mA	199 Ω	25°

First, determine the inductive reactance:

$$X_L = 2\pi fL$$
$$= 2 \times 3.14 \times 3 \times 10^2 \text{ Hz} \times 2.50$$
$$\times 10^{-3} \text{ H}$$
$$= \mathbf{471 \ \Omega}$$

Next, calculate the branch currents:

$$I_L = \frac{V_T}{X_L} = \frac{22 \text{ V}}{471 \ \Omega} = \mathbf{46.7 \ mA}$$

$$I_R = \frac{V_T}{R} = \frac{22 \text{ V}}{220 \ \Omega} = \mathbf{100 \ mA}$$

Now, calculate the total current:

$$I_T = \sqrt{I_R^2 + I_L^2}$$
$$= \sqrt{(46.7 \times 10^{-3})^2 + (100 \times 10^{-3})^2} \text{ A}$$
$$= \sqrt{(2181 \times 10^{-6}) + (10{,}000 \times 10^{-6})^2} \text{ A}$$
$$= \sqrt{12{,}181 \times 10^{-6}} \text{ A}$$
$$= 110.4 \times 10^{-3} \text{ A} = \mathbf{110.4 \ mA}$$

Next, calculate the impedance:

$$Z = \frac{V_T}{I_T} = \frac{22 \text{ V}}{110.4 \times 10^{-3} \text{ A}} = \mathbf{199 \ \Omega}$$

Finally, determine the phase angle:

$$\text{cosine } \theta = \frac{I_R}{I_T} = \frac{100 \times 10^{-3} \text{ A}}{110.4 \times 10^{-3} \text{ A}}$$
$$= 0.906$$
$$\theta = \mathbf{25°}$$

Table 24–1 shows a comparison of the results from the earlier series connection and from the parallel connection. We can see significant differences in the circuit characteristics.

SERIES *RL* CIRCUIT MEASUREMENTS

The primary objective of this experiment is to determine the impedance of an *RL* circuit. Other objectives are to become familiar with the voltage and phase relationships that exist in a series *RL* circuit. Your first step is to connect the circuit in Figure 24–13A.

Once the circuit has been connected and inspected, you should energize the generator and measure the voltages across the inductor and resistor with an oscilloscope. This information will be used later for determining current, reactance, impedance, and phase, so you should enter it in a table as follows:

V_T	V_R	V_L
10 V	7.8 V	6.1 V

The next step is to connect the oscilloscope as shown in Figure 24–13B. You must also switch the horizontal sweep for an external source.

Now, adjust the gain controls to produce a waveform similar to that in Figure 24–14. Since you are not measuring voltages but are comparing them, the inputs do not need to be calibrated. In fact, you should adjust the horizontal and vertical controls to produce a waveform size that is convenient for viewing.

The waveform you will see is the phase relationship between total voltage and total current. The total voltage appears because it is connected directly to the horizontal input of the oscilloscope. Current phase can be repre-

FIGURE 24-13
SERIES *RL* CIRCUIT MEASUREMENT

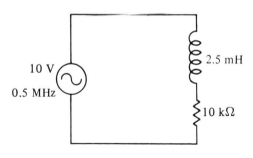

A. Circuit to be evaluated

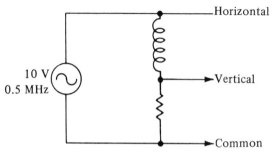

B. Oscilloscope connections for
phase measurements

FIGURE 24-14
OSCILLOSCOPE PATTERN FOR PHASE
MEASUREMENTS

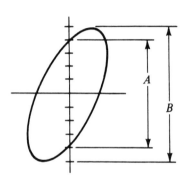

sented by resistor voltage because resistors do not affect phase. Its voltage and current are in phase with one another, so the resistor voltage represents the current phase at the oscilloscope vertical input.

With a waveform like the one in Figure 24–14, you can determine the phase shift of a circuit. Just determine the size of *A* and *B* on the oscilloscope, divide, and find the angle in the sine table. For example,

$$\text{sine } \theta = \frac{A}{B} = \frac{7.8 \text{ divisions}}{10 \text{ divisions}} = 0.78$$

Therefore, $\theta = 51°$.

The measured values you entered earlier in a table can now be used for determining current, reactance, impedance, and phase shift of the experimental circuit:

$$I_T = \frac{V_R}{R} = \frac{7.8 \text{ V}}{10,000 \text{ }\Omega}$$

$$= \textbf{0.00078 A}$$

$$X_L = \frac{V_L}{I_T} = \frac{6.1 \text{ V}}{0.00078 \text{ A}} = \textbf{7820 }\boldsymbol{\Omega}$$

$$Z = \frac{V_T}{I_T} = \frac{10 \text{ V}}{0.00078 \text{ A}}$$

$$= \textbf{12,820 }\boldsymbol{\Omega}$$

$$\text{tangent } \theta = \frac{O}{A} = \frac{V_L}{V_R} = \frac{6.1 \text{ V}}{7.8 \text{ V}} = 0.7820$$

$$\theta = \textbf{38°}$$

You can also use the voltage measurements to produce a graph like the one in Figure 24–9.

SUMMARY

Inductance is the property of a circuit that opposes change in the flow of current. Inductance exists in many electronic components, whether you want it or not. It usually is produced by winding a coil of wire around some form of core. Variations in current through the coil cause a magnetic field to expand and contract. This action, in turn, produces a counter electromotive force that opposes the current change.

Inductive reactance is the opposition by an inductor to the flow of alternating current. While alternating current is continually changing, the opposition to it by an inductor is constant. The amount of opposition is expressed in ohms and is directly proportional to both inductance and frequency. When inductors are connected in series, their total inductance — and therefore, reactance — increases.

The opposition to current change causes a phase shift in an inductive circuit. The current through an inductor lags the voltage across it by 90°. Resistors do not cause a phase shift. When inductors and resistors are combined in a circuit, there will be a phase shift lag of between 0° and 90°.

Since the oppositions caused by resistors and reactors are not identical, they cannot be added directly. Instead, they are added vectorially, which means that both their magnitudes and phase angles must be added. The vector sum of resistance and reactance is called impedance, and it is also measured in ohms.

Four methods can be used for determining the impedance of a series *RL* circuit. One approach is to calculate reactance and draw a vector graph of resistance and reactance, similar to the one in Figure 24–10. Impedance can be measured on it with a scale, and phase can be measured on it with a protractor. Another method is to calculate the impedance value by using the Pythagorean theorem. This formula states that impedance equals the square root of reactance squared plus resistance squared.

A third method involves the use of the sine, cosine, and tangent functions of basic trigonometry. With these, we can compare resistances with reactance and mathematically determine phase angle and impedance. The fourth method of determining impedance is to measure voltage and current and then divide voltage by current.

Voltages and oppositions are used when evaluating series *RL* circuits. Currents and oppositions are used with parallel circuits. In fact, the easiest way to evaluate a parallel circuit is to determine branch oppositions, branch currents, total current, and then impedance. This method is called the total current method of evaluation and is far easier than other methods.

In conclusion, we note a few important points. Impedance is an AC circuit characteristic with magnitude and phase angle. Resistances and inductors in series produce an impedance greater than the resistance or reactance but less than their sum. This impedance causes the current to lag the voltage somewhere between 0° and 90°. Resistors and inductors in parallel produce an impedance less than the smallest resistance or reactance and a current phase shift of between 0° and 90°.

CHAPTER 24

REVIEW TERMS

choke: common name for an inductor used as a filter

filter: inductor used to reduce the amount of AC in a circuit with combined AC and DC

impedance: vector sum of all oppositions in an AC circuit

inductance: property of a circuit that opposes change in the current of the circuit

inductive reactance: opposition of an inductor to the flow of alternating current

mutual inductance: inductance produced by the interaction of adjacent inductors

phase: relationship between the zero-voltage crossings of two signals, expressed in degrees

total current method: method for determining impedance in parallel *RL* circuits

vector: description of a signal in terms of both magnitude and direction

vector sum: sum of two signals including both magnitudes and directions

REVIEW QUESTIONS

1. Define inductance.

2. Describe mutual inductance.

3. Define inductive reactance.

4. Name the units of inductance and inductive reactance.

5. Describe the relationship between inductive reactance and frequency.

6. Describe the relationship between inductive reactance and inductance.

7. Define impedance.

8. Define phase.

9. What is the effect of an increase in frequency on the current in an *RL* circuit?

10. What is the effect of an increase in frequency on the phase of an *RL* circuit?

11. What is the relationship between current and voltage in an inductive circuit?

12. Describe the relationship between inductor voltage and resistor voltage in a series AC *RL* circuit.

13. What is the effect on total inductance when two inductors are connected in series? In parallel?

14. Why is phase angle an important consideration in an AC *RL* circuit?

15. Describe how to multiply impedances expressed in terms of vector magnitudes and phase angles.

16. Describe how to divide impedances expressed in terms of vector magnitudes and phase angles.

FIGURE 24-15

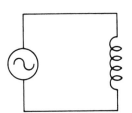

REVIEW PROBLEMS

1. Determine the inductive reactance of a 1.5-henry inductor at 60 hertz. At 120 hertz.

2. Determine the inductive reactance of a 2.5-millihenry inductor at 60 hertz. At 120 hertz.

FIGURE 24-16

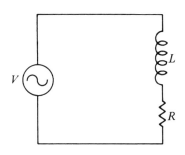

3. Determine the current in the circuit shown in Figure 24–15 if the voltage is 18 volts peak to peak at 60 hertz across a 3.5-henry inductor. Sketch the voltage and current waveshapes.

4. Determine the current in the circuit shown in Figure 24–15 if the voltage is 24 volts peak to peak at 400 hertz across a 500-millihenry inductor. Sketch the voltage and current waveshapes.

5. Determine the impedance, current, voltages, and phase of the circuit shown in Figure 24–16. The parameters are $V = 48$ volts at 8 kilohertz, $R = 6.8$ kilohms, and $L = 0.15$ henry. Use a graphical approach.

6. Determine the impedance, current, voltages, and phase of the circuit in Figure 24–16. The parameters are $V = 12$ volts at 1.5 megahertz, $R = 750$ kilohms, and $L = 50$ millihenrys. Use a graphical approach.

7. Determine the impedance, current, voltages, and phase of the circuit described in Problem 5. Use a mathematical approach.

8. Determine the impedance, current, voltages, and phase of the circuit described in Problem 6. Use a mathematical approach.

9. Determine the currents, impedance, and phase of the circuit shown in Figure 24–17. The voltage is 36 volts at 1 kilohertz, $L = 8$ henrys, and $R = 68$ kilohms. Use a graphical approach.

10. Determine the currents, impedance, and phase of the circuit in Figure 24–17 if the voltage is 24 volts at 800 hertz, $L = 8$ henrys, and $R = 68$ kilohms. Use a graphical approach.

11. Determine the currents, impedance, and phase of the circuit described in Problem 9. Use a mathematical approach.

FIGURE 24-17

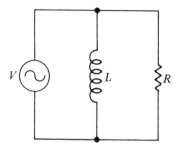

FIGURE 24-19

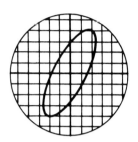

FIGURE 24-18

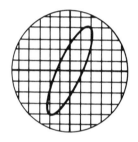

12. Determine the currents, impedance, and phase of the circuit described in Problem 10. Use a mathematical approach.

13. In Figure 24–18, what phase angle is indicated?

14. In Figure 24–19, what phase angle is indicated?

25

CAPACITIVE REACTANCE

Define capacitive reactance.

State the relationship among capacitance, frequency, and capacitive reactance.

Calculate the capacitive reactance and impedance of series and parallel AC *RC* circuits.

Perform a lab experiment to determine the capacitive reactance of an AC circuit.

Prepare a lab report describing an *RC* circuit experiment.

INTRODUCTION

The objective of this chapter is to introduce the theory and measurement of capacitive reactance. Both vector analysis and mathematical analysis are used in evaluating series and parallel AC *RC* circuits.

Review of Capacitance

Capacitance is defined as the property of a circuit that opposes change in the voltage of the circuit. The key words here are *change* and *voltage*. In Chapter 13, we explained that the capacitor in a series *RC* circuit does not charge or discharge instantly but does so after a delay based on the *RC* time constant of the circuit. Capacitors in AC circuits act in a similar manner. Since AC means that there is a continual change in voltage, capacitors in AC circuits provide continuous opposition to it. This opposition is called capacitive reactance.

We begin this topic by briefly reviewing capacitor theory. Recall that capacitors are electrostatic devices. In the basic capacitor construction, two conductors are separated by an insulator. While any two conductors separated by an insulator have some capacitance, the amount is insignificant and can usually be ignored. Components constructed as capacitors have large conductors, or plates, separated by a carefully selected insulator, or dielectric. Aluminum is a common plate material, while paper, ceramic, mica, and air are often used as dielectrics. Capacitance is expressed in farads (F); most capacitors have capacitance values in the microfarad (μF) or picofarad (pF) ranges. The amount of capacitance depends on the area of the plates, the distance between the plates, and the quality of the dielectric.

When a voltage is applied across a capacitor in a circuit, current begins to flow. Free electrons move from the negative source to the nearest capacitor plate. They do not continue beyond that point because of the insulating dielectric, so they begin to collect on the plate. At the same time, free electrons on the other plate move toward the positive source, causing an electron deficiency to develop at that plate. The time it takes for this process to occur is directly proportional to the size of the resistance in the circuit and the capacitance — that is, $t = RC$. In fact, the time it takes for a capacitor to charge to 63% of the applied voltage or fall to 37% of it is called the *RC* time constant of the circuit.

A charged capacitor will maintain its charge when the charging source is removed unless there is an external path for the electrons to follow. If an external path is provided, electrons from the plate with the excess will flow to the deficient plate, if allowed, until both plates have equal charges. Therefore, you must always be careful when working with capacitors.

> **Safety tip:** Even if power is no longer being applied, before handling capacitors, be sure they are not still charged.

Capacitance increases when capacitors are connected in parallel because the total area of the plates is increased. Capacitance decreases when they are connected in series because the total distance between the plates is increased. The voltage distribution across capacitors connected in series is inversely proportional to their capacitance. Since this rule is the opposite of the rule for resistances in series, capacitor voltage ratings must be carefully observed. The relationship between voltage, capacitance, and charge is described by the formula $Q = CV$.

Capacitive Reactance

Capacitive reactance is defined as the opposition from a capacitor to the flow of alternating current. Its symbol is X_C, and it is measured in ohms. One way to understand capacitive reactance is to consider a capacitor in series with a resistor. When voltage is applied to this circuit, a large amount of current flows through the resistor, although it quickly drops to zero as the capacitor becomes charged. The initial opposition from a capacitor to a voltage change is low, but it increases as the capacitor charges. If the capacitor value is increased, it will take longer to become charged. That is, it will take longer for the opposition to become significant. In other words, capacitive reactance decreases as capacitance increases. It also decreases as frequency increases.

As just mentioned, the opposition of a capacitor begins to become significant as the capacitor charges to the same level as the applied voltage. But in AC circuits, the applied voltage is always changing, so the capacitor does not have a chance to become charged before the voltage changes. And as the frequency of the applied voltage increases, the opposition by the capacitor is even less likely to become significant. These relationships can be stated this way: Capacitive reactance decreases with an increase of capacitance or an increase in frequency. Thus, the formula for capacitive reactance is as follows, where f is the frequency and C is the capacitance:

$$X_C = \frac{1}{2\pi f C}$$

For instance, the amount of reactance caused by a 0.1-microfarad capacitor at 60 hertz is

$$X_C = \frac{1}{2\pi f C}$$

$$= \frac{1}{2 \times 3.14 \times 60 \text{ Hz} \times 0.1 \times 10^{-6} \text{ F}}$$

FIGURE 25-1
CAPACITIVE REACTANCE VERSUS FREQUENCY

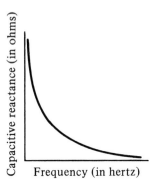

$$= \frac{1}{37.68} \times 10^6 \ \Omega = 0.02654 \times 10^6 \ \Omega$$

$$= \mathbf{26{,}540 \ \Omega}$$

Doubling the frequency to 120 hertz reduces the capacitive reactance by a half:

$$X_C = \frac{1}{2\pi f C}$$

$$= \frac{1}{2 \times 3.14 \times 1.2 \times 10^2 \text{ Hz} \times 0.1 \times 10^{-6} \text{ F}}$$

$$= \frac{1}{75.36 \times 10^{-6}} \ \Omega = \mathbf{13{,}270 \ \Omega}$$

The inverse relationship between capacitive reactance and frequency is shown in Figure 25–1. At very low frequencies or at DC, capacitive reactance is so high that the capacitor essentially acts as an open circuit. At very high frequencies, the reactance of a capacitor becomes very small.

As mentioned in the preceding chapter, most circuits do not have just DC or just AC but have, instead, a combination of both. Capacitors are useful in these circuits because

FIGURE 25-2
AC CAPACITIVE CIRCUIT

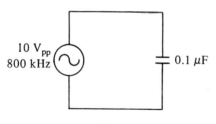

A. Circuit

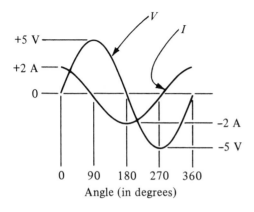

B. Phase relationships

they respond differently to the two types of signals. Whereas inductors offer little opposition to DC signals, capacitors oppose them greatly. And while inductors do offer opposition to AC signals, capacitors do not. If you compare Figures 24–1 and 25–1, you will see that inductors and capacitors act in opposite ways as frequency varies from low to high.

Phase in a Capacitive Circuit

Capacitors, like inductors, cause a phase shift between voltage and current. However, the direction of the shift is not the same.

Rule: In a capacitor, the voltage always follows, or lags, the current by 90°.

This effect occurs because the capacitor is opposing the voltage change.

Consider the circuit in Figure 25–2A. The capacitive reactance of this circuit is

$$X_C = \frac{1}{2\pi f C}$$

$$= \frac{1}{6.28 \times 0.8 \text{ MHz} \times 0.1 \text{ μF}}$$

$$= \frac{1}{6.28 \times 0.8 \times 10^6 \text{ Hz} \times 0.1 \times 10^{-6} \text{ F}}$$

$$= \mathbf{2\ \Omega}$$

From Ohm's law, the circuit has the following current:

$$I = \frac{V}{X_C} = \frac{10 \text{ V}_{pp}}{2\ \Omega} = \mathbf{5\ A_{pp}}$$

The magnitude and phase of the voltage and current are shown in Figure 25–2B. You can see the phase shift caused by the capacitor when you compare the points where voltage and current cross zero in the positive direction. As expected, the voltage is behind the current. The time it takes the electrostatic field to charge or discharge causes the capacitor voltage to lag behind the current phase in the circuit.

SERIES *RC* CIRCUIT CALCULATIONS

As we saw in the preceding chapter, the vector sum of all oppositions in an AC circuit is called impedance. Impedance *Z* has a magnitude in ohms and a phase angle in degrees. In a series RC circuit, the magnitude is greater than the sum of the resistances or the sum of the reactances but less than the sum of all them. Since resistors cause no phase shift but a capacitor causes a voltage lag, an RC circuit will have a voltage phase lag between 0° and 90°.

The two methods of circuit evaluation outlined in Chapter 24 can be used here for an RC circuit, with a little revision. One difference involves the reactance formulas. The inductive reactance formula involves a direct proportion, while the capacitive reactance formula involves an inverse proportion. The other difference is in what lags what. Current lags voltage in inductors, but voltage lags current in capacitors. The rest of the evaluation techniques are similar.

Vector Analysis

The circuit to be evaluated is shown in Figure 25–3A. Since much of the technique for vector analysis was explained in Chapter 24, it will not be explained again here. Instead, we will outline the work. The reactance for the circuit is

$$X_C = \frac{1}{2\pi f C}$$

$$= \frac{1}{6.28 \times 50 \text{ Hz} \times 0.33 \times 10^{-6} \text{ F}}$$

$$= 9650 \ \Omega$$

The determination of capacitive reactance is the only calculation required at this point. Once this calculation is done, we can draw a graph showing resistance and reactance in their proper magnitudes and directions. Note that while the inductive reactance line is drawn at $+90°$, the capacitive reactance line is drawn at $-90°$. See Figure 25–3B.

Once again, two dashed lines are drawn

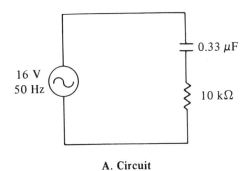

FIGURE 25-3
AC *RC* CIRCUIT

16 V
50 Hz

0.33 μF

10 kΩ

A. Circuit

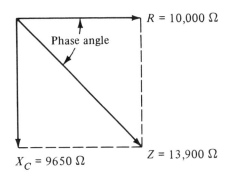

$R = 10,000 \ \Omega$

Phase angle

$X_C = 9650 \ \Omega$

$Z = 13,900 \ \Omega$

B. Determining impedance and phase angle

on the graph. One line begins at the end of the resistance line and is parallel to the reactance line. The other begins at the end of the reactance line and is parallel to the resistance line. The impedance line is then drawn from the graph origin to the intersection of the two dashed lines. We can now measure impedance and phase from this graph with the aid of a scale and a protractor. When we do so, we find a magnitude of about 13,900 ohms and an angle of about $-44°$. The impedance, then, is 13,900 $\angle -44°$ ohms.

Current is equal to the total voltage divided by the total opposition. Phase angles must be included along with magnitudes.

FIGURE 25-4
VECTOR SUM OF CIRCUIT VOLTAGES

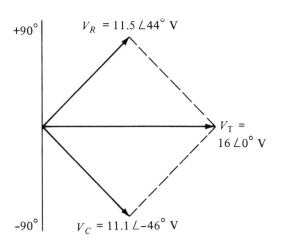

Since the applied voltage is our starting point, its phase angle is 0°. Thus, the current is

$$I = \frac{V}{Z} = \frac{16 \angle 0° \text{ V}}{13,900 \angle -44° \text{ }\Omega}$$

$$= \textbf{0.00115} \angle \textbf{44° A}$$

In this calculation, current magnitude was determined by dividing the numerator magnitude by the denominator magnitude. The phase angle was determined by subtracting the denominator angle from the numerator angle. From mathematics rules, we know that subtracting a value that is already negative makes the answer positive.

Voltage magnitudes are determined by multiplication, and phase angles are determined by addition. As shown in Figure 25–3B, the resistor has a phase angle of 0°; the capacitor phase angle is −90°. Therefore, we have the following calculations:

$$V_R = IR$$

$$= 0.00115 \angle 44° \text{ A} \times 10,000 \angle 0° \text{ }\Omega$$

$$= \textbf{11.5} \angle \textbf{44° V}$$

$$V_C = IX_C$$

$$= 0.00115 \angle 44° \text{ A} \times 9650 \angle -90° \text{ }\Omega$$

$$= \textbf{11.1} \angle \textbf{−46° V}$$

These voltages can be checked by adding them vectorially on a graph, as shown in Figure 25–4. The resistor voltage leads the applied voltage at the same phase as the total circuit current. The capacitor voltage lags this current by 90°.

Mathematical Analysis

The mathematical approach is more accurate than the graphical method, and it is faster if you use a calculator. We will consider the circuit in Figure 25–3A used in the graphical approach. Figure 25–3B shows what we already know.

Since the opposite and adjacent sides of the triangle are known, we can begin by using the tangent function. Keep in mind that the phase of the impedance is negative because the capacitive reactance points downward. The tangent function is

$$\text{tangent } \theta = \frac{O}{A} = \frac{X_C}{R} = \frac{9650 \text{ }\Omega}{10,000 \text{ }\Omega}$$

$$= 0.965$$

From the tables, we can conclude that the phase angle is −44°.

We now use the sine function to determine the length of the hypotenuse, or the impedance magnitude:

$$\text{sine } \theta = \frac{O}{H} = \frac{X_C}{Z}$$

$$Z = \frac{X_C}{\text{sine } \theta} = \frac{9650 \text{ }\Omega}{0.6947} = \textbf{13,891 }\Omega$$

Therefore, $Z = 13,891 \angle -44° \text{ }\Omega$.

Impedance can also be determined by using the Pythagorean theorem, as follows:

$$Z = \sqrt{R^2 + X_C^2}$$
$$= \sqrt{10,000^2 + 9650^2} \ \Omega$$
$$= \sqrt{100,000,000 + 93,122,500} \ \Omega$$
$$= \sqrt{193,122,500} \ \Omega = \mathbf{13,897 \ \Omega}$$

The difference between the two values of impedance is caused by earlier rounding off.

Once we have these values, we can continue the process of determining current and voltages in the manner explained earlier in the chapter.

As another example, consider a circuit with the following values:

$$V_T = 22 \ V \qquad R = 470 \ \Omega$$
$$f = 0.75 \ MHz \qquad C = 400 \ pF$$

First, determine the capacitive reactance:

$$X_C = \frac{1}{2\pi f C}$$

$$= \frac{1}{2 \times 3.14 \times 0.75 \times 10^6 \ Hz \times 400 \times 10^{-12} \ F}$$

$$= \frac{1}{1884 \times 10^{-6}} \ \Omega = \mathbf{531 \ \Omega}$$

Next, calculate the impedance:

$$Z = \sqrt{R^2 + X_C^2}$$
$$= \sqrt{(4.7 \times 10^2)^2 + (5.31 \times 10^2)^2} \ \Omega$$
$$= \sqrt{22.09 \times 10^4 + 28.20 \times 10^4} \ \Omega$$
$$= \sqrt{50.29 \times 10^4} \ \Omega = \mathbf{709 \ \Omega}$$

Now, determine the current:

$$I = \frac{V_T}{Z} = \frac{22 \ V}{709 \ \Omega} = \mathbf{31 \ mA}$$

Next, determine the voltage drops:

$$V_R = I \times R = 0.031 \ A \times 470 \ \Omega$$
$$= \mathbf{14.6 \ V}$$

$$V_C = I \times X_C = 0.031 \ A \times 531 \ \Omega$$
$$= \mathbf{16.5 \ V}$$

Finally, determine the phase angle:

$$\text{cosine } \theta = \frac{R}{Z} = \frac{470 \ \Omega}{709 \ \Omega} = 0.6629$$

$$\theta = \mathbf{48.5°}$$

Since this circuit is capacitive, the current leads the voltage by 48.5°.

PARALLEL *RC* CIRCUIT CALCULATIONS

In the preceding chapter, we considered resistor-inductor combinations in both series and parallel. So far in this chapter, we have considered only the combined effects of resistors and capacitors in series connection. Now, we look at parallel connection. Our same two methods will be used, vector analysis and mathematical analysis. These methods will be used again in the following two chapters when we investigate the combined effects of resistors, inductors, and capacitors in series and parallel connections.

Vector Analysis

Once again, we begin parallel circuit evaluation by considering voltage as constant. Thus, the various values of current must be determined. These values, in turn, depend on the branch oppositions, which, for the circuit shown in Figure 25–5A, are resistance and capacitive reactance. The final step in the evaluation is to determine the circuit impedance. At this point, we will still use the total current method for determining parallel circuit impedance. The circuit is shown in Figure 25–5A, and the calculations are as follows:

FIGURE 25-5
PARALLEL AC *RC* CIRCUIT

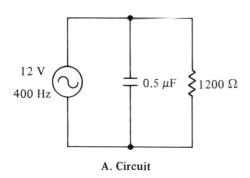

A. Circuit

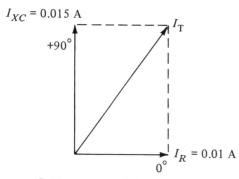

B. Vector sum of circuit currents

$$X_C = \frac{1}{2\pi fC}$$

$$= \frac{1}{6.28 \times 4 \times 10^2 \text{ Hz} \times 0.5 \times 10^{-6} \text{ F}}$$

$$= 0.0796 \times 10^4 \ \Omega = \mathbf{796 \ \Omega}$$

$$I_{XC} = \frac{V}{X_C} = \frac{12 \text{ V}}{796 \ \Omega} = \mathbf{0.01507 \ A}$$

$$I_R = \frac{V}{R} = \frac{12 \text{ V}}{1200 \ \Omega} = \mathbf{0.010 \ A}$$

Now, we draw a graph showing the branch currents in this circuit, as pictured in Figure 25–5B. Resistor current is in phase with the

total voltage, which is our reference. Capacitor current leads resistor current by 90°. Parallel dashed lines are used, once again, for locating one end of the hypotenuse. The other end of the hypotenuse is at the origin. Total current magnitude is indicated by the length of the hypotenuse. The phase is the angle between the total current and resistor current. When we measure the magnitude and angle, we find the total current to be about 0.018 ∠56° ampere.

Circuit impedance can now be found by using the AC vector version of Ohm's law. Its magnitude will be less than that of the smallest branch opposition. The impedance is

$$Z = \frac{V_T}{I_T} = \frac{12 \ \angle 0° \text{ V}}{0.018 \ \angle 56° \text{ A}}$$

$$= \mathbf{667 \ \angle -56° \ \Omega}$$

Mathematical Analysis

The circuit in Figure 25–5A can also be evaluated by using trigonometric functions. With this approach, our answers should be a bit more accurate. Figures 25–5A and 25–5B show what we already know.

We can determine phase angle by using the tangent function, as follows:

$$\text{tangent } \theta = \frac{O}{A} = \frac{I_{XC}}{I_R} = \frac{0.01507 \text{ A}}{0.010 \text{ A}}$$

$$= 1.507$$

Therefore, $\theta = 56.4°$. (The number of digits in your answer, 56.4 or 56, will depend on the number of digits available in the table or calculator you use.)

With the phase angle known, we can use the sine function to determine total current, as follows:

$$\text{sine } \theta = \frac{O}{H} = \frac{I_{XC}}{I_T}$$

$$I_T = \frac{I_{XC}}{\text{sine } \theta} = \frac{0.01507 \text{ A}}{\text{sine } 56.4°}$$

$$= \frac{0.01507 \text{ A}}{0.8329} = \mathbf{0.01809 \text{ A}}$$

Combining this magnitude with the phase angle, we get our final expression for total current:

$$I_T = \mathbf{0.01809 \angle 56.4° \text{ A}}$$

The current magnitude can also be found with the Pythagorean theorem, as follows:

$$I_T = \sqrt{I_R^2 - I_{XC}^2}$$

$$= \sqrt{0.010^2 - 0.01507^2} \text{ A}$$

$$= \sqrt{0.0001 - 0.0002271} \text{ A}$$

$$= \sqrt{0.0003271} \text{ A} = \mathbf{0.01809 \text{ A}}$$

As another example, consider the following values, which we used earlier in a series circuit problem:

$$V_T = 22 \text{ V} \qquad R = 470 \text{ } \Omega$$
$$f = 0.75 \text{ MHz} \qquad C = 400 \text{ pF}$$

First, determine the capacitive reactance:

$$X_C = \frac{1}{2\pi f C}$$

$$= \frac{1}{2 \times 3.14 \times 0.75 \times 10^6 \text{ Hz} \times 400 \times 10^{-6} \text{ F}}$$

$$= \frac{1}{1884 \times 10^{-6}} \text{ } \Omega = \mathbf{531 \text{ } \Omega}$$

Next, calculate the currents:

$$I_C = \frac{V_T}{X_C} = \frac{22 \text{ V}}{531 \text{ } \Omega} = \mathbf{41.4 \text{ mA}}$$

$$I_R = \frac{V_T}{R} = \frac{22 \text{ V}}{470 \text{ } \Omega} = \mathbf{46.8 \text{ mA}}$$

$$I_T = \sqrt{I_R^2 + I_{XC}^2}$$

TABLE 25-1
COMPARISON OF SERIES AND PARALLEL CONNECTIONS

Connection	I_T	Z	θ
Series	31 mA	709 Ω	48.5°
Parallel	62.5 mA	352 Ω	41.5°

$$= \sqrt{(46.8 \times 10^{-3})^2 + (41.4 \times 10^{-3})^2} \text{ A}$$

$$= \sqrt{2190 \times 10^{-6} + 1714 \times 10^{-6}} \text{ A}$$

$$= \sqrt{3904 \times 10^{-6}} \text{ A} = \mathbf{62.5 \text{ mA}}$$

Now, calculate the impedance:

$$Z = \frac{V_T}{I_T} = \frac{22 \text{ V}}{64.5 \times 10^{-3} \text{ A}} = \mathbf{352 \text{ } \Omega}$$

Finally, determine the phase angle:

$$\text{cosine } \theta = \frac{I_R}{I_T} = \frac{46.8 \times 10^{-3} \text{ A}}{62.5 \times 10^{-3} \text{ A}}$$

$$= 0.7488$$

$$\theta = \mathbf{41.5°}$$

Table 25–1 shows a comparison of the results for the series connection and the parallel connection. We see that there are significant differences in the two circuits' characteristics.

RC CIRCUIT MEASUREMENTS

This chapter has provided you with a few ways of looking at AC capacitive circuits. Understanding their rules and calculations, on paper, is important. But so is understanding them for a bench circuit. Therefore, the experiments offered in this section provide other ways of viewing *RC* circuits.

FIGURE 25-6
RC SERIES CIRCUIT FOR EVALUATION

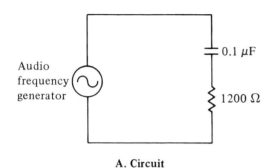

Audio frequency generator

0.1 μF

1200 Ω

A. Circuit

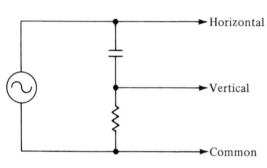

→ Horizontal

→ Vertical

→ Common

B. Oscilloscope connections

Series Circuit

Begin by connecting the circuit in Figure 25–6A. Then, follow this procedure:

1. Adjust the source for 10 volts at 1 kilohertz.

2. Measure and record all voltages. Use a table that includes columns for frequency, V_T, V_C, V_R, and I_T.

3. Change the frequency to 2 kilohertz. Measure and record the voltages again.

4. Measure and record the circuit current at both frequencies with 10 volts applied to the circuit.

5. Connect an oscilloscope as shown in Figure 25–6B.

6. Determine the phase angle between V_R and V_T by using the procedure shown in Figure 24–14. That is, measure the two dimensions indicated, and use the sine function. Record your data in a table that includes columns for frequency, A, B, A/B, and phase angle.

7. Disconnect the circuit.

Parallel Circuit

This experiment gives you an opportunity to compare series and parallel connections as well as measured and calculated values. Begin by reconnecting the components used in the preceding experiment according to the schematic in Figure 25–7. Now, follow the experimental procedure outlined next:

1. Adjust the source for 10 volts at 1 kilohertz.

2. Measure and record all currents. Use a table that includes columns for frequency, I_T, I_C, and I_R.

3. Change the frequency to 2 kilohertz. Measure and record the currents again.

4. Disconnect the circuit.

Reporting Results

Prepare your report in the following manner:

1. Provide a title page with the experiment name, your name, the names of anyone who worked with you, and the date of the experiment.

2. Write a statement of your objective, which would indicate that you investigated the operation of series and parallel *RC* circuits with AC applied.

3. List the components and equipment used. Include brand, model, and serial numbers, as appropriate.

4. Draw schematics and prepare tables. Data for both methods of connection should be included.

5. Provide calculations for the expected values of voltage and current for the series connection, and compare those values with what you measured. Enter all values in a table for the convenience of readers.

6. Provide calculations for the phase angle of the series circuit by using both voltage measurements and the Lissajous patterns. Show the calculation for the expected value of the phase.

7. Provide calculations for the expected impedance at both frequencies.

8. Show calculations for the measured impedance by using V_T and I_T at both frequencies.

9. Provide calculations for the impedance of the parallel circuit at both frequencies by using V_T and I_T.

10. Show calculations for the expected impedance of the parallel circuit at both frequencies.

11. Write a statement of your conclusions.

Consider these questions as you write your conclusion: What happened to the current in these circuits as the frequency increased? What does this result say about impedance? What effect did frequency have on phase? Does the impedance of a parallel *RC* circuit change in the same direction as it does in a series circuit when frequency increases? How close were your measurements to your calculations? Why?

SUMMARY

Capacitance is defined as the property of a circuit that opposes change in voltage. It exists

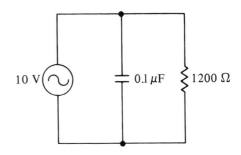

FIGURE 25-7
RC PARALLEL CIRCUIT FOR EVALUATION

whenever two conductors are close together. Most of the time, though, capacitance exists because a capacitor is included in the circuit. A capacitor is made with two large conductors separated by an insulator. The conductors are often aluminum and are called plates. The insulator is called the dielectric and can be made of paper, ceramic, or an electrolytic paste.

Capacitors in a circuit allow electrons to be stored on one plate and removed from the other. In this situation, the capacitor is said to be charged. The plate with an excess is negatively charged; the other plate is positively charged. This ability to store energy in an electrostatic field is what makes a capacitor oppose voltage change. A charged capacitor provides electrons to the circuit connected to it when the applied voltage starts to fall. When the applied voltage begins to rise, energy is absorbed by the capacitor as it begins to collect electrons.

When the voltage is constantly and rapidly changing, the capacitor continuously stores and returns electrons. This process allows current to flow in the circuit when the voltage charge is out of phase. In fact, the voltage across a capacitor lags its charging current by 90°. The opposition of a capacitor to the flow of alternating current is called capacitive reactance. This opposition decreases with an increase in capacitance or frequency.

When a capacitor is connected in series with a resistor in an AC circuit, phase shift results. Although the resistor does not change phase, the capacitor does. The result is a circuit in which the total current leads the applied voltage by more than 0° but less than 90°. This shift, in turn, causes the resistor to be at this same phase angle, but the capacitor voltage leads it by 90°. A vector diagram of this situation shows that the capacitor voltage leads the applied voltage and the resistor voltage lags it. The vector sum of these voltages equals the applied voltage at a 0° phase shift.

Parallel *RC* circuits can be evaluated by measuring branch currents, since the total voltage will be a constant, or a reference point. The branch currents can be added vectorially on a graph or compared with trigonometric functions in order to determine total current. Then, total current is used to calculate impedance. This method is used with series and parallel circuits because it avoids more complex mathematical approaches.

The calculation of impedance for current and voltage in AC circuits requires more work than that for DC circuits. Vector voltages and currents have both magnitude and angle. Thus, when dividing vectors, you divide the magnitudes and subtract the denominator angle from the numerator angle. When multiplying vectors, you multiply the magnitudes and add the angles. This process will be used again in the next few chapters.

This chapter outlined two experiments for measuring parameters in *RC* series and parallel circuits. Upon completion of experiments like these, a report is necessary. A lab report is the most effective method for communicating what you intended to do, what you did, and what your results were. Your complete report needs a statement of your objectives, a list of the materials and equipment you used, and your procedure. Schematics and block diagrams usually are included, as are your data and sample calculations. The final part of your report is a statement of your conclusions — your opinion of the data in light of the objective of the experiment.

CHAPTER 25

REVIEW TERMS

capacitance: property of a circuit that opposes change in the voltage of the circuit

capacitive reactance: opposition from a capacitor to the flow of alternating current

REVIEW QUESTIONS

1. Define capacitance.

2. Define capacitive reactance.

3. Name the units of capacitance and capacitive reactance.

4. Describe the relationship between capacitive reactance and frequency.

5. Describe the relationship between capacitive reactance and capacitance.

6. Compare capacitive reactance with inductive reactance.

7. What is the relationship between voltage and current in a capacitive circuit?

8. How does the current-voltage relationship of a capacitor compare with that of an inductor?

9. What is the effect on capacitive reactance when two capacitors are connected in series? In parallel?

10. How does an increase in frequency affect the current in an *RC* circuit?

11. How does an increase in frequency affect the phase in an *RC* circuit?

12. Explain how current can flow in an AC *RC* circuit if the dielectric of a capacitor is an insulator.

13. Describe how to measure phase shift in a series *RC* circuit by using an oscilloscope.

14. Describe and compare three methods of determining phase in an AC *RC* circuit.

REVIEW PROBLEMS

1. Determine the capacitive reactance of a 1.5-microfarad capacitor at 60 hertz and at 120 hertz.

FIGURE 25-8

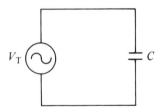

FIGURE 25-9

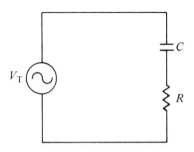

2. Determine the capacitive reactance of a 47-picofarad capacitor at 500 kilohertz and 1.5 megahertz.

3. Determine the current in the circuit shown in Figure 25–8 if the total voltage is 12 volts peak to peak at 60 hertz and C = 0.47 microfarad. Sketch the voltage and current waveforms, and indicate values.

4. Determine the current in the circuit described in Problem 3 if the frequency is changed to 1.5 kilohertz. Sketch the voltage and current waveforms.

5. Determine the impedance, current, voltages, and phase of the circuit shown in Figure 25–9 if the total voltage is 4 volts at 90 hertz, C = 0.33 microfarad, and R = 4.7 kilohms. Use a graphical approach.

FIGURE 25-10

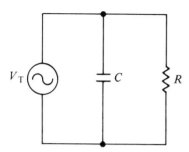

6. Determine the impedance, current, voltages, and phase of the circuit shown in Figure 25–9 if the total voltage is 12 volts at 1.2 kilohertz, $C = 0.1$ microfarad, and $R = 2.2$ kilohms. Use a graphical approach.

7. Determine the impedance, current, voltages, and phase of the circuit described in Problem 5, using a mathematical approach.

8. Determine the impedance, current, voltages, and phase of the circuit described in Problem 6, using a mathematical approach.

9. Determine the impedance, current, voltages, and phase of the circuit shown in Figure 25–10 if $V_T = 4$ volts at 90 hertz, $C = 0.33$ microfarad, and $R = 4.7$ kilohms. Use a graphical approach.

10. Determine the impedance, current, voltages, and phase of the circuit in Figure 25–10 if $V_T = 12$ V volts at 1.2 kilohertz, $C = 0.1$ microfarad, and $R = 2.2$ kilohms. Use a graphical approach.

11. Determine the impedance, current, voltages, and phase of the circuit described in Problem 9, using a mathematical approach.

12. Determine the impedance, current, voltages, and phase of the circuit described in Problem 10, using a mathematical approach.

SERIES *RLC* CIRCUITS

FIGURE 26-1
AC SERIES *RLC* CIRCUIT

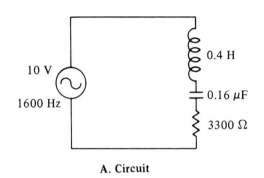

A. Circuit

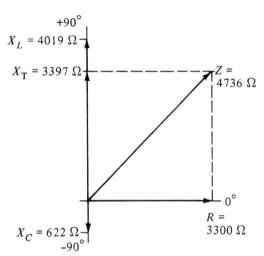

B. Vector sum of circuit oppositions

CALCULATIONS FOR SERIES *RLC* CIRCUITS

The objective of this chapter is to extend your knowledge of AC circuits to those involving series *RLC* connections. We have discussed quite a bit about how inductors and resistors work in AC and DC circuits. We have also discussed circuits involving capacitors and resistors. Two important points should now be clear. When the frequency increases, inductive reactance also increases but capacitive reactance decreases. In addition, inductors cause current to lag, but capacitors cause it to lead. In other words, these two components affect circuits in opposite ways, and the effects change with frequency. In this chapter, we will examine what happens when these different components are connected in series with each other in an AC circuit.

Impedance

The impedance of a series *RLC* circuit depends on capacitance, inductance, frequency, and resistance. Specifically, **impedance** is the vector sum of oppositions in a circuit.

To begin, we consider the circuit pictured in Figure 26–1A. The first step in evaluating this circuit is to determine the impedance. Although we have determined impedance before, there is one additional consideration here. Since capacitors and inductors act in opposite ways, we have the following rule.

> **Rule:** The total reactance of a series *RLC* circuit is the difference between X_C and X_L.

Thus, the impedance calculations are as follows:

$$X_C = \frac{1}{2\pi f C}$$

$$= \frac{1}{6.28 \times 1600 \text{ Hz} \times 0.16 \times 10^{-6} \text{ F}}$$

$$= 622 \ \Omega$$

$$X_L = 2\pi f L = 6.28 \times 1600 \text{ Hz} \times 0.4 \text{ H}$$

$$= 4019 \ \Omega$$

$$Z = \sqrt{R^2 + (X_L - X_C)^2}$$

$$= \sqrt{3300^2 + (4019 - 622)^2} \ \Omega$$

$$= \sqrt{3300^2 + 3397^2}\ \Omega$$
$$= \sqrt{10,890,000 + 11,539,609}\ \Omega$$
$$= \sqrt{22,429,609}\ \Omega = \textbf{4736}\ \Omega$$

Trigonometric functions can be used for determining phase and impedance. Figure 26–1B provides a visual description of the values used in the following calculations, where $X_T = X_L - X_C$:

$$\text{tangent } \theta = \frac{O}{A} = \frac{X_T}{R} = \frac{3397\ \Omega}{3300\ \Omega} = 1.029$$

$$\theta = \textbf{45.8°}$$

$$\text{sine } \theta = \frac{O}{H} = \frac{X_T}{Z}$$

$$Z = \frac{X_T}{\text{sine } \theta} = \frac{3397\ \Omega}{0.7169} = \textbf{4738}\ \Omega$$

Or $Z = 4736\angle45.8°$ ohms. The first impedance value is used in our final answer since it did not result from rounding off.

Current

The current in this circuit can be determined in the usual manner. That is, current is equal to the total voltage divided by the total opposition. This calculation, of course, involves magnitude and phase:

$$I = \frac{V}{Z} = \frac{10\ \angle0°\ V}{4736\ \angle45.8°\ \Omega}$$
$$= \textbf{0.00211}\ \angle-\textbf{45.8°}\ \textbf{A}$$

As shown in Figure 26–2, the voltage leads the current in this circuit. That is, the voltage crosses the 0° point before the current does. Remember that time goes from left to right in the phase diagrams. Since voltage leads, or current lags, this circuit is said to be inductive. Stated another way, this result gives the following point.

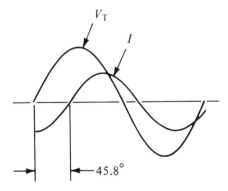

Key point: The largest reactance in a series circuit dominates.

Voltages

As with all series circuits, voltage drops here are equal to oppositions multiplied by current, as the following calculations show:

$$V_C = IX_C$$
$$= 0.00211\ \angle-45.8°\ A$$
$$\times\ 622\ \angle-90°\ \Omega$$
$$= \textbf{1.31}\ \angle-\textbf{135.8°}\ \textbf{V}$$
$$V_L = IX_L$$
$$= 0.00211\ \angle-45.8°\ A$$
$$\times\ 4019\ \angle90°\ \Omega$$
$$= \textbf{8.48}\ \angle\textbf{44.2°}\ \textbf{V}$$
$$V_R = IR$$
$$= 0.00211\ \angle-45.8°\ A$$
$$\times\ 3300\ \angle0°\ \Omega$$
$$= \textbf{6.96}\ \angle-\textbf{45.8°}\ \textbf{V}$$

A graphical representation of these voltage values, as shown in Figure 26–3, illus-

FIGURE 26-3
RELATIVE VOLTAGE DROPS IN A CIRCUIT

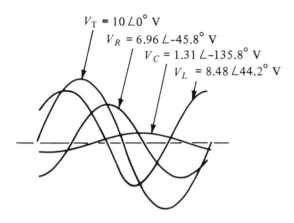

$V_T = 10 \angle 0°$ V

$V_R = 6.96 \angle -45.8°$ V

$V_C = 1.31 \angle -135.8°$ V

$V_L = 8.48 \angle 44.2°$ V

FIGURE 26-4
VECTOR SUM OF THE VOLTAGE DROPS

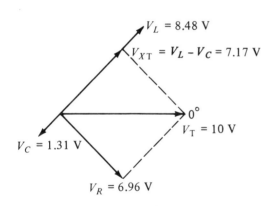

$V_L = 8.48$ V

$V_{XT} = V_L - V_C = 7.17$ V

$0°$

$V_T = 10$ V

$V_C = 1.31$ V

$V_R = 6.96$ V

trates some surprising but important effects of this circuit. It is common for the sum of the voltage drops in a series *RLC* circuit to exceed the applied voltage. This effect will be especially apparent during resonance, which is described later in the chapter. However, since these voltages are out of phase with each other, the vector sum of their instantaneous values always equals the instantaneous value

of the applied voltage. This result is indicated in Figure 26–3.

As shown in Figure 26–3, some voltages are high when others arc low. Some also may be positive when another is negative. The important points to note are the instantaneous values of V_R, V_C, and V_L.

> **Key point:** The sum of the instantaneous values of the voltage drops at any point on the graph is equal to the instantaneous value of V_T at that point.

Figure 26–4 shows vectorially how this effect occurs. Voltage V_R is in phase with the current, which lags the applied voltage by 45.8°. The inductor voltage leads this current by 90°. This shift makes the inductor and capacitor voltages 180° out of phase with each other. Their voltages are always of opposite polarity, so the net voltage across them is the difference between their voltage drops. The applied voltage, then, is equal to the vector sum of resistor voltage and net inductor voltage. Don't forget, though, that there may be larger voltages across individual components than the voltage applied to the circuit.

Frequency Response

We mentioned earlier that the opposition of inductors and capacitors to alternating current flow changes with frequency. The best way to illustrate this change is to change the frequency of the circuit just evaluated from 1600 to 180 hertz. The results for the circuit in Figure 26–1A are as follows:

$$X_C = \frac{1}{2\pi f C}$$

$$= \frac{1}{6.28 \times 180 \text{ Hz} \times 0.16 \times 10^{-6} \text{ F}}$$

$$= \textbf{5529} \ \Omega$$

$$X_L = 2\pi fL = 6.28 \times 180 \text{ Hz} \times 0.4 \text{ H}$$
$$= \textbf{452 } \boldsymbol{\Omega}$$

This change in frequency has caused a large increase in capacitive reactance and a decrease in inductive reactance. Thus, the circuit has shifted from inductive to capacitive.

There are other changes also, as shown by the following calculations:

$$Z = \sqrt{R^2 + (X_C - X_L)^2}$$
$$= \sqrt{3300^2 + (5529 - 452)^2} \ \Omega$$
$$= \sqrt{3300^2 + 5077^2} \ \Omega$$
$$= \sqrt{10,890,000 + 25,775,929} \ \Omega$$
$$= \sqrt{36,665,929} \ \Omega = \textbf{6055 } \boldsymbol{\Omega}$$

Figure 26–5 shows the resistance in phase with a value of 3300 ohms and the vector sum of reactance of 5077 ohms at $-90°$. Phase can now be calculated by dividing the opposite side, or reactance, by the adjacent side, or resistance, and finding the angle that has that value for its tangent. Impedance is then determined by dividing the reactance by the sine. The phase and impedance calculations are as follows:

$$\text{tangent } \theta = \frac{O}{A} = \frac{X_T}{R} = \frac{5077 \ \Omega}{3300 \ \Omega} = 1.538$$

$$\theta = \boldsymbol{-57°}$$

$$\text{sine } \theta = \frac{O}{H} = \frac{X_T}{Z}$$

$$Z = \frac{X_T}{\text{sine } \theta} = \frac{5077 \ \Omega}{0.8387} = \textbf{6053 } \boldsymbol{\Omega}$$

Since the process of obtaining this answer involved some slight rounding off, the first value will be used as the final answer:

$$Z = 6055 \angle -57° \ \Omega$$

There is not much difference in impedance magnitude after this frequency change, but the phase has changed by 102.8°.

FIGURE 26-5
OPPOSITION AT THE SECOND FREQUENCY

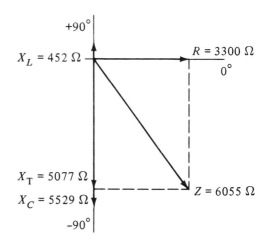

The total current is

$$I = \frac{V_T}{Z} = \frac{10 \angle 0° \text{ V}}{6055 \angle -57° \ \Omega}$$
$$= \textbf{0.00165 } \boldsymbol{\angle 57°} \textbf{ A}$$

The current magnitude is less than it was before, and its phase is leading rather than lagging. Thus, the circuit is now capacitive.

Since we know the current value, we can now calculate the three voltage drops. These drops show us how the applied voltage is distributed throughout the circuit. The calculations follow.

$$V_C = IX_C$$
$$= 0.00165 \angle 57° \text{ A} \times 5529 \angle -90° \ \Omega$$
$$= 9.12 \angle -33° \text{ V}$$
$$V_L = IX_L$$
$$= 0.00165 \angle 57° \text{ A} \times 452 \angle 90° \ \Omega$$
$$= \textbf{0.74 } \boldsymbol{\angle 147°} \textbf{ V}$$
$$V_R = IR$$
$$= 0.00165 \angle 57° \text{ A} \times 3300 \angle 0° \ \Omega$$
$$= \textbf{5.44 } \boldsymbol{\angle 57°} \textbf{ V}$$

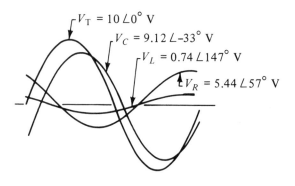

$V_T = 10 \angle 0°$ V

$V_C = 9.12 \angle -33°$ V

$V_L = 0.74 \angle 147°$ V

$V_R = 5.44 \angle 57°$ V

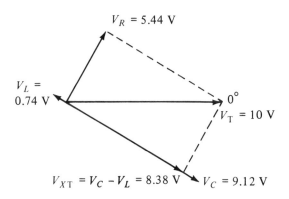

$V_R = 5.44$ V

$V_L = 0.74$ V

$0°$

$V_T = 10$ V

$V_{XT} = V_C - V_L = 8.38$ V

$V_C = 9.12$ V

This voltage distribution, shown in Figures 26–6 and 26–7, can be compared with the voltage distribution for the same circuit at a lower frequency, shown in Figures 26–3 and 26–4. We see that increasing the frequency has changed both the phase shift and the voltage distributions.

The effects of frequency change are only partly explained by these calculations. They indicate phase and voltage changes, but there are other changes, too. In addition, these cal-culations may give the impression that imped-ance and current change only slightly. That is not true. The next section describes the other major changes that occur.

RESONANCE

We begin our discussion by considering two points discussed earlier. First, inductive reac-tance increases with frequency, and capacitive reactance decreases. Second, the total reac-tance in a series circuit is the difference be-tween these two reactances. These points lead to the idea of resonance.

Resonant Frequency

At very low frequencies, the impedance of the circuit in Figure 26–8A is high because X_C is high. At very high frequencies, it is high be-cause X_L is high. Impedance becomes lower between these two points. In fact, it reaches its lowest point when X_C equals X_L. The con-dition that occurs when X_C equals X_L is called **resonance,** and the frequency at which it oc-curs is called the **resonant frequency** of a cir-cuit. Thus, by definition, resonance occurs when capacitive reactance equals inductive re-actance. The formula for resonant frequency f_r is derived as follows:

$$X_L = X_C$$

$$2\pi f_r L = \frac{1}{2\pi f_r C}$$

$$f_r^2 = \frac{1}{4\pi^2 LC}$$

$$\sqrt{f_r^2} = \sqrt{\frac{1}{4\pi^2 LC}}$$

$$f_r = \frac{1}{2\pi\sqrt{LC}}$$

Thus, the resonant frequency of the circuit in Figure 26–1A is

$$f_r = \frac{1}{2\pi\sqrt{LC}}$$

$$= \frac{1}{6.28 \sqrt{0.4 \text{ H} \times 0.16 \times 10^{-6} \text{ F}}}$$

$$= \frac{1}{6.28 \sqrt{0.064 \times 10^{-6}} \text{ Hz}}$$

$$= \frac{1}{6.28 \times 0.2530 \times 10^{-3} \text{ Hz}}$$

$$= \textbf{629.4 Hz}$$

This frequency is indicated in Figure 26–8B at the point where the two reactances cross.

A few more calculations will help explain resonance. These calculations are as follows:

$$X_L = 2\pi fL = 6.28 \times 629.4 \text{ Hz} \times 0.4 \text{ H}$$

$$= \textbf{1581 } \boldsymbol{\Omega}$$

$$X_C = \frac{1}{2\pi fC}$$

$$= \frac{1}{6.28 \times 629.4 \text{ Hz} \times 0.16 \times 10^{-6} \text{ F}}$$

$$= \textbf{1581 } \boldsymbol{\Omega}$$

$$X_T = X_C - X_L$$

$$= 1581 \angle -90° \ \Omega - 1581 \angle 90° \ \Omega$$

$$= \textbf{0 } \boldsymbol{\Omega}$$

$$Z = \sqrt{R^2 - X_T^2} = \sqrt{3300^2 - 0^2} \ \Omega$$

$$= \sqrt{3300^2} \ \Omega = \textbf{3300 } \boldsymbol{\Omega}$$

In other words, we have the following rule.

Rule: The impedance of a series circuit at resonance is equal to the resistance of the circuit.

Now, let's consider the current and voltage distributions at resonance:

FIGURE 26-8
RESONANCE

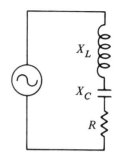

A. Circuit

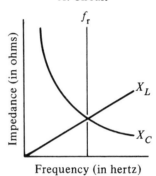

B. X_L and X_C versus frequency

$$I = \frac{V_T}{Z} = \frac{10 \angle 0° \text{ V}}{3300 \angle 0° \ \Omega}$$

$$= \textbf{0.00303} \angle 0° \textbf{ A}$$

$$V_C = IX_C$$

$$= 0.00303 \angle 0° \text{ A} \times 1581 \angle -90° \ \Omega$$

$$= \textbf{4.79} \angle -90° \textbf{ V}$$

$$V_L = IX_L$$

$$= 0.00303 \angle 0° \text{ A} \times 1581 \angle 90° \ \Omega$$

$$= \textbf{4.79} \angle 90° \textbf{ V}$$

$$V_R = IR$$

$$= 0.00303 \angle 0° \text{ A} \times 3300 \angle 0° \ \Omega$$

$$= \textbf{10} \angle 0° \textbf{ V}$$

FIGURE 26-9
OPPOSITIONS AND VOLTAGES AT RESONANCE

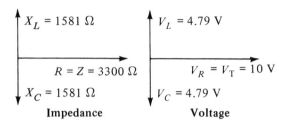

$X_L = 1581\ \Omega$ $V_L = 4.79\ V$

$R = Z = 3300\ \Omega$ $V_R = V_T = 10\ V$

$X_C = 1581\ \Omega$ $V_C = 4.79\ V$

Impedance   Voltage

FIGURE 26-10
RELATIVE VOLTAGES AT RESONANCE

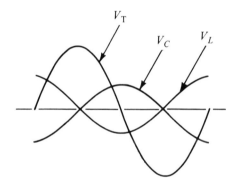

V_T V_C V_L

The results of these calculations are summarized in Figure 26–9. The figure and the calculations show that a series circuit at resonance has the following features:

— A phase shift of 0°;
— An impedance equal to the resistance;
— A resistor voltage equal to the applied voltage;
— Equal and opposite voltages across X_L and X_C;

— Maximum current and minimum impedance.

We can explain what occurs in a circuit at resonance in many ways. For instance, in one method, we consider inductors and capacitors as having the ability to store electric energy. Inductors do so electromagnetically with their coils; capacitors do so electrostatically with their plates and dielectric. Inductors oppose current change by developing a large counter emf. Capacitors, in turn, charge slowly while current flows elsewhere in the circuit. At resonance, the counter emf drops at the same rate that the capacitor charges. When the applied polarity changes, the capacitor discharges, and the inductor counter emf rises.

We can compare this effect to that of a person on a swing. When the swing is pulled back, the *potential energy* from the force of gravity causes the swing to move forward. When the swing reaches its lowest point, *kinetic energy* from the force of motion keeps it going up to another high point. Then, once again, potential energy starts it back, and the cycle is completed. The swing has a natural, or resonant, frequency at which it will *oscillate* (move back and forth). That frequency depends on the weight on the swing and the length of the rope. A shorter rope causes a higher resonant frequency, in the same way that a smaller inductance or capacitance does.

The transfer of energy back and forth from potential to kinetic to potential in a swing is similar to the transfer between electromagnetic and electrostatic energy in an *RLC* circuit. The effect can be seen in the equal but opposite voltages of the inductor and capacitor, as shown in Figure 26–10. Finally, the energy loss caused by the person on the swing is similar to resistance losses in the series circuit. Swing losses are replaced by a push from an energy source, just as electrical losses are replaced by an energy source.

Effects around Resonance

We have looked at a circuit at one point below resonance, at one point above resonance, and at resonance. If calculations of reactance, impedance, phase, current, and voltage were made for a number of frequencies, we would be able to draw a more complete graph of phase, impedance, and current. Figure 26–11 is such a graph, and it summarizes the key points about series *RLC* circuits.

As shown in the figure, below resonance, the circuit has a high impedance and a low current that leads the applied voltage. When the applied frequency is increased, the dominating X_C will decrease. Reactance X_L will increase at the same time but at a much lower rate. The decreasing size of X_C also decreases its effect on phase shift, which begins to get smaller. When both reactances are equal, the phase shift is zero and current is at its maximum. Increasing the frequency causes X_L to begin to dominate and cause a lagging current. As the frequency becomes even higher, phase lag approaches almost $-90°$, and the circuit appears to be inductive. Current is again at a minimum.

As indicated earlier, the impedance at resonance is equal to the resistance. Because of this feature, a lower resistance will cause the impedance curve — and therefore, the current curve — to be sharper. This relationship is called the **quality factor** Q of a circuit, which is calculated in the following way:

$$Q = \frac{X_L}{R}$$

Thus, for the circuit in Figure 26–1A, the quality factor is

$$Q = \frac{X_L}{R} = \frac{1581 \ \Omega}{3300 \ \Omega} = 0.479$$

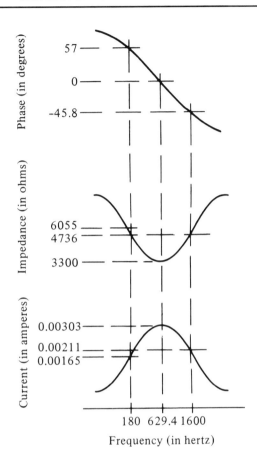

FIGURE 26-11
RELATIONSHIPS IN A SERIES *RLC* CIRCUIT

Reducing the resistance of this circuit will cause an increase in Q and more current at resonance:

$$Q = \frac{X_L}{R} = \frac{1581 \ \Omega}{270 \ \Omega} = 5.85$$

Figure 26–12 shows the effects of resistance on the impedance — and therefore, current — of a series *RLC* circuit. A lower resistance produces a higher Q, as indicated by the sharper curve. However, there is little change

FIGURE 26-12
EFFECTS OF RESISTANCE IN A SERIES *RLC* CIRCUIT

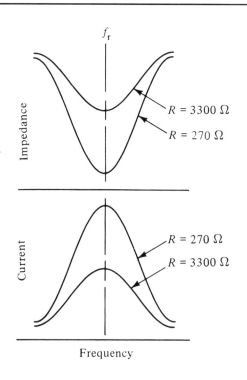

FIGURE 26-13
BANDWIDTH OF A SERIES *RLC* CIRCUIT

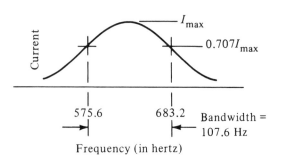

in impedance or current far from resonance since the reactances dominate there.

Uses for Series Resonant Circuits

Here is the most significant characteristic of a series *RLC* circuit with regard to its applications.

> **Key point:** A series *RLC* circuit allows current to flow freely at one frequency and oppose it at other frequencies.

This effect around resonance, shown in Figure 26–11, is further explained this way: The circuit is said to have a bandwidth from one frequency to another since it allows current between those limits to flow, opposing the flow at other frequencies. The **bandwidth** of a circuit is the frequency range between the points where the current exceeds 0.707 of its maximum value, I_{max}. The lower and upper limits of bandwidth are indicated in Figure 26–13. Note that the phase shift at these points is 45°.

The formula for the bandwidth *BW* is

$$BW = \frac{f_r}{Q}$$

Thus, for the circuit in Figure 26–1A with *R* changed to 270 ohms the bandwidth is

$$BW = \frac{f_r}{Q} = \frac{629.4 \text{ Hz}}{5.85} = \textbf{107.6 Hz}$$

This characteristic allows the circuit to be used as a bandpass filter. It can be connected in series with a source of many frequencies and adjusted to pass the range desired while opposing others beyond that range. The adjustment is made by selecting *L* and *C* so that they resonate at the desired frequency. More will be said about bandpass filters in the next chapter.

A CB transceiver often uses a series *RLC* circuit. The antenna and ground are separated by a space, which tends to make them function as a capacitor. An adjustable inductor is connected in series with the antenna and output stage to form the series *RLC* circuit, as pictured in Figure 26–14. A licensed technician adjusts the inductor so that the circuit resonates in the middle of the desired band. This adjustment ensures maximum current flow, allowing maximum power to transfer to the antenna.

FIGURE 26-14
TRANSMITTER OUTPUT USING A SERIES RESONANT CIRCUIT

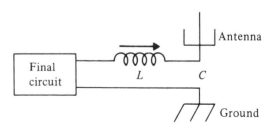

POWER IN AC SERIES CIRCUITS

AC circuits use power in much the same way that DC circuits do. But there is a significant difference that must be considered. Much of the current that flows in an AC *RLC* circuit does not originate at the source; rather, it comes from an inductor or capacitor. As mentioned earlier in the chapter, these components often pass energy back and forth from their stored fields. Other components such as motors generate a current flow as their magnetic field rotates. Thus, to determine the true power of an AC circuit, we must be more specific than we were with DC circuits. Our first step is to determine the power factor of the circuit.

Power Factor and True Power

The percentage of apparent power that is dissipated as true power is called the **power factor** of a circuit. It can be expressed as a decimal or a percentage. Note that pure inductance and pure capacitance do not *dissipate* energy; they only *store* energy. Resistance, though, does dissipate energy in the form of heat. Therefore, the resistance portion of energy must be found when true power is determined.

The following power factor can be used for this calculation:

power factor = cosine θ

The **true power** of a circuit is then found by multiplying the apparent power *VI* by the power factor, as follows:

$P = VI \cos \theta$

In circuits with only resistors, there is no phase shift θ. The phase angle is 0°, and the power factor equals 1, or 100%, since the cosine of 0° equals 1. Thus, in resistive circuits, the actual power dissipated (true power) equals voltage times the current (apparent power). This result holds for all DC and AC resistive circuits.

An AC circuit with a perfect inductor is a much different case, however. The current in a perfect inductive circuit lags the voltage by 90°. Since the cosine of 90° equals 0, the power factor is 0, or 0%. While current does flow in the circuit, no power is dissipated because there is no resistance in the circuit. This result is not typical since most circuits have some resistance. In the following section, we show how true power can be calculated for a typical circuit.

Determining Circuit Power

A version of the circuit in Figure 26–1A was evaluated earlier in the chapter. It was described in Figure 26–5 and is summarized here:

$$V_T = 10 \text{ V at } 180 \text{ Hz}$$
$$L = 0.4 \text{ H}$$
$$C = 0.16 \text{ μF}$$
$$R = 3300 \text{ Ω}$$
$$X_L = 452 \text{ Ω}$$
$$X_C = 5529 \text{ Ω}$$
$$I = 0.00165 \angle 57° \text{ A}$$

The power factor for this circuit equals the cosine of 57°, and it can be written as 0.545 or 54.5%. The true power of this circuit, then, is

$$P = VI \text{ cosine } \theta$$
$$= 10 \text{ V} \times 0.00165 \text{ A} \times 0.545$$
$$= \mathbf{0.00899 \text{ W}}$$

There is another approach we can use to find true power. The only parameter dissipating power is resistance, so the following formula can be used:

$$P = I^2R$$

Thus, using this formula, we have the following result:

$$P = I^2R$$
$$= 0.00165 \text{ A} \times 0.00165 \text{ A} \times 3300 \text{ Ω}$$
$$= \mathbf{0.00898 \text{ W}}$$

Once again, rounding off numbers has led to some difference in the answers.

Note that since DC circuits as well as AC resistive circuits have a 0° phase shift, and since the cosine of 0° equals 1, the formula $P = VI$ can be used only with those circuits. The power factor must be included with it if any phase shift exists. However, the formula $P =$

I^2R can be used for power determination in all circuits.

SUMMARY

The impedance of a series RLC circuit depends on capacitance, inductance, frequency, and resistance. Specifically, the impedance of a series RLC circuit is equal to the vector sum of the resistance and the net (total) reactance. The total reactance is equal to the difference between capacitive and inductive reactance. The difference is used because these two components affect circuits in opposite ways. Inductive reactance increases with frequency, but capacitive reactance decreases. Also, inductive reactance causes current to lag, but capacitive reactance causes it to lead. When these components are connected in series, the impedance can vary greatly. It is, however, always greater than the circuit resistance.

The current in a series RLC circuit varies inversely with the impedance. Its phase can also vary, depending on which reactance dominates. At low frequencies, the capacitive reactance will be large, causing a small current that will lead the applied voltage by up to 90°. At high frequencies, the inductive reactance will be large, causing a small current that can lag the applied voltage by up to 90°. These reactances are equal and opposite at resonance, where they allow maximum current with a 0° phase shift.

The voltage drops vary with frequency, also, since both current and oppositions can change. Yet the vector sum of all voltage drops equals the applied voltage. However, some large voltages do develop, especially around resonance. At this frequency, large, equal and opposing voltages can develop across the reactances, and they can far exceed the applied voltage. One will be at +90° and the other at −90°, while the applied voltage appears in full across the resistance at a 0° phase shift.

All these characteristics change as the frequency changes. Frequency determines the value of each reactance and the point where they will be equal. Doubling the frequency of a series *RLC* circuit can cause more difference than a change in any one component. This result is especially true as resonance is approached.

Resonance occurs when inductive reactance and capacitive reactance are equal. The formula for calculating the resonant frequency of a circuit is

$$f_r = \frac{1}{2\pi\sqrt{LC}}$$

At the resonant frequency, impedance is at its lowest point, and it equals the circuit resistance. Stored energy is transferred back and forth between inductors and capacitors, and little is supplied by the source. Large current and large voltages can also exist.

The quality factor Q of a circuit is determined by dividing X_L by R. A high number, such as 100, indicates that the circuit has little resistance. This result would be indicated by a steep curve for impedance versus frequency. A flat curve indicates a low Q and a high resistance. The bandwidth of this curve can be determined by dividing the resonant frequency by the circuit Q.

Bandwidth is the frequency range between the points where the current is greater than 0.707 times its maximum value. The area between these points is called the bandpass since it represents the range of frequencies given easy access through the circuit.

AC circuit power can be described in two ways: Apparent power equals voltage times current, while true power equals the power actually dissipated. Voltage times current is not a true power indicator since the energy in an AC *RLC* circuit is not always coming from the source. Inductors and capacitors transfer energy back and forth between their respective fields. Thus, to determine true power, we must first determine the power factor, which is equal to the cosine of the circuit phase angle. True power is then equal to the apparent power (VI) times the power factor.

DC circuits have no phase shift, and since the cosine of 0° equals 1, the power factor can be ignored. An AC reactive circuit can have a leading or lagging phase shift of up to 90°, though, so the power factor cannot be ignored. If the circuit were perfectly reactive, the phase shift would be 90° and the factor would be 0, indicating no power. However, current would flow in the circuit.

One other approach will always work for calculating the power in an AC or DC circuit. That approach uses the formula $P = I^2R$.

CHAPTER 26

REVIEW TERMS

bandwidth: number, equal to resonant frequency divided by quality factor, that describes the frequency range between the upper and lower half-power points of a resonant circuit

impedance: vector sum of all oppositions in a circuit

power factor: percentage of power dissipated as true power in an AC circuit

quality factor: number equal to inductive reactance divided by resistance; used to describe the quality of an inductor

resonance: condition of a circuit that occurs when capacitive reactance equals inductive reactance; the behavior includes transfer of energy between reactances

resonant frequency: frequency at which inductive reactance equals capacitive reactance

true power: product of voltage, current, and power factor in an AC circuit

REVIEW QUESTIONS

1. Define resonance.

2. Define resonant frequency.

3. Define bandwidth.

4. Define the quality factor Q.

5. What is the relationship between Q and the bandwidth of a series RLC circuit?

6. Define power factor.

7. What is the power factor of a DC circuit? Explain.

8. At what frequency is power at its maximum in a series RLC circuit?

9. Why is impedance determined by the difference between inductive and capacitive reactance?

10. Describe the characteristics of a series RLC circuit at resonance.

11. Describe the characteristics of a series RLC circuit above and below resonance.

12. Explain the relationship between resistance and impedance in a series resonant circuit.

13. Explain how the measured voltages across a capacitor, inductor, and resistor in a series RLC circuit can add to more than the applied voltage.

14. Describe an application for series *RLC* circuits.

15. Describe how a series *RLC* circuit can be used to filter out undesired frequencies.

REVIEW PROBLEMS

1. Calculate the resonant frequency of the circuit shown in Figure 26–15 if $V_T = 24$ volts, $L = 150$ millihenrys, $C = 47$ picofarads, and $R = 6.8$ kilohms.

2. Calculate the resonant frequency of the circuit in Figure 26–15 if $V_T = 48$ volts, $L = 1.2$ henrys, $C = 0.01$ microfarad, and $R = 470$ ohms.

3. What value capacitor will be required if the circuit described in Problem 1 is to be resonant at 100 kilohertz?

4. What value inductor will be required if the circuit described in Problem 2 is to be resonant at 500 hertz?

5. Calculate the impedance, current, voltages, and phase, at resonance, for the circuit described in Problem 1.

6. Calculate the impedance, current, voltages, and phase, at resonance, for the circuit described in Problem 2.

7. Calculate Q and *BW* for the circuit described in Problem 1.

8. For Problem 7, what effect will changing R to 680 ohms have on Q and bandwidth?

9. Calculate Q and *BW* for the circuit described in Problem 2.

FIGURE 26-15

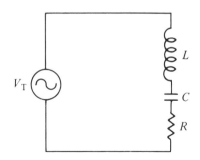

10. For Problem 9, what effect will changing R to 330 ohms have on Q and bandwidth?

11. Calculate the impedance, current, voltages, and phase at the upper and lower bandwidth limits of the circuit described in Problem 1.

12. Calculate the impedance, current, voltages, and phase at the upper and lower bandwidth limits of the circuit described in Problem 2.

13. Calculate the power factor and power of the circuit described in Problem 11.

14. Calculate the power factor and power of the circuit described in Problem 12.

15. Perform all calculations necessary in order to draw a graph showing impedance, current, and phase of the circuit described in Problem 1. Plot the values for at least five frequencies.

16. Perform all calculations necessary in order to draw a graph showing impedance, current, and phase of the circuit described in Problem 2. Plot the values for at least five frequencies.

27

PARALLEL *RLC* CIRCUITS

Calculate the impedance, the phase, and all currents of a parallel *RLC* circuit.

Determine the resonant frequency, bandwidth, and quality factor of a parallel *RLC* circuit.

Describe the relationship among frequency, impedance, and phase in a parallel *RLC* circuit.

Calculate the power factor and true power of a parallel *RLC* circuit.

Evaluate the performance of series and parallel *RLC* circuits.

CALCULATIONS FOR PARALLEL *RLC* CIRCUITS

The objective of this chapter is to extend your knowledge of AC circuits to those involving parallel *RLC* connections. We begin with the calculation of circuit parameters.

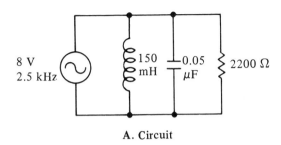

A. Circuit

Current

The easiest way of evaluating parallel *RLC* circuits is by the total current method. That is, we calculate the current in each branch and then determine their vector sum. The circuit we will consider is shown in Figure 27–1A. We begin by finding the opposition in each branch, as follows:

$$X_L = 2\pi f L$$
$$= 6.28 \times 2.5 \times 10^3 \text{ Hz} \times 150$$
$$\times 10^{-3} \text{ H}$$
$$= 2355 \ \Omega$$

$$X_C = \frac{1}{2\pi f C}$$
$$= \frac{1}{6.28 \times 2.5 \times 10^3 \text{ Hz} \times 0.05 \times 10^{-6} \text{ F}}$$
$$= \frac{1 \times 10^3}{0.785} \ \Omega = 1274 \ \Omega$$

Now, the current in each branch must be calculated. Keep in mind that we must use both magnitude and phase. The calculations are

$$I_L = \frac{V_T}{X_L} = \frac{8 \ \angle 0° \text{ V}}{2355 \ \angle 90° \ \Omega}$$
$$= \textbf{0.00340} \ \angle \textbf{-90° A}$$

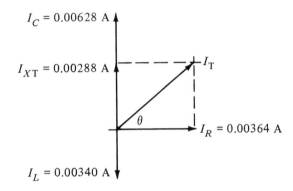

B. Vector sum of circuit currents

$$I_C = \frac{V_T}{X_C} = \frac{8 \ \angle 0° \text{ V}}{1274 \ \angle -90° \ \Omega}$$
$$= \textbf{0.00628} \ \angle \textbf{90° A}$$

$$I_R = \frac{V_T}{R} = \frac{8 \ \angle 0° \text{ V}}{2200 \ \angle 0° \ \Omega} = \textbf{0.00364} \ \angle \textbf{0° A}$$

The total current equals the vector sum of all branch currents, as shown in Figure 27–1B. The calculations for this sum are as follows:

$$I_{XT} - I_C - I_L$$
$$= 0.00628 \ \angle 90° \text{ A}$$
$$- 0.00340 \ \angle -90° \text{ A}$$
$$= \textbf{0.00288} \ \angle \textbf{90° A}$$

$$I_T = \sqrt{I_R^2 + I_{XT}^2} \text{ A}$$
$$= \sqrt{0.00364^2 + 0.00288^2} \text{ A}$$
$$= \sqrt{0.0000132 + 0.00000829} \text{ A}$$
$$= \sqrt{0.0000215} \text{ A} = \mathbf{0.00464 \text{ A}}$$

$$\text{tangent } \theta = \frac{O}{A} = \frac{I_{XT}}{I_R} = \frac{0.00288 \text{ A}}{0.00364 \text{ A}}$$
$$= 0.7912$$

$$\theta = \mathbf{38.3°}$$

$$\text{sine } \theta = \frac{O}{H} = \frac{I_{XT}}{I_T}$$

$$I_T = \frac{I_{XT}}{\text{sine } \theta} = \frac{0.00288 \text{ A}}{0.6198}$$
$$= \mathbf{0.00464 \text{ A}}$$

Therefore, $I_T = 0.00464 \angle 38.3°$ ampere.

This circuit is capacitive because the current leads the voltage. Note the following important point.

> **Key point:** The smallest reactance in a parallel circuit dominates, since it has the greatest current flow.

Impedance

Once the value of total current is known, impedance can be determined. As always, the impedance of a circuit equals the total voltage divided by the total current. The total current method continues to be our basic method for evaluating AC circuits. (There are other methods, such as the use of rectangular coordinates, that you may learn about later, but they are not included in this text.) The calculations continue as follows:

$$Z = \frac{V_T}{I_T} = \frac{8 \angle 0° \text{ V}}{0.00464 \angle 38.3° \text{ A}}$$
$$= \mathbf{1724 \angle -38.3° \ \Omega}$$

Frequency Response

The best way to consider the circuit's response to different frequencies is to change the frequency from 2500 hertz to, say, 1070 hertz and repeat all calculations. If no other circuit values are changed, we will see what effect the frequency change has. The calculations at 1070 hertz, then, are as follows:

$$X_L = 2\pi f L$$
$$= 6.28 \times 1.07 \times 10^3 \text{ Hz} \times 150$$
$$\times 10^{-3} \text{ H}$$
$$= \mathbf{1008 \ \Omega}$$

$$X_C = \frac{1}{2\pi f C}$$
$$= \frac{1}{6.28 \times 1.07 \times 10^3 \text{ Hz} \times 0.05 \times 10^{-6} \text{ F}}$$
$$= \frac{1 \times 10^3}{0.336} \ \Omega = \mathbf{2976 \ \Omega}$$

$$I_L = \frac{V_T}{X_L} = \frac{8 \angle 0° \text{ V}}{1008 \angle 90° \ \Omega}$$
$$= \mathbf{0.00794 \angle -90° \text{ A}}$$

$$I_C = \frac{V_T}{X_C} = \frac{8 \angle 0° \text{ V}}{2976 \angle -90° \ \Omega}$$
$$= \mathbf{0.00269 \angle 90° \text{ A}}$$

$$I_R = \frac{V_T}{R} = \frac{8 \angle 0° \text{ V}}{2200 \angle 0° \ \Omega}$$
$$= \mathbf{0.00364 \angle 0° \text{ A}}$$

$$I_{XT} = I_L - I_C$$
$$= 0.00794 \angle -90° \text{ A}$$
$$- 0.00269 \angle 90° \text{ A}$$
$$= \mathbf{0.00525 \angle -90° \text{ A}}$$

Figure 27–2 shows the capacitor current at $+90°$, the inductor current at $-90°$, and the vector difference also at $-90°$, since the inductor current is larger. Remember that the larger current dominates in a parallel circuit. The total current is the vector sum of the resistor current at 0° and the net inductor current. Total current and phase are determined as follows:

$$I_T = \sqrt{I_R^2 + I_{XT}^2} \text{ A}$$
$$= \sqrt{0.00364^2 + 0.00525^2} \text{ A}$$
$$= \sqrt{0.0000132 + 0.0000276} \text{ A}$$
$$= \sqrt{0.00004081} \text{ A}$$
$$= \mathbf{0.00639 \text{ A}}$$
$$\text{tangent } \theta = \frac{O}{A} = \frac{I_{XT}}{I_R} = \frac{0.00525 \text{ A}}{0.00364 \text{ A}}$$
$$= 1.442$$
$$\theta = \mathbf{-55.3°}$$
$$\text{sine } \theta = \frac{O}{H} = \frac{I_{XT}}{I_T}$$
$$I_T = \frac{I_{XT}}{\text{sine } \theta} = \frac{0.00525 \text{ A}}{0.8221}$$
$$= \mathbf{0.00639 \text{ A}}$$

Therefore, $I_T = 0.00639 \angle -55.3°$ ampere.

In response to the frequency change, capacitive reactance has become larger and its current smaller. Inductive reactance has become smaller and its current larger. Thus, the circuit is now inductive. This result is also evident from the fact that the current is lagging the voltage.

Another change occurs in the circuit impedance, as the following calculation shows:

$$Z = \frac{V_T}{I_T} = \frac{8 \angle 0° \text{ V}}{0.00639 \angle -55.3° \text{ A}}$$
$$= \mathbf{1252 \angle 55.3° \ \Omega}$$

More will be said about the effects of frequency change later in this chapter.

Power

The first step in calculating the value of the true power for this circuit is to determine the

FIGURE 27-2
VECTOR SUM OF CURRENTS AT THE
SECOND FREQUENCY

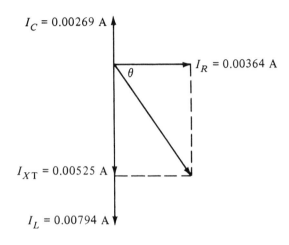

$I_C = 0.00269$ A

$I_R = 0.00364$ A

θ

$I_{XT} = 0.00525$ A

$I_L = 0.00794$ A

power factor. As before, it is equal to the cosine of the phase angle. So the power factor is

$$\text{power factor} = \text{cosine } 55.3° = \mathbf{0.5693}$$

Thus, 56.93% of the energy supplied to the circuit is dissipated in some form, such as heat.

The true power can be determined two ways, as shown next:

$$P = VI \text{ cosine } \theta$$
$$= 8 \text{ V} \times 0.00639 \text{ A} \times 0.5693$$
$$= \mathbf{0.0291 \text{ W}}$$
$$P = I^2R$$
$$= 0.00364 \text{ A} \times 0.00364 \text{ A} \times 2200 \ \Omega$$
$$= \mathbf{0.0291 \text{ W}}$$

The total current is not used in the second formula because that current does not pass through the resistor. Only the resistor branch current is used. However, total current is used in the formula $P = VI$ cosine θ.

RESONANCE

Parallel and series circuits do not behave the same way at and around resonance. However, the method for determining their resonant frequency is the same.

Resonant Frequency

Our next step in evaluating the circuit in Figure 27–1A is the calculation of resonant frequency, as follows:

$$f_r = \frac{1}{2\pi \sqrt{LC}}$$

$$= \frac{1}{6.28 \sqrt{0.150 \text{ H} \times 0.05 \times 10^{-6} \text{ F}}}$$

$$= \frac{1}{6.28 \sqrt{0.0075 \times 10^{-6}}} \text{ Hz}$$

$$= \frac{1}{6.28 \times 0.0866 \times 10^{-3}} \text{ Hz}$$

$$= 1.839 \times 10^3 \text{ Hz} = \textbf{1839 Hz}$$

A parallel circuit at resonance has the following characteristics:

— A phase shift of 0°;
— An impedance equal to resistance;
— A resistor current that equals the total current;
— Equal and opposite currents through the inductive and capacitive branches;
— Maximum impedance and minimum current.

Effects of Resonance

The best way to consider the effects of resonance on a parallel circuit is to calculate all values and construct a current graph. The reactance calculations, which are done first, are

$$X_L = 2\pi fL$$

$$= 6.28 \times 1839 \text{ Hz} \times 150 \times 10^{-3} \text{ H}$$

$$= \textbf{1732 } \Omega$$

$$X_C = \frac{1}{2\pi fC}$$

$$= \frac{1}{6.28 \times 1839 \text{ Hz} \times 0.05 \times 10^{-6} \text{ F}}$$

$$= \frac{1 \times 10^3}{0.5774} \Omega = \textbf{1732 } \Omega$$

As the calculations show, one characteristic of resonance exists: Inductive reactance equals capacitive reactance.

We can now calculate current. Since both reactances are equal, both currents will be equal. However, the phase angles will be in opposition. The calculations follow:

$$I_L = \frac{V_T}{X_L} = \frac{8 \angle 0° \text{ V}}{1732 \angle 90° \Omega}$$

$$= \textbf{0.00462} \angle -\textbf{90° A}$$

$$I_C = \frac{V_T}{X_C} = \frac{8 \angle 0° \text{ V}}{1732 \angle -90° \Omega}$$

$$= \textbf{0.00462} \angle \textbf{90° A}$$

$$I_{XT} = I_L - I_C$$

$$= 0.00462 \angle -90° \text{ A}$$
$$- 0.00462 \angle 90° \text{ A}$$

$$= \textbf{0 A}$$

$$I_R = \frac{V_T}{R} = \frac{8 \angle 0° \text{ V}}{2200 \angle 0° \Omega}$$

$$= \textbf{0.00364} \angle \textbf{0° A}$$

$$I_T = \sqrt{I_R^2 + I_{XT}^2} \text{ A}$$

$$= \sqrt{0.00364^2 + 0^2} \text{ A}$$

$$= \sqrt{0.00001325} \text{ A} = \textbf{0.00364 A}$$

As shown in Figure 27–3, the currents of the two reactors are equal and opposite, so their vector sum is zero. The net total current in the circuit is the current in the resistor. That does not mean that the reactors have no current. What it does mean is that current flows through one reactor in one direction but flows through the other reactor in the opposite direction. In other words, at resonance, energy transfers back and forth between the two reactors, with any energy loss replaced by the voltage source.

Another characteristic of parallel resonance is present: The total current equals the resistor current. Since this circuit is a parallel circuit, you may be mislead by this result. That is, the total current in a parallel circuit equals the sum of the separate branch currents. But since the total current here is equal to the resistor branch current, what happens to inductive and capacitive branch currents? Something quite interesting is going on.

We begin the explanation by recalling that, in a parallel circuit, the voltage is the same across all components. The resistor current is in phase with the applied voltage, so they rise and fall together. Current flowing through the resistor comes from the source and alternates in direction as the source voltage polarity alternates.

The current through the inductor lags the applied voltage by 90°, while the capacitor current leads it by 90°. That result means that the inductor and capacitor branch currents are 180° out of phase with each other. In other words, when one has a great supply of electrons, the other has a demand, and vice versa. The net result is that, once resonance occurs, these two reactors transfer electrons back and forth, as indicated in Figure 27–4. This flow is 90° out of phase with the resistor current, which is in phase with the applied voltage. The only energy required from the source is the energy needed to replace resistive losses in a branch.

Several conditions must exist before this oscillation, or resonance, can occur. The reactances must be equal for the timing of their fields to be the same. The frequency of the ap-

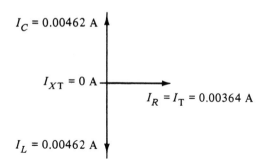

FIGURE 27-3
CIRCUIT CURRENTS AT RESONANCE

$I_C = 0.00462$ A

$I_{XT} = 0$ A

$I_R = I_T = 0.00364$ A

$I_L = 0.00462$ A

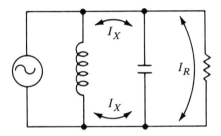

FIGURE 27-4
CURRENT PATHS AT RESONANCE

plied source must be the same as the resonant frequency for oscillation to occur. Otherwise, it will hinder rather than support the effect. Consider, again, the analogy of a person on a swing. Rope length and body weight determine resonant frequency. If the person providing an occasional push does so at the correct frequency and phase, little energy is required to keep the swing moving. A similar situation exists in a parallel *RLC* circuit at resonance. Since little energy is taken from the source at resonance, the circuit has a high impedance.

While the source current at resonance may be at its lowest level, large currents can exist between reactances. This situation can be compared with the push you would give a

FIGURE 27-5
RELATIONSHIPS IN A PARALLEL *RLC*
CIRCUIT

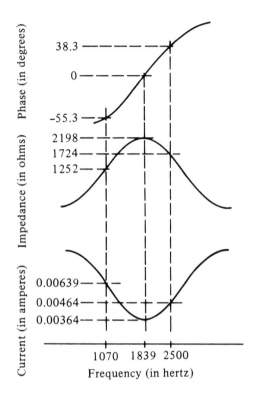

resonant swing to keep it going and the push it would give you if you got in the way at the wrong time. Remember what happened in the series resonant circuit, also: Voltages far in excess of the applied voltage were developed across the reactances at resonance.

Our next step is to calculate impedance. Since little energy is taken from the source at resonance, we expect impedance to be high. In fact, the impedance of a parallel *RLC* circuit is at its highest level during resonance. Remember, also, that the phase angle at resonance is 0°. The impedance calculations are

$$Z = \frac{V_T}{I_T} = \frac{8 \angle 0° \text{ V}}{0.00364 \angle 0° \text{ A}} = \mathbf{2198 \angle 0° \ \Omega}$$

We have evaluated this circuit at frequencies above, below, and at resonance. The frequency characteristics of this circuit are summarized in the graph in Figure 27–5.

A comparison of Figure 27–5 with Figure 26–11 gives a good idea of the differences between series and parallel *RLC* circuits. Basically, their impedance, current, and phase react in opposite ways as frequency varies. An understanding of these figures is essential for progress in electronics.

FIGURE 27-6
COIL RESISTANCE

Quality Factor and Bandwidth

As with series circuits, the quality factor *Q* is an indication of the relationship between reactance and resistance in an *RLC* parallel circuit. Practically speaking, *Q* for a parallel circuit is determined by the reactance of the inductor and its internal resistance. For the circuit in Figure 27–6, R_L represents the 100-ohm internal resistance of the inductor, while X_L is its 2400-ohm reactance. Reactances X_L and X_C are equal since the circuit is resonant. Then, *Q* is calculated as follows:

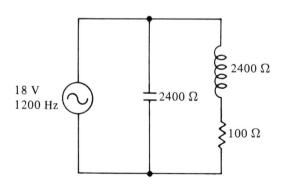

$$Q = \frac{X_L}{R_L} = \frac{2400 \ \Omega}{100 \ \Omega} = 24$$

Bandwidth *BW* is inversely related to *Q* and gives an indication of the frequency response of a circuit. High-*Q* circuits have a narrow bandwidth; low-*Q* circuits have a wide bandwidth. The bandwidth for this circuit is

$$BW = \frac{f_r}{Q} = \frac{1200 \ \text{Hz}}{24} = 50 \ \text{Hz}$$

The half-power points of a circuit are the two frequencies where the voltage drops to 0.707 times the peak, or resonant frequency, value. Since voltage causes current in a circuit, the current also drops to 0.707 of its peak value. As we know, power equals voltage times current. If current and voltage both drop by 0.707, then the power, or 0.707 times 0.707, is now 0.5 of what it was at resonance. In addition, the phase shift is normally 45° at the half-power points.

The half-power points for this circuit are

$$1200 \ \text{Hz} - \frac{50}{2} \ \text{Hz} = 1175 \ \text{Hz}$$

$$1200 \ \text{Hz} + \frac{50}{2} \ \text{Hz} = 1225 \ \text{Hz}$$

If a 100-ohm resistor R_S is placed in series with the 100 ohms of internal resistance that this inductor has, *Q* and *BW* change as follows:

$$Q = \frac{X_L}{R_L + R_S} = \frac{2400 \ \Omega}{200 \ \Omega} = 12$$

$$BW = \frac{f_r}{Q} - \frac{1200 \ \text{Hz}}{12} = 100 \ \text{Hz}$$

The effects of this change can be seen in Figure 27–7. A lower *Q* results in a broader bandwidth. Values of *Q* can range from 20 to 100 in a typical *RLC* circuit.

FIGURE 27-7
CIRCUIT *Q* AND BANDWIDTH

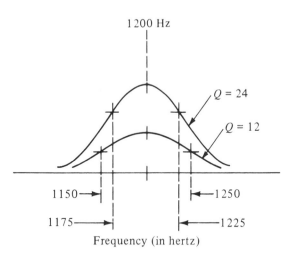

Frequency (in hertz)

Uses for Parallel Resonant Circuits

The most common uses for parallel resonant circuits are in radio and television equipment. Their ability to discriminate among different frequencies makes them useful in the signal selection and rejection process. Inductors and capacitors can be placed in parallel to form a network that allows most frequencies to pass except for those that are close to the resonant frequency. This connection is much like the series connection described in the previous chapter. The difference is that series circuits provide their least opposition at resonance, but parallel circuits provide their greatest opposition at resonance.

A radio schematic has many circuits that look like the one in Figure 27–8. It shows a transformer that is tuned to resonate at a particular frequency. A complex signal of many frequencies is applied to the transformer primary. The *LC* circuit on the primary side resonates only at the desired frequency, with an amplitude proportional to the size of the

FIGURE 27-8
ADJUSTABLE TRANSFORMER IN A RADIO CIRCUIT

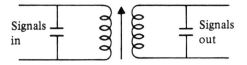

FIGURE 27-9
SERIES *RLC* CIRCUIT FOR EVALUATION

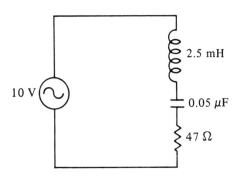

desired portion of the incoming signal. Transformer action causes this signal to be reproduced at the secondary side of the transformer. The desired signal is then developed across the high impedance of the secondary parallel resonant circuit. Undesired signals at frequencies beyond the half-power points of the bandwidth are further suppressed. This selection-discrimination process allows a radio or television circuit to select the desired station and block all others.

EVALUATING *RLC* CIRCUITS

In this section, we evaluate two *RLC* circuits. Keep the following point in mind.

Key point: Three parameters are of interest in tests of *RLC* circuits: resonant frequency, impedance, and bandwidth.

Two experiments to determine these values are outlined in this section. Before we begin, however, we emphasize one point: The preceding sections involve ideal components; experiments, however, are conducted with actual components. Thus, your measured results may be somewhat different from the calculated values.

Series Connection

Real inductors have internal series resistance because of their windings. Real capacitors have internal parallel resistance because of their leakage. Capacitor leads have inductance, and inductor leads have capacitance. The point is, therefore, that the three components in an actual series *RLC* circuit do not each have only one form of opposition. Rather, each component has complex oppositions. This feature will produce results that are not what you might expect.

Figure 27–9 shows the series circuit you will evaluate. Your objective is to determine the resonant frequency, impedance, and bandwidth. While you may not have the exact values indicated in the figure, you should try to come close to them. If you did use these exact values, and if your components were ideal, you would expect to get the following results:

$$f_r = \frac{1}{2\pi\sqrt{LC}}$$

$$= \frac{1}{6.28\sqrt{2.5 \times 10^{-3}\ H \times 0.05 \times 10^{-6}\ F}}$$

$$= \frac{1}{6.28\sqrt{0.0125 \times 10^{-8}}}\ Hz$$

$$= \frac{1}{6.28 \times 0.1118 \times 10^{-4}}\ Hz$$

$$= \textbf{14,243 Hz}$$

$$I_\text{T} = \frac{V_\text{T}}{Z} = \frac{V_\text{T}}{R} = \frac{10\ \text{V}}{47\ \Omega} = \textbf{0.2128 A}$$

$$Q = \frac{X_L}{R}$$

$$= \frac{6.28 \times 14.243 \times 10^3\ \text{Hz} \times 2.5 \times 10^{-3}\ \text{H}}{47\ \Omega}$$

$$= \frac{223.6}{47} = \textbf{4.757}$$

$$BW = \frac{f_\text{r}}{Q} = \frac{14{,}243\ \text{Hz}}{4.757} = \textbf{2944 Hz}$$

The half-power points indicating the bandwidth limits are

$$14{,}243\ \text{Hz} - \frac{2944}{2}\ \text{Hz} = \textbf{12{,}771 Hz}$$

$$14{,}243\ \text{Hz} + \frac{2944}{2}\ \text{Hz} = \textbf{15{,}715 Hz}$$

These results are expected results, but it is unlikely that you will get them since you are using real components. For instance, the resonant frequency may be different because of the tolerance of your reactances. Try to determine their tolerances so that you can predict how much different the actual resonant frequency will be. Remember that tolerance errors must be added.

Bandwidth and impedance are affected by resistance, and the resistor is not the only source of resistance. Inductor winding resistance will be a significant factor. Try to determine what it is so that you can predict its effect. Finally, make sure that you do not exceed the current capacity of the inductor. Excess current will saturate an inductor and cause incorrect results.

Now that you have done the calculations, you should proceed with the experiment. The best way to determine resonance is by measuring the resistor voltage. Just vary the frequency until the resistor voltage reaches maximum. This technique determines resonance because, at resonance, series circuits have their lowest impedance, highest current, and highest resistor voltage. This voltage relationship can be expressed by the following formula:

$$V_R = \text{maximum at } f_\text{r}$$

Stated another way, resistor voltage equals its maximum at the resonant frequency.

Since your basic data will be resistor voltages and frequencies, it is a good idea to make a table. You will need data at resonance, at the half-power points, and at about ten other places each side of resonance in order to construct a graph. Remember that the best way to describe an *RLC* circuit is with a graph of frequency versus current, resistor voltage, or impedance.

Impedance can be determined by dividing the measured value of total voltage by the measured value of total current. Once again, if you know the actual value of the resistor, you can use resistor voltage to calculate circuit current. Use the following equation to calculate impedance:

$$Z = \frac{V_\text{T}}{I_\text{T}} = \frac{V_\text{T}}{V_R/R}$$

Bandwidth can also be determined by measuring the resistor voltage. Recall that the bandwidth limits are reached when the circuit current drops to 0.707 of its maximum value, which, of course, occurs at resonance. Since the resistor voltage is determined by circuit current, you can use resistor voltage at resonance as a reference. Increase the frequency until the resistor voltage drops to 0.707 times its resonant value. At that point, you are at the

FIGURE 27-10
PARALLEL *RLC* CIRCUIT FOR EVALUATION

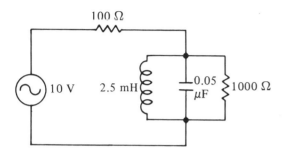

upper limit. Lower the frequency through resonance to find the lower limit. Monitor the generator output voltage to ensure that it remains constant. The varying impedance of the *RLC* circuit places a changing load across the generator. And the output voltage of some generators changes under changing loads. Remember, again, that you are looking for the bandwidth limits that occur at the points where the resistor voltage drops to 0.707 times its maximum value. The following equation will serve as a reminder:

$$BW \text{ limits at } V_R = 0.707 V_{R \text{ max}}$$

Once you have completed your measurements, compare your measured values with the calculated values. Then, answer these two questions: Were your measured and calculated results similar? Can you explain the cause of the differences beyond normal tolerance limits?

Parallel Connection

Once again, you will evaluate an *RLC* circuit to determine resonant frequency, impedance, and bandwidth. You can use the same inductor and capacitor that you used before, but two changes are needed for the parallel connec-

tion. The resistance branch should be larger, and a series resistance should be added to limit inductor current. The parallel circuit you will evaluate is shown in Figure 27–10.

The resonant frequency of this circuit should be the same as the resonant frequency of the series circuit. Since we have not yet studied series-parallel circuits, the overall impedance will be difficult to calculate. However, the impedance of the parallel section of this circuit, if it were ideal, would be equal to the resistive branch of 1000 ohms. The resistance of the inductor will upset that value as well as the circuit *Q*. Therefore, just measure these values.

One way to determine resonance is to measure the parallel circuit voltage. The parallel circuit voltage is not the generator voltage but the voltage across the parallel network. Since impedance in a parallel *RLC* circuit is maximum at resonance, the voltage there will also be maximum. Use an oscilloscope to measure the voltage, and change the frequency until the voltage is maximum. Record your two numbers in a table, and then change the frequency until you find the upper and lower half-power points. The network voltage will be 0.707 times the resonant value at those points.

Now that you know the resonant frequency and bandwidth, you can make measurements for determining impedance. To do so, measure the current at resonance, or measure the voltage across the series resistor and then calculate the current value. Data for a graph of frequency versus current or impedance can be obtained by measuring the series resistor voltage at ten points above and ten points below resonance. You can then calculate current or impedance.

Once you have all your data, consider these questions: Was the curve shaped the same above and below resonance? If not, why?

SUMMARY

The easiest approach to calculating parallel AC circuit parameters is the total current method. In this method, reactance, current, and then impedance are determined. The current in each branch of a parallel circuit is equal to the voltage divided by the opposition of the branch. Resistor branch current is in phase with the voltage, capacitor branch current leads the voltage by 90°, and inductor branch current lags by 90°. The total current of the circuit is equal to the vector sum of all branch currents.

The impedance of a parallel *RLC* circuit is equal to the voltage divided by the total current. When the inductor has the lowest reactance, the largest current passes through it. This result produces a total current that is lagging, which, in turn, produces a leading impedance. When the capacitor has the lowest reactance, the total current is leading, and the impedance lags.

The capacitive reactance decreases and the inductive reactance increases when the applied frequency increases. At one point, both reactances are equal, and resonance occurs. The main characteristics of a parallel resonant circuit are minimum current, maximum impedance, and 0° phase shift. A change in the frequency to one above or below resonance will cause an increase in current, a decrease in impedance, and a change in the phase angle. The frequencies above and below resonance where the impedance drops to 0.707 of its resonant frequency value are called the half-power points. The range between them is called the bandwidth. The phase at the half-power points is 45°.

The quality factor *Q* of an *RLC* circuit is an indication of the relationship between the reactance and resistance of the circuit. Values for *Q* range from 20 to 100 in a typical *RLC* circuit, with a high-*Q* circuit having a steep curve and a narrow bandwidth. As the series resistance decreases, *Q* increases, producing a circuit that is more selective. Circuits such as these find application in radio and television systems, where the product must pass a narrow range of frequencies and reject all others.

The performance of an *RLC* circuit can be evaluated by taking several measurements. The three parameters of interest are impedance, resonant frequency, and bandwidth. Voltage is measured across a series resistance, and its value as a function of frequency is noted. Resonance is determined by observing the frequency of maximum current (indicated by resistor voltage) for series circuits and of minimum current for parallel circuits. Half-power points are noted for the determination of bandwidth.

REVIEW QUESTIONS

1. How are L and C in a resonant circuit like the kinetic and potential energy of a mechanical system?

2. Explain why the smallest reactance dominates in a parallel RLC circuit.

3. What is the difference between the formula used to determine resonant frequency in a parallel circuit and the formula used for a series circuit?

4. Describe the characteristics of a parallel RLC circuit at resonance.

5. Compare the characteristics of a parallel RLC circuit at resonance with those of a series RLC circuit at resonance.

6. At what frequency is impedance maximum in a parallel RLC circuit?

7. At what frequency is current maximum in a parallel RLC circuit?

8. Compare the characteristics of a parallel RLC circuit above and below resonance with those of a series RLC circuit above and below resonance.

9. Explain why the upper and lower bandwidth frequencies are called the half-power points.

10. What is the phase angle at the upper cutoff frequency? At the lower cutoff frequency?

11. What is the relationship between coil resistance and Q in a parallel RLC circuit?

12. What is the relationship between coil resistance and bandwidth in a parallel RLC circuit?

13. At what frequency is power maximum in a parallel RLC circuit?

14. Describe an application of a parallel RLC circuit.

15. Explain how a parallel RLC circuit can be used to filter out undesired frequencies.

REVIEW PROBLEMS

1. Determine the resonant frequency of the circuit shown in Figure 27–11 if $V_T = 24$ volts, $L = 150$ millihenrys, $R_L = 6.8$ kilohms, $C = 47$ picofarads, and $R = 0.15$ megohm.

2. Determine the resonant frequency of the circuit shown in Figure 27–11 if $V_T = 48$ volts, $L = 1.2$ henrys, $R_L = 470$ ohms, $C = 0.01$ microfarad, and $R = 0.22$ megohm.

3. How will changing C to 100 picofarads in Problem 1 affect the operation of the circuit?

4. How will changing L to 0.2 henry in Problem 1 affect the operation of the circuit?

5. What effect will changing C to 0.005 microfarad have on the circuit in Problem 2?

6. Calculate the impedance, currents, and phase at resonance for the circuit described in Problem 1.

7. Calculate the impedance, currents, and phase at resonance for the circuit described in Problem 2.

8. Calculate Q and *BW* for the circuit described in Problem 1.

9. Calculate Q and *BW* for the circuit described in Problem 2.

10. Calculate the currents, impedance, and phase at the upper and lower bandwidth limits for the circuit described in Problem 1.

11. Calculate the currents, impedance, and phase at the upper and lower bandwidth limits for the circuit described in Problem 2.

12. Calculate the power factor and true power of the circuit described in Problem 10.

13. Calculate the power factor and true power of the circuit described in Problem 11.

14. Consider the circuit in Figure 27–11, with V_T = 14 volts, L = 3.5 millihenrys, R_L = 47 ohms, C = 0.047 microfarad, and R = 470 kilohms. Perform all calculations necessary in order to draw a graph showing impedance, current, and phase of this circuit as a function of frequency. Plot at least five points for each curve.

FIGURE 27-11

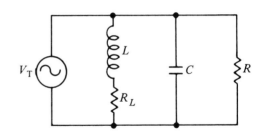

15. Consider the circuit in Figure 27–11, with V_T = 26 volts, L = 2.5 millihenrys, R_L = 100 ohms, C = 0.003 microfarad, and R = 2.2 megohms. Perform all calculations necessary in order to draw a graph showing impedance, current, and phase of this circuit as a function of frequency. Plot at least five points for each curve.

16. Write a summary paper describing the characteristics of series and parallel *RLC* circuits. Compare their changes in impedance, current, and phase as frequency changes.

SEMICONDUCTOR DIODES

OBJECTIVES

Define semiconductor, p type, n type, pn junction, and doping.

Describe forward and reverse biasing.

Identify semiconductor diodes by their appearance and symbols.

State the rules for the care and handling of semiconductor diodes.

Determine the condition of a semiconductor diode by measuring the front-to-back ratio with a multimeter.

Determine the anode and cathode of a semiconductor diode by using a multimeter.

DIODE THEORY

The objective of this chapter is to introduce the basic theory and operation of semiconductor diodes. Up to this point, your studies have been about electric circuits designed around *passive* components such as resistors, inductors, and capacitors. Now, you are about to begin your study of electronics and the use of *active* components such as transistors and integrated circuits.

Most active components today are solid-state components. They come in small sealed packages that are very durable, and they have no moving parts. Inside the package is a semiconductor material that has been chemically modified so that it will act in a particular manner. This chapter introduces the characteristics of semiconductor elements and some components made from them.

Semiconductors

The resistivity scale in Figure 28–1 shows the current-carrying characteristics of a wide range of materials. At one end are the good conductors, such as gold, silver, and copper. At the other end are the poor conductors, such as rubber, glass, and ceramic. Midway between these two groups are the two basic elements used for making semiconductors, silicon and germanium. In their pure state, these elements are neither good conductors nor good insulators. Thus, **semiconductors** are defined as materials with a resistivity midway between that of conductors and that of insulators. They will conduct under some conditions and will not conduct under others. The name *semiconductor* means half-time conductor. For a semiconductor to behave this way, it must first be chemically modified. An overview of this process is given next.

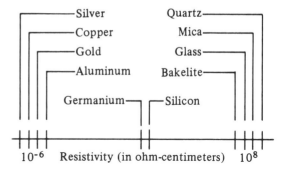

FIGURE 28-1
RESISTIVITY OF VARIOUS MATERIALS

Silver
Copper
Gold
Aluminum
Germanium
Quartz
Mica
Glass
Bakelite
Silicon

10^{-6} Resistivity (in ohm-centimeters) 10^8

Doping

Like all atoms, a normal silicon atom has an equal number of protons and electrons. This equality of positive and negative charges means that the atom is balanced. It is not charged. A large piece of pure silicon is likewise uncharged. In pure silicon (abbreviated Si), a crystal-like structure develops, with the four electrons in the outer orbit of one atom forming a bond with electrons from other atoms, as pictured in Figure 28–2.

Through a process called **doping,** a small number of atoms of another element are added to part of the silicon crystal, as shown in Figure 28–3. These added atoms, less than one for each one million silicon atoms, are called impurities since, because of them, the crystal is no longer pure silicon. One impurity often used, the element phosphorus (P), has five electrons in its outer orbit compared with the four that silicon has. It has five protons also, so it is electrically balanced. However, the fifth electron does not fit in the shared bond structure that exists between the other atoms. Also, it is not tightly bound to its own atom and can easily be freed. Because of the available electron, this section of the crystal is

FIGURE 28-2
LATTICE STRUCTURE OF A SILICON
CRYSTAL

FIGURE 28-2
LATTICE STRUCTURE OF A SILICON
CRYSTAL

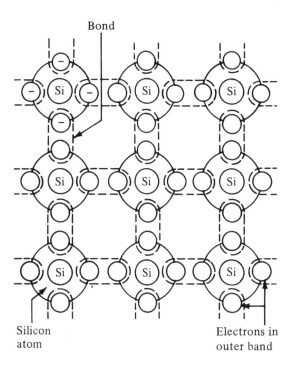

Silicon
atom

Electrons in
outer band

FIGURE 28-3
SILICON CRYSTAL WITH A DONOR ATOM

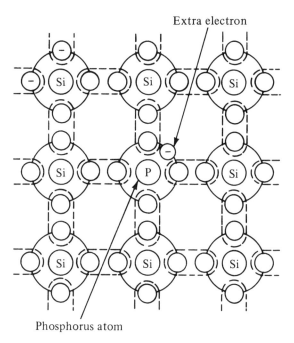

Phosphorus atom

called **n type** (the n stands for "negative"). The phosphorus is called a donor atom because it brought this electron to the silicon crystal.

Another section of the same semiconductor is doped with a different element, such as boron (B). This element has three outer orbit electrons. It provides a space called a hole that electrons can move into, as shown in Figure 28–4. The boron atom is called an acceptor atom, since it will accept electrons, and this section of the semiconductor is called **p type** (p means "positive").

It should be noted that many procedures and elements can be used for doping. The basic point is that the semiconductor elements germanium and silicon are modified so that they have a p type section with available holes and an n type section with available electrons. These two sections are not two pieces that have been attached. They are included within the one solid semiconductor chip.

Leads are attached to the two ends of the crystal so that voltage can be applied. The result is called a semiconductor diode, shown in Figure 28–5. A **diode** is an active electronic component with two electrodes.

pn Junctions

The border between the p type and n type sections of a semiconductor diode is called the **pn junction,** which is shown in Figure 28–6. On one side of the junction are the silicon and acceptor atoms. On the other side are more silicon atoms with the donor atoms. Although

FIGURE 28-4
SILICON CRYSTAL WITH AN ACCEPTOR ATOM

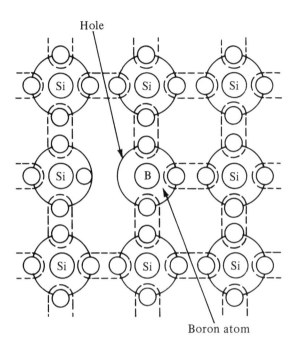

FIGURE 28-5
SEMICONDUCTOR DIODE

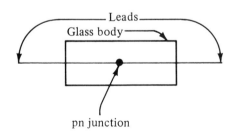

FIGURE 28-6
pn JUNCTION

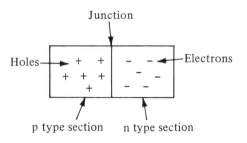

some electrons at the boundary do cross over into the nearby holes, the number doing so is very small. There is essentially no flow from one side to the other. An external force is required to make any noticeable flow occur.

Biasing

The word *bias* is usually used to indicate a leaning in a particular direction that already exists before a person considers a situation. In electronics, **bias** is a fixed DC voltage that is applied to a circuit before a signal is added. The process of adding the DC bias voltage and AC signal voltage will be described in Chapter 29. At this point, we will be concerned only with biasing — that is, the process of applying bias.

There are two ways in which bias can be applied. The first method is called reverse biasing. **Reverse bias** occurs when negative is connected to the p type section and positive is connected to the n type section, as shown in Figure 28–7A. In reverse biasing, free electrons in the n type are drawn out toward the positive terminal of the source. A very small and brief flow occurs, but there is essentially no current flow because the removed electrons are not replaced. Some electrons from the negative source also drift into holes in the p type, but, again, the flow is insignificant. The most significant result is that the barrier at the junction becomes wider since the holes and electrons have been drawn apart.

The second method of bias connection is called foward biasing. **Forward bias** occurs when positive is connected to the p type section and negative is connected to the n type

FIGURE 28-7
REVERSE-BIASED AND FORWARD-BIASED pn JUNCTIONS

Holes and electrons separate

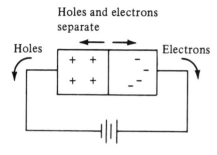

A. Reverse-biased junction

Holes and electrons combine

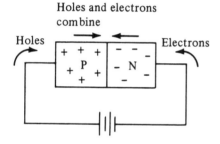

B. Forward-biased junction

FIGURE 28-8
MOTION OF HOLES

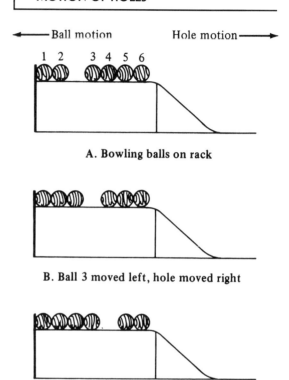

A. Bowling balls on rack

B. Ball 3 moved left, hole moved right

C. Ball 4 moved left, hole moved right

section, as shown in Figure 28–7B. This connection causes current to flow in the semiconductor. The electrons in the n type area are pushed across the junction by the negative source voltage. Once across, they move through holes toward the positive source. As they move, their replacements follow from the negative source. The approximate voltage necessary to overcome the junction barrier is about 0.2 volt for germanium and 0.6 volt for silicon.

A pn junction must be forward-biased for current to flow through it. Reverse biasing stops the current flow. Throughout this text, we have considered current as moving electrons. Therefore, current flows through a diode from the n type section to the p type,

just the way the electrons flow. But recall that, earlier in this text, we mentioned that many people prefer to speak of conventional current as flowing from positive to negative. In that case, we just speak of holes moving rather than electrons.

Imagine some bowling balls on a rack, as pictured in Figure 28–8A. Suppose ball 3 moves to the left, as in Figure 28–8B. The hole moves to the right. Moving ball 4 to the left, as in Figure 28–8C, moves the hole even further to the right. Thus, when the balls move right to left, the holes move left to right.

You can consider electrons and holes in the same way. As electrons move one way, the

FIGURE 28-9
FORWARD-BIASED AND REVERSED-BIASED
JUNCTIONS

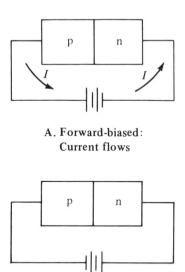

A. Forward-biased:
Current flows

B. Reverse-biased:
No current flow

FIGURE 28-10
DIODE CHARACTERISTIC CURVE

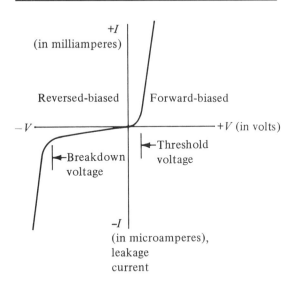

holes move another. Forward biasing pushes the positive holes and negative electrons together, causing them to combine at the junction. In this case, current flows, as shown in Figure 28–9A. Reverse biasing has the opposite effect. The positive source draws the few free electrons away from the junction, and the negative source draws the few holes. The junction gets wider as the carriers that could combine get farther apart. In this case, current does not flow, as indicated in Figure 28–9B.

Diode Curves

One curve can describe the operating characteristics of a semiconductor diode, as shown in Figure 28–10. It shows the current that oc-

curs for different junction voltages and polarities.

The upper right, or forward, region of the curve shows that a positive (forward) bias voltage will cause a positive (forward) current to flow, once the voltage is above the threshold, or barrier, level. Notice that this current is in milliamperes. The lower left, or reverse, region shows that a negative (reverse) bias voltage causes only a slight leakage current to occur. This current is in microamperes. If the reverse voltage is increased to the zener, or breakdown, level, current will flow in the reverse direction. Some diodes are designed to operate this way, but others will be ruined by this operation.

As we see, a diode acts like a switch. It can be turned on with forward bias and turned off with reverse bias. There is a difference, however, between a switch and a diode. A diode is not a perfect switch since it has some resistance when it is conducting and some leakage when it is not conducting. It is very

FIGURE 28-11
DIODES

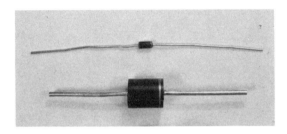

FIGURE 28-12
DIODE SYMBOLS

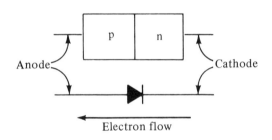

A. Electron flow in diode

B. Zener diode

fast, though, and it is electrically operated by its bias.

DIODE CHARACTERISTICS

When you work with diodes, you will not see a pn junction. Instead, you will see small components with two leads, or terminals, similar to those in Figure 28–11. The upper component is a small-signal diode and the lower one, a diode for moderate currents, is a large-current, or power, diode. In both cases, the cathode (described later) is on the right. Because you must be able to recognize diode types and their leads, we describe diode symbols, types, and specifications in this section.

Symbols

The two sections of a semiconductor diode are called the cathode and the anode. A **cathode** is made of n type material, and an **anode** is made of p type. Electrons flow from n type to p type, or from cathode to anode, as shown in Figure 28–12A. The diode symbol uses an arrow, which points in the direction of *conventional current,* or hole flow. Electrons flow in a direction opposite that of the arrow. The bar in the symbol represents the cathode, and the arrow represents the anode.

Diodes designed to operate beyond the zener, or breakdown, level are called **zener diodes.** Their symbol is slightly different, as shown in Figure 28–12B.

Types

Diodes can be made of germanium or silicon, as mentioned previously. Diodes can be described another way, too. The small ones are generally called signal, or switching, diodes. The large ones are called power diodes. Power diodes are designed to tolerate the hundreds of amperes that flow in high-power circuits. They are often made of threaded metal so that they can be attached to a heat sink, which helps drain off the heat.

Signal and switching diodes are often small glass components with colored stripes. Manufacturers usually indicate the cathode end with a stripe, spot, angular cut on the lead, or some other distinctive mark. These diodes can switch on or off in microseconds and are intended for currents in the milliampere range. Diodes designed to serve as voltage references operate at the reverse breakdown point and are called zener diodes, as noted previously.

Various methods are used for identifying diodes. For example, since diodes tend to be quite small, a number may be printed on it. Another common method of identification is a color code, as shown in Figure 28–13. Many diodes have a number that begins with 1N. We read the colors as we read them for resistors and attach the proper numbers after the 1N at the beginning. Thus, the diode at the top of Figure 28–13 is a 1N235 — that is, the usual 1N followed by the 235 specified by the colors. The lower diode is a 1N1725.

Signal, switching, and power diodes are used in many applications. Power supply rectifier circuits are their most common use; these circuits are described in Chapter 30. Diodes are also used to block signals from entering sections where they are not wanted. Another use of diodes is in circuits that produce complex waveshapes. Diode applications will be considered in detail in your next electronics course.

Another type of diode is the **light-emitting diode** (LED), which glows when it is forward-biased. The LED is often used for the digital display of numbers on products such as stereos, test instruments, and cash registers. Figure 28–14 shows how eight diode segments are used to produce a number between 0 and 9 as well as a decimal point. Typically, an integrated circuit precedes the digital display and sends forward-biasing voltages to the appropriate segments. These voltages cause the segments to glow red, green, or some other color according to their chemical makeup. Note that liquid crystal displays (LCDs), the black-on-silver displays used on watches and meters, are not diodes but, rather, are low-power, AC-operated devices.

FIGURE 28-13
DIODE IDENTIFICATION MARKINGS

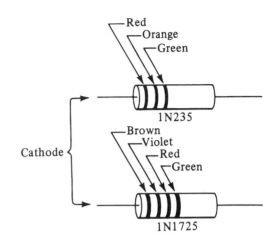

FIGURE 28-14
LED DIGITAL DISPLAY

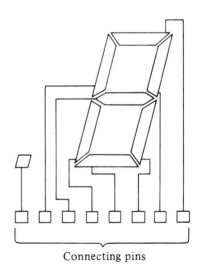

Connecting pins

Specifications

Just as resistors have ohm and watt ratings, and capacitors have microfarad and volt ratings, diodes also have ratings. The ratings of diodes can be described in a long list of specifications. There are, however, a few basic ones that are generally used to describe a diode. These basic specifications are as follows:

— Maximum forward current (I_F) is the highest current the diode can safely allow when forward-biased.

— Peak reverse voltage (PRV) is the highest reverse bias voltage the diode can tolerate before it breaks down.

— Maximum power dissipation $(P_{D\ max})$ is the highest forward current times forward voltage that the diode can tolerate (increasing current decreases the safety margin of power).

— Forward voltage (V_F) is the voltage dropped across the junction of a forward-biased diode.

— Reverse current (I_R) is the leakage current of a reverse-biased diode.

— Maximum operating temperature (T_a) is the highest temperature at which a diode can safely operate.

WORKING WITH DIODES

Your primary work with diodes will involve installing, removing, and checking them. These tasks include lead identification, soldering without causing damage, desoldering, and resistance checking. These are critical skills, but they are easy to learn and will become second nature to you with experience.

Installation and Removal

There are three questions you must ask yourself before installing a semiconductor diode. Do you have the specified diode or an equivalent? Are you installing it in the proper direction? Are you protecting it from damage by the heat of your soldering iron? We'll consider the first question first.

The diode you install must be the one specified or an equivalent. You can tell if it is the specified diode by comparing its identifying marks with information on the schematic or parts lists. There are, however, acceptable substitutes for all diodes. Semiconductor substitution booklets list diodes that can be used to replace others. These booklets indicate which diodes have specifications similar to those of the diode you wish to replace.

Diodes with different numbers are different, so before you use another diode, you should consider the differences. If the difference is in physical size and the replacement fits, then there is no problem. If the difference is in temperature rating, though, that difference may be very important. Generally, diodes indicated as equivalent have equal or better I_F, PRV, and $P_{D\ max}$ ratings, and those specifications are the ones you should consider.

The direction in which you install a diode is also very important. While current passes through a resistor in either direction, it does not do so in a diode. Cathode markings are given on diodes to ensure that you install them properly. A diode installed in the wrong direction will cause the circuit to malfunction and probably cause damage to the equipment. Remember that the striped end is the cathode.

Heat is another problem for diodes. They are temperature-sensitive, and their operating characteristics change considerably with temperature. As temperature rises, their internal resistance drops and they become more prone to damage. You can also damage a diode with excessive heat when you are soldering it. Keep in mind that the junction is a very small item inside the case. By using a heat sink and an iron under 35 watts, as indicated in Figure 28–15, you can be sure that enough heat is applied at the connection without an excessive amount reaching the junction. Heat that goes up the lead toward the diode case is absorbed by the heat sink.

The removal of a diode requires attention to two points. First, you should make a note of its direction in the circuit so that you will reinstall it or its replacement properly. The second concern is the use of a heat sink. You may be removing the diode for testing. There is no point in damaging it if it is good when you remove it. So use a heat sink. The use of a desoldering tool is also a good idea.

Checking Conditions

A wide range of tests can be performed on a diode. At this point, we are concerned only about two. Is the diode conducting when it should? Is it not conducting when it should not conduct? The assumption is that the diode was good once. If it has a defect now, then it is probably a short or an open. That is, it conducts both ways (a short), or it does not conduct at all (an open). In a later course, you will learn about leakage and other, more advanced tests.

The easiest way to test a diode is to use the range on a multimeter marked "Diode Test." If the instrument does not have such a range, measure resistance one way and then the other. That is, apply a positive voltage to the cathode, which makes the diode reverse-biased. The multimeter will show a high resistance reading, indicating little current flow. Switch the leads, and apply a negative voltage to the cathode. Current flows in this case, and the meter will show a low resistance reading.

FIGURE 28-15
HEAT SINK USED WHEN SOLDERING DIODES

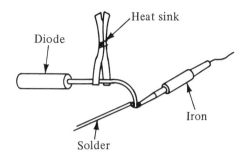

In other words, there is a high resistance in one direction and a low resistance in the other direction.

Because specific resistances vary, it is the ratio of one resistance to the other that is important. The ratio between the two resistances is called the **front-to-back ratio** (FB ratio). It should be more than 100 : 1 for power diodes and 1000 : 1 for signal diodes. The formula for the ratio is

$$\text{FB ratio} = \frac{\text{reverse resistance}}{\text{forward resistance}}$$

The $R \times 100$ range is suggested for this test. Using very high or very low resistance ranges may cause too much current to flow through the diode or not enough voltage to make the diode to conduct. Also, do not be concerned about the actual resistance of the diode, because you will get different answers. Differing answers are due to nonlinearity in the diode, a subject you will learn about in your next course.

Identifying Leads

There are times when the markings on a diode are worn and you are not sure which end is the

FIGURE 28-16
CHECKING A DIODE FOR LEAD
IDENTIFICATION

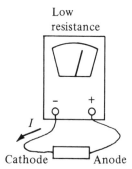

cathode. A simple test with a multimeter will help you determine the cathode end. Just place the meter on the $R \times 100$-ohm range and measure the resistance of the diode. It should be low one way and high the other way. Once you have determined the direction for the low resistance, connect the meter so that the resistance is at the lower of the two values. Since the resistance is at the lower value, the diode is conducting. If electrons are coming out of the negative terminal of the multimeter and entering the left side of the diode, that side must be n type and it must be the cathode, as shown in Figure 28–16.

SUMMARY

Semiconductors are elements that are neither excellent conductors nor excellent insulators. On a resistivity scale, they are found midway between copper and rubber. The semiconductor elements used in electronic circuits are germanium and silicon. Through a process called doping, elements such as boron and phosphorus are added to semiconductors. These impurities cause the semiconductors to conduct

under some circumstances and not conduct under others.

The doping of half of the semiconductor crystal produces some electrons that can be easily moved from one location. This section is called n type. The other half of the semiconductor is doped so that there are holes in the outer orbit of the atoms; here, electrons can move. This section is called p type. The boundary between these two sections of the single crystal is called a pn junction.

Applying a positive voltage to the p type section and a negative voltage to the n type section is called forward biasing. This procedure causes the electrons and holes to move toward the junction. Combinations of electrons and holes occur, allowing current to flow through the crystal. If the polarity is changed, reverse biasing occurs. Electrons and holes are drawn apart and the junction gets wider. Negligible current flows.

Diodes come in a variety of shapes and sizes, ranging from small signal and switching diodes to large power diodes. Most small diodes have a glass enclosure with an identification number or colored stripes. The stripes indicate the cathode end and diode number. Diodes using identification stripes have numbers that begin with 1N followed by numbers that correspond to the same color code used for resistors. Power diodes are large metal components that often have a threaded mounting stud. Because of the heat they develop and their sensitivity to it, they are usually mounted on a heat sink.

The basic diode symbol is used for all types of diodes. It consists of an arrow pointing to a solid bar. The bar represents the cathode, and the arrow represents the anode. Electrons flow in a direction opposite that of the arrow. Diode specifications include voltage, current, and power ratings. The specifications that are most common are maximum forward current (I_F), peak reverse voltage

(PRV), and maximum power dissipation ($P_{D \text{ max}}$).

When working with diodes, you should observe some basic rules. Be sure you have the correct diode or its equivalent, install it in the proper direction, and use a heat sink when installing or removing a diode.

You can perform a few basic tests on a diode with a multimeter. By using the $R \times$ 100-ohm range, you can measure the resistance in one direction and the resistance in the other. The larger number divided by the smaller number is called the front-to-back ratio. It should be greater than 100 : 1 for power diodes and 1000 : 1 for signal diodes.

A very high resistance in both directions indicates that a diode is not conducting when it should be. It is open. If the resistance is very low in both directions, it is conducting when it should not be. It is shorted.

One other test is possible with a multimeter: You can identify the leads. Connect a diode that checks out as good in the direction that indicates low resistance. Since current leaves the multimeter at the negative terminal and flows into the diode, the end connected to the negative terminal must be the cathode.

CHAPTER 28

REVIEW TERMS

anode: p type section of a diode

bias: fixed DC voltage applied to a circuit before a signal is added

cathode: n type section of a diode

diode: two-terminal semiconductor component that conducts in one direction but not in the other

doping: process of adding a small amount of impurities to a semiconductor material to alter its properties

forward bias: application of a positive voltage to the p type side of a diode and a negative voltage to the n type side, a situation that causes conduction

front-to-back (FB) ratio: ratio of reverse resistance to forward resistance in a diode

light-emitting diode (LED): semiconductor device that glows when it is forward-biased; used to form the segments of a number in a digital display

n type section: section of a semiconductor material that has been doped with donor atoms; section that has available electrons

pn junction: area in a semiconductor diode where the p type and n type sections meet

p type section: section of a semiconductor material that has been doped with acceptor atoms; section that has available holes (accepts electrons)

reverse bias: application of a negative voltage to the p type side of a diode and a positive voltage to the n type side, a situation that prevents conduction

semiconductor: material with a resistivity midway between that of conductors and that of insulators

zener diode: diode designed to operate beyond the zener, or breakdown, level

REVIEW QUESTIONS

1. Define semiconductor.

2. Define front-to-back ratio.

3. Define doping.

4. Define p type and n type sections of a semiconductor.

5. Define pn junction.

6. Describe the steps that must be taken when working with semiconductor diodes.

7. Describe the process called doping.

8. Describe how a diode responds when it is reverse-biased. Use a sketch.

9. Describe how a diode responds when it is foward-biased. Use a sketch.

10. Describe how to check the condition of a diode with a multimeter.

11. Describe how to identify diode leads with a multimeter.

12. Compare the resistivity of insulators, conductors, and semiconductors.

FIGURE 28-17

A.   B.

FIGURE 28-18

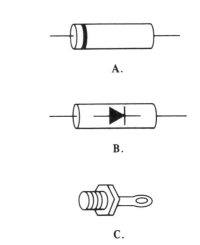

A.

B.

C.

REVIEW PROBLEMS

1. Identify the type of each diode shown in Figure 28–17.

2. Name the leads (cathode and anode) of the diodes shown in Figure 28–18. Indicate the direction of electron flow.

3. Draw a typical diode characteristic curve, and name the three sections.

FIGURE 28-19

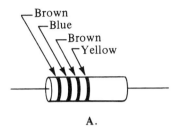

A.

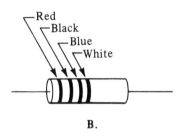

B.

FIGURE 28-20

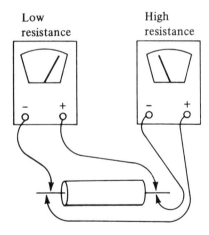

FIGURE 28-21

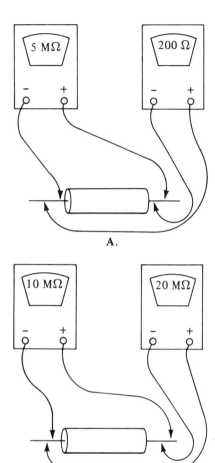

A.

B.

6. Indicate the condition of the diodes shown in Figure 28–21. Explain your answer.

4. What are the identification numbers of the diodes shown in Figure 28–19?

5. Name the leads of the diode shown in Figure 28–20. Explain your answer.

TRANSISTORS AND INTEGRATED CIRCUITS

OBJECTIVES

Identify transistors and integrated circuits by their appearance and symbols.

State the rules for the care and handling of transistors and integrated circuits.

Determine the condition of a transistor by testing it with a multimeter.

Identify the emitter, base, and collector of a transistor by using symbols or a multimeter.

Describe the characteristics of a transistor switch and a transistor amplifier.

TRANSISTOR THEORY

The objective of this chapter is to introduce the basic theory and operation of transistors and integrated circuits. The basic operation of transistors will be discussed first.

Basic Operation

The heart of a basic **transistor** is a doped semiconductor crystal, much like the one in a diode. However, the transistor crystal has three sections rather than two. These sections are called **emitter** (E), which is identified by an arrow; **base** (B), which is identified by a bar; and **collector** (C). Their doping can be either npn or pnp, as shown in Figure 29–1. Leads are attached to each of the sections and brought out of the protective case for future connections, as shown in Figure 29–2.

Like diodes, transistors must be biased in order to operate.

> **Key point:** The normal method for biasing a transistor is to forward bias the emitter-base junction and reverse bias the collector-base junction.

Resistors are used to limit the current flow, a process that will be described in your next electronics course. Notice the biasing polarities of the two transistors in Figure 29–1.

In this discussion, we will consider the npn transistor, but the same concepts also apply to pnp transistors. Just reverse the batteries and replace the word *electrons* with the word *holes*.

We begin with the **npn transistor** represented in Figure 29–3A. The emitter and collector have been doped as n type. The base is p type and is only a few thousandths of an inch thick. Since the emitter-base junction is forward-biased, it conducts like a simple diode.

The emitter resistor (R_E) limits the current produced by the emitter-base battery (V_{BB}), which is only a few volts. The collector-base battery (V_{CC}) is larger and will typically range from 6 to 30 volts. This junction is reverse-biased, so it does not conduct — not yet, anyway. Figure 29–3B shows what happens.

Electrons (I_E in the figure) leave the negative side of V_{BB} and enter the emitter. Because the first junction is forward-biased, they continue on to the base. Once in the base, they have a choice of paths. They can leave the base and go to the positive side of V_{BB}. Or they can follow the attraction of the much larger force of V_{CC} and go to it through the collector. Approximately 95% of the electrons that enter the emitter take the second choice: They go right on through the base and out the collector. It is easy for them to do so because they have considerable speed by the time they reach the base and because the base is so thin. Only about 5% of the electrons leave through the base.

Note that although the base current (I_B) may be small, it is important. If the base current is interrupted, the emitter-base junction will no longer be forward-biased. Electrons can no longer enter the base; and if they do not get there, they cannot go to the collector. Thus, we have the following key point.

> **Key point:** The base current (I_B) controls the collector current (I_C).

The process by which a small current, or voltage, controls a larger one is called **amplification.** Under normal conditions, a transistor can be used as an amplifier because variations in the base current cause proportional variations in the collector current.

As mentioned earlier, the concepts just discussed also apply, with modifications, to a **pnp transistor,** which has n type base and p type collector and emitter. The modifications

FIGURE 29-1
JUNCTION TRANSISTOR

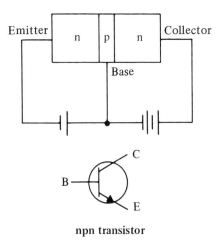

npn transistor

pnp transistor

FIGURE 29-2
TRANSISTOR

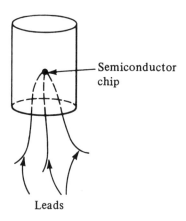

are that the batteries are reversed and the word *electron* is replaced by the word *hole*.

Types of Transistors

We have already introduced two types of transistors, npn and pnp. There is no way to distinguish one from the other except by an identification number or through tests. The most

common method of identification of transistors is a number, such as 2N234. A transistor with this number is generally similar to all others with the same number, regardless of manufacturer. Sometimes, a manufacturer has its own part number for a given transistor. In this case, the manufacturer may need to be contacted when a replacement is needed.

Another method for grouping transistors is by application. Switching, signal, and power transistors are a few examples. More will be said about these transistors later in the chapter.

The transistors described so far are called **bipolar-junction transistors** (BJT). Their construction and method of operation were discussed in the previous subsection. Briefly, a BJT is a transistor in which the current between two terminals is controlled by the current through a third terminal. A few applications for the BJT will be presented shortly.

Other families of transistors include the field effect transistor (FET) and the metal-oxide semiconductor field effect transistor (MOSFET). Their care and handling are discussed later in this chapter, but their construction, operation, and applications are reserved for an advanced electronics course.

Transistor Switch

A transistor is a component used for control since its small base current controls a larger collector current. There are two primary applications for transistors: switching and amplification. **Switching** is a process by which one current turns another on or off. Only two states are possible here, on or off; there is no halfway. Amplification is a process in which one current varies another over a continuous range, although on and off are also possible here. In addition, all the possible values between on and off are possible.

We will consider a switching circuit first. When the switch is open, as illustrated in Figure 29–4A, the emitter-base battery is disconnected so that there is no bias voltage. Without this voltage, base current does not flow, and there is no collector current. Since there is no collector current, there is no voltage drop across the lamp. It is off. The transistor collector-emitter is like an open switch. Therefore, it has the applied voltage of 12 volts across it.

Closing the switch causes the emitter-base junction to become forward-biased. Thus, base current flows, as shown in Figure 29–4B. Since the collector-base junction is reverse-biased, current now flows through the lamp, causing it to light. The transistor, acting like a closed switch, has just about 0 volts dropped across it. The 12 volts applied by V_{CC} are dropped across the lamp. If the switch is opened again, the lamp will go off.

You may be asking yourself, why not just connect the switch to the lamp? There are various reasons for using a transistor. For example, this circuit has a base current that is only a twentieth of the size of the collector current, so the switch is handling a smaller current. Thus, a smaller switch with smaller contacts

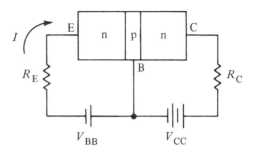

A. Biasing

B. Current through transistor

can be used, and contact wear is less of a concern when a transistor is employed.

Another advantage arises if the device being operated with the circuit is an inductor rather than a light. With an inductor, arcing can be a problem. You may have noticed this problem when removing the power cord of an operating motor. Arcing reduces the life of switches. In the circuit shown in Figure 29–5, the transistor carries the inductor current. Since it has no moving parts, the transistor does not arc. The problem of contact wear and pitting is reduced since the switch carries only a fraction of the current. There is another advantage to this circuit. It is also possible to operate the circuit from a distance. To do so, we

FIGURE 29-4
TRANSISTOR SWITCHING CIRCUIT

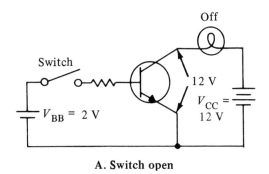

A. Switch open

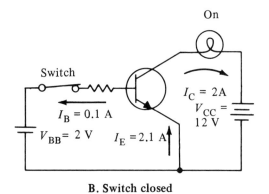

B. Switch closed

FIGURE 29-5
SMALL CURRENT CONTROLLING A LARGE CURRENT

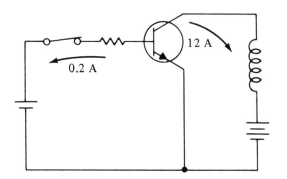

FIGURE 29-6
TRANSISTOR AMPLIFIER CIRCUIT

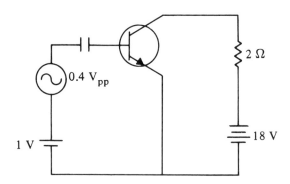

just extend the low-current wires of the switch rather than the larger wires required for the higher current.

The ratio between the base current (I_B) and the collector current (I_C) it controls is called the **amplification factor,** or **beta** (β), of a transistor. Beta is calculated as follows:

$$\beta = \frac{I_C}{I_B}$$

For the circuit in Figure 29–5, the amplification factor is

$$\beta = \frac{I_C}{I_B} = \frac{12\ A}{0.2\ A} = 60$$

Transistor Amplifier

The next circuit we will discuss, shown in Figure 29–6, is an amplifier circuit. Note that, while the switching circuit had just DC signals, the amplifier has both AC and DC signals. Also, while the switching circuit was either on or off, the amplifier is always on.

In an amplifier, operation begins with a DC bias voltage (see Figure 29–7) that places

FIGURE 29-7
COMPOSITE TRANSISTOR INPUT VOLTAGE

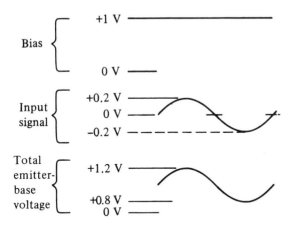

the transistor midway between fully on and off. These points are called **saturation** and **cut-off,** respectively. Then, an AC voltage is superimposed on it. As we know, an AC voltage goes through both positive and negative values. Thus, the voltage combination seen by the emitter-base junction is a varying voltage, but it is always greater than 0 volts, as shown in Figure 29–7. Therefore, the transistor is on more or on less, but it is always on.

When the total emitter-base voltage (bias and signal) increases, the base current increases and causes the collector current to increase. A decrease in total emitter-base voltage causes a drop in base current and collector current. Keeping these ideas in mind, we now can see what occurs in the collector circuit.

Imagine that the transistor, load resistor (R_L), and battery (V_{CC}) form a series circuit. The battery voltage and R_L resistance values are constant. But the collector current and transistor internal resistance change. And because the transistor turns on and off, it appears as a variable resistor to the rest of the collector circuit loop.

Let's consider what happens now during one cycle of the input signal. The input signal is the AC applied at the base-emitter, and the output signal is the AC that appears at the collector-emitter. We start our analysis at 0° of the input signal, as shown in Figure 29–8A. With a 1-volt forward bias across the emitter-base junction, the transistor is turned on. A collector current of 3 amperes causes a voltage drop of 6 volts across R_L. Since the collector-emitter is in series with V_{CC} and R_L, the drop across the transistor is 12 volts.

At 90° of the input signal, as shown in Figure 29–8B, the emitter-base junction is forward-biased with 1.2 volts. This voltage causes an increase in base current, which, in turn, increases collector current. The exact amount of increase depends on the specifications of the transistor. For convenience, we will assume that I_C is now up to 5 amperes. The voltage drop across R_L is increased to 10 volts, and the signal from collector to emitter drops to 8 volts. At 180°, the circuit is the same as it was at 0°, which was shown in Figure 29–8A.

Finally, we consider the circuit when the negative input signal peak occurs, at 270°, as shown in Figure 29–8C. Here, the emitter-base junction voltage is reduced to 0.8 volt, causing lower base and collector currents. Voltage across R_L drops to only 2 volts, so the collector-emitter voltage rises to 16 volts.

Figure 29–9 summarizes the results of transistor operation for emitter-base voltage (V_{BE}), collector current (I_C), and collector-emitter voltage (V_{CE}). As shown in the figure, an input voltage change of 0.4 volt peak to peak causes an output voltage change of 8 volts peak to peak. The ratio of these two voltages is called the **voltage gain** of the transistor, and it is calculated as follows:

$$\text{gain} = \frac{V_{out}}{V_{in}}$$

FIGURE 29-8
TRANSISTOR OPERATION

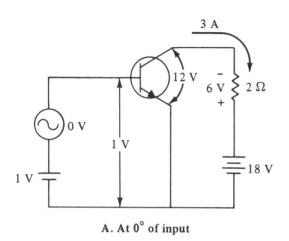

A. At 0° of input

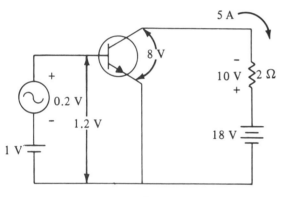

B. At 90° of input

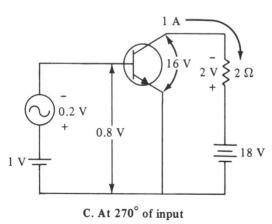

C. At 270° of input

FIGURE 29-9
TRANSISTOR INPUT VOLTAGE AND
RESULTING OUTPUT VOLTAGE

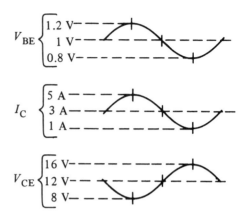

Thus, for the circuits in Figure 29–8, the voltage gain is

$$\text{gain} = \frac{V_{\text{out}}}{V_{\text{in}}} = \frac{8 \text{ V pp}}{0.4 \text{ V pp}} = 20$$

WORKING WITH TRANSISTORS

Once again, your contact with components will involve the outside rather than the inside. Thus, you must be able to identify the different leads of a transistor. You must be able to install and remove a transistor without damaging it or any adjacent components. Finally, you should be able to check its quality with simple tests.

Identification of Leads

The basic junction transistor has three places for connection: emitter, base, and collector. Normally, then, there are three leads; but

FIGURE 29-10
TRANSISTOR LEAD IDENTIFICATION

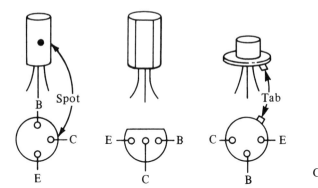

in some cases, there are not. The best way to learn where the emitter, base, and collector connections are to be made is to check the manufacturer's specifications, because so many possibilities exist. A number of examples are shown in Figure 29–10.

Many of the other transistor types do not physically look much different from the basic junction transistor. However, the schematic drawings and lead identifications are quite different. While their operation and application are part of your next electronics course, we can introduce their appearance and symbols here. We mentioned two types, the FET and the MOSFET, earlier. Some other semiconductor devices are the unijunction transistor (UJT), the triac, and the silicon-controlled rectifier (SCR). The symbols for these components are numerous, as are the names for the leads. Some common symbols are shown in Figure 29–11. Keep in mind that there are many variations for each of these types.

Installation and Removal

Considerations for the care and handling of transistors are much like those for other semiconductors, such as diodes. Make sure you

FIGURE 29-11
VARIOUS TRANSISTOR SYMBOLS

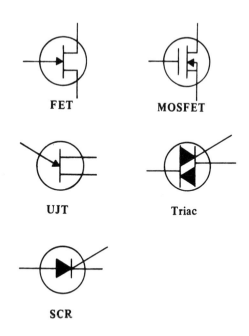

FET

MOSFET

UJT

Triac

SCR

FIGURE 29-12
PROPERLY SOLDERING A TRANSISTOR

FIGURE 29-13
REMOVING SOLDER FROM A TRANSISTOR
WITH A WICK

have the correct transistor, identify the proper leads, and protect the component from excessive heat and, if it is an FET or MOSFET, from static electricity. Some transistors are installed by just plugging them into a socket. But before you can do so, you must trim the leads. Cut them to a length of about 1 centimeter with small diagonal cutters, and align them with the socket by using needle-nose pliers. If they are to be soldered, use an iron of no more than 30 watts and a heat sink. (When soldering FETs and MOSFETs, use a grounded iron.) If there is a grounding ring, leave it on until the transistor is installed. As always, make a good mechanical connection, and use as little solder as possible. The correct soldering technique is shown in Figure 29–12.

Removing a transistor can be as easy as pulling it out or as difficult as unsoldering it. Before you remove it, make sure it is the corret one, and make a sketch of where each lead

is connected. Remember that you have another one to install. Do not damage the transistor when you are removing it because it may still be good.

If the transistor is soldered in, the first step is to remove the solder. You can use a soldering iron and a desoldering tool or wick, as shown in Figure 29–13. Before removing the solder, be sure to install a heat sink. Now, place the braid over the solder to be removed. Then, place a small iron over the braid. The iron will heat the braid, which, in turn, will melt the solder and draw it away from the terminal. As the braid fills with solder, it should be moved so that an unused section is available. When all solder is removed from one connection, the heat sink, braid, and iron should be moved to another connection. Once the solder has been removed, you can undo the mechanical connection. Just straighten the lead ends with needle-nose pliers and carefully

FIGURE 29-14
TESTING A TRANSISTOR

High
resistance

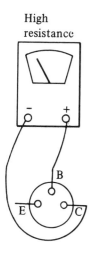

A. Check of emitter-
base junction

Low
resistance

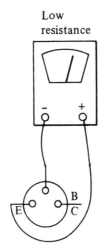

B. Reversing the emitter-
base connection

High
resistance

C. Check of collector-
base junction

Low
resistance

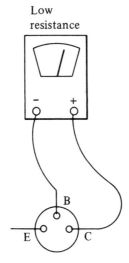

D. Reversing the collector-
base connection

remove the transistor. Clean the terminals for future installation.

Transistor Testing

A wide range of tests can be performed on transistors. Here, we will consider only a few checks that can be made with a multimeter. As we did for semiconductor diodes, we can determine whether the component is shorted or opened by measuring the front-to-back ratios. We also can determine whether it is npn or pnp.

The test for defects requires a multimeter on diode check or on the $R \times 100$ range. The procedure for step 1, checking the emitter-base junction, is shown in Figure 29–14A. The emitter-base junction should indicate a high resistance in one direction and a low resistance in the other. While we do not know the exact ratio that indicates that the transistor is perfect, we do know that it was good once and that if it was bad, it is probably short or open. In either case, this check will leave little question in your mind.

Once you determine whether the resistance is low or high, proceed with step 2, switching the leads, as shown in Figure 29–14B. A high reading on step 1 should be followed by a low reading on step 2. If both are low, the junction is shorted. If both are high, the junction is open. If one is high and the other low, the transistor is probably good.

This test can be repeated with the collector-base junction. It also should indicate high in one direction and low in the other. Step 3, checking the collector-base junction in one direction, is shown in Figure 29–14C. Switching the leads, step 4, is shown in Figure 29–14D. As before, there is a short if both readings are low and an open if both readings are high. Both junctions must be good for the transistor to be good.

The information from these four steps can also indicate if the transistor is npn or pnp. As

with diodes, you must be sure of the meter terminal polarities during resistance checks. Once you are sure that the negative (−) terminal does have a negative polarity during resistance measurements, you can proceed.

Look at your results from steps 1 and 2. Under which set of connections did you get the lower resistance? For the situation pictured in Figures 29–14A and 29–14B, the lower resistance occurred when negative was connected to the base and positive was connected to the emitter. Now, a low resistance indicates conduction, which, in turn, indicates forward bias, as illustrated in Figure 29–15. Also, forward bias means positive to p type and negative to n type. So, the base must be n type and the emitter p type. Our transistor is pnp.

INTEGRATED CIRCUITS

Integrated circuits (ICs) were developed from an expansion of the semiconductor transistor concept. Their small size and huge capacity are almost beyond imagination. While their fabrication is not within the scope of this text, a brief overview is possible.

Introduction

An **integrated circuit** is a collection of electronic components formed by doping a small slice of material such as silicon. These devices are designed by engineers who are aided by computers. Large, wall-size detailed drawings indicate all features of the component. Then, photographs are taken of the drawings and reduced for the production stage. Figure 29–16 shows the detail of an IC chip surface.

Automatic processing systems perform a number of operations on the silicon crystal. Some sections are etched away, while others remain. The net result is that one block is

FIGURE 29-15
DETERMINING TRANSISTOR TYPE

FIGURE 29-16
IC CHIP DETAIL

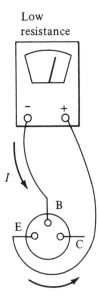

Low
resistance

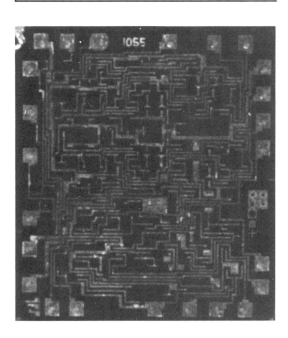

modified and then cut into thousands of small chips. A cross section of such a chip is shown in Figure 29–17.

It is not uncommon for an integrated circuit to be a tenth of the size of a postage stamp yet include thousands of circuits. Some, though, are much larger. But no matter what their size, most ICs are mounted in holders or chip carriers for connection to other parts of a circuit. For instance, a microprocessor IC, like the one in Figure 29–18, has 40 pins on the outside of its protective case. The pins connect internally to an array of transistors, diodes, resistors, and other elements. This IC is a complete central-processing unit for a computer and is capable of handling a wide range of data-processing procedures or routines.

While the development cost of ICs is very high, the final manufacturing cost is quite low. When you consider the number of discrete

FIGURE 29-17
IC LAYERS

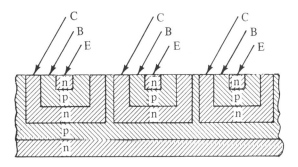

FIGURE 29-18
MICROPROCESSOR IC

FIGURE 29-19
SEMICONDUCTOR DIODE ARRAY

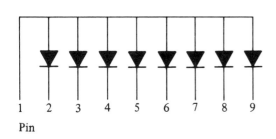

Pin

FIGURE 29-20
INTEGRATED CIRCUITS

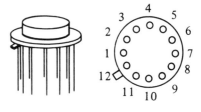

Bottom view

A. TO-5 integrated circuit

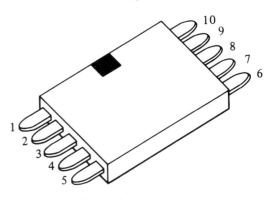

B. Flat-pack integrated circuit

components, the space, and the assembly time it would take to hand-wire an electronic pocket calculator or a radio, you realize the tremendous cost and space savings made by ICs.

Types and Symbols

Some of the more simple forms of ICs consist of one package of diodes called a diode array. These components are then available for connection to other components on a circuit board. The schematic for such an IC is illustrated in Figure 29–19. However, most ICs contain far more complex circuitry that we

have not discussed (it is part of an advanced course).

Early ICs were packaged in round cases, similar to those used for junction transistors, and had 8, 10, or 12 pins. Figure 29–20A shows an example. The pins are identified by number and can be plugged into a socket or soldered to a board.

As more complex ICs were developed, more pins were needed and another package style was introduced. In-line IC packages come in plug-in or solder-in forms. Their pins are numbered counterclockwise as seen from the top, as shown in Figure 29–20B. A notch is centrally located on the end between the

FIGURE 29-21
IC OPERATIONAL AMPLIFIER

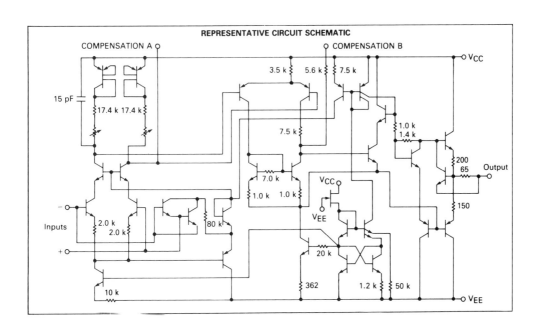

lowest- and highest-numbered pins. The most common package, though, is the dual in-line pin (DIP) configuration, shown earlier, in Figure 3–2.

As mentioned earlier, the schematic drawing of an IC is very large and complex. For this reason, simplified block diagrams are more often used. Furthermore, there is really no need to know all elements within an IC, since they cannot be checked anyway. Thus, what is most useful is a block diagram indicating the functions performed within or the signals expected at each pin. Figure 29–21A shows an IC op amp (operational amplifier). Its schematic in Figure 29–21B gives an indication of the complexity possible in such a small space.

Care and Handling

Some ICs are expensive, and many are very delicate. To reduce the likelihood of too much handling, which can lead to possible damage, you should solder in the IC socket and then plug the IC into it. This technique can also prevent another problem: damage from heat.

Integrated circuits have an additional problem: static electricity. They, like FETs and MOSFETs, are so sensitive that the energy transferred when you touch the leads can burn out some internal sections. However, once they are installed, static electricity is not a problem. It is also not a problem when they are stored on the conductive foam they came in. So, when it is time to install an IC, carefully remove it from its package and plug it in. Grounded personnel, bench, and tools may also be necessary.

There are a few reasons for removing an IC from a circuit. In-circuit testers make it possible to evaluate an IC without removing it. If, however, removal becomes necessary, it can be done simply with a small prying tool made specifically for that purpose. If the IC is

good and is to be kept out of the circuit, mounting it on conductive foam is a good idea.

SUMMARY

The basic transistor is similar to a diode, except that there are two junctions rather than one. Sections called the emitter, base, and collector are doped npn or pnp, with the very thin base in the center. For normal operation, the emitter-base junction is forward-biased, allowing current carriers to enter the base. These carriers are electrons in npn transistors and holes in pnp transistors. Once the carriers enter the base, they are subjected to the attraction of the much stronger collector voltage. This voltage causes the majority of the carriers, about 95%, to continue on through the collector to the power source. In this situation, the collector-base junction must be reverse-biased.

The small base current controls a collector current, which can be 20 to 50 times larger. Without base current, however small, there is no collector current. This relationship of currents in the transistor is called beta, which equals the collector current divided by the base current.

Transistors can be described by the way they are doped: npn or pnp. They can also be described by their designed use: signal, switching, or power. These are the basic junction models. Many other types also exist, including unijunction transistors (UJTs), silicon-controlled rectifiers (SCRs), field effect transistors (FETs), and metal-oxide semiconductor field effect transistors (MOSFETs).

In a typical application, a transistor is used as a switch. The device to be turned on or off is placed in series with the collector, and a switch is placed in series with the base.

Closing the switch causes base current to flow, which, in turn, allows collector current to flow. The device operates. Opening the switch interrupts both currents, and the device turns off. The primary advantages here are the reduced switch current, which prolongs switch life, and the capability of operating the circuit remotely.

Amplifier circuits are a bit more complicated. To begin with, DC voltage sources are connected to forward bias the emitter-base junction and to reverse bias the collector-base junction. The level of bias at the emitter-base is midway between cutoff (zero bias) and saturation (maximum bias). An AC input signal is then added in series with the emitter-base bias, causing the transistor to first conduct more and then less. This operation eventually leads to a collector-emitter voltage that varies at a much greater level, which is the purpose of a voltage amplifier. One additional point should be noted: Since an increase in voltage at the base causes a decrease in collector voltage, the input and output waveshapes are 180° out of phase.

The most common method of identifying transistor leads is to remember the names: emitter, base, and collector. A great number of transistor leads are placed so that their order is emitter, base, and collector going clockwise from the space or tab, as seen from the bottom. However, there are so many base configurations that the best way to be sure of each lead's name is to use a manual or the manufacturer's specifications.

A few basic rules should be followed when installing a transistor. Make sure you have the correct one. Identify all leads. Trim and straighten the leads. Make a secure mechanical connection. Solder with an iron under 30 watts, and use a heat sink. Remember, also, that grounded facilities may be required.

The best way to remove a transistor is to first desolder the connection with a small iron, heat sink, and a desoldering tool or braid. Placing the braid on the solder and the iron on the braid will cause the solder to flow into the braid. Once desoldering is complete, the mechanical connection can be undone and the component removed.

Two simple tests can be performed on a transistor by using a multimeter. By measuring the emitter-base front-to-back ratio, you can determine if it is short, open, or good. The same test should then be made with the collector-base junction. You can also determine if the transistor is npn or pnp by noting which connection produces the lower resistance. A low resistance indicates conduction, which means forward bias. Thus, the transistor lead with the negative multimeter terminal connected to it must be n type.

Integrated circuits (ICs) are solid semiconductor components that have been doped and etched to produce hundreds or thousands of internal segments. Transistors, diodes, resistors, and capacitors of microscopic size can exist within a silicon chip of postage stamp size. Integrated circuits can range from a few diodes to a radio, amplifier, electronic watch, or computer microprocessor.

Because of their complexity, most IC drawings show only block diagrams and pin numbers. Most IC pins are numbered counterclockwise as seen from the top. They are usually installed by plugging them into a soldered-in socket while grounded, to reduce the likelihood of damage. While they are solid and physically strong, some ICs are electrically sensitive and can be destroyed by touching the leads with your finger. For this reason, new ICs are often stored on conductive foam holders until they are ready for installation. Once installed, they are safe from static discharge.

This chapter completes the introduction to solid-state active components. More will be said about their operation and application in your next course.

CHAPTER 29

REVIEW TERMS

amplification: process in which one current (or voltage) varies another through a transistor or a similar device

amplification factor: ratio between the base current and the collector current in a transistor

base: BJT terminal identified by a bar

beta: see *amplification factor*

bipolar-junction transistor (BJT): semiconductor device in which the current between two terminals is controlled by the current through a third

collector: BJT terminal

cutoff: transistor condition of minimum current

emitter: BJT terminal identified by an arrow

integrated circuit: collection of electronic components and their connections formed by doping a small slice of material such as silicon

npn transistor: transistor with p type base and n type collector and emitter

pnp transistor: transistor with an n type base and p type collector and emitter

saturation: transistor condition of maximum current

switching: process by which one current turns another on or off

transistor: semiconductor device with three sections — an emitter, a base, and a collector

voltage gain: ratio of output voltage to input voltage in a transistor

REVIEW QUESTIONS

1. Define bipolar junction transistor.

2. Define beta.

3. State what the abbreviations FET, UJT, and MOSFET mean.

4. Describe an integrated circuit.

5. Define amplification.

6. Describe the normal biasing of a junction transistor.

7. Describe the basic advantages of an integrated circuit over discrete circuits.

8. Name three advantages of a transistor switch over a mechanical switch.

9. Outline the safety considerations to be used when working with transistors.

10. Outline the safety considerations to be used when working with integrated circuits.

11. Describe how integrated circuit leads are identified.

REVIEW PROBLEMS

1. Draw the schematic symbols of four types of transistors.

2. Name the transistor types shown in Figure 29–22.

3. Draw the schematic symbols of npn and pnp transistors, and name the leads.

4. Name the leads of the transistors shown in Figure 29–23.

5. Sketch an npn transistor, show how it is biased, and indicate relative current amounts.

6. Looking at Figure 29–24, determine whether the transistor is good or bad.

7. Is the transistor of Figure 29–24 npn or pnp? Why?

8. Sketch the voltage waveforms that appear across the base-emitter and across the collector-emitter of the transistor in Figure 29–25 if the source voltage is 25 millivolts peak to peak and the voltage gain is 150. Indicate values.

9. Sketch the voltage waveforms that appear across the base-emitter and across the collector-emitter of the transistor in Figure 29–25 if the source voltage is 50 millivolts peak to peak and the voltage gain is 100. Indicate values.

FIGURE 29-23

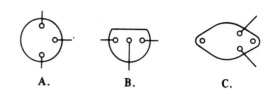

A. B. C.

FIGURE 29-24

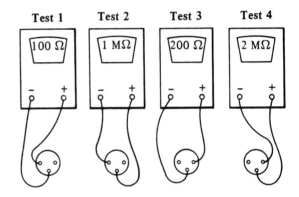

Test 1   Test 2   Test 3   Test 4

100 Ω   1 MΩ   200 Ω   2 MΩ

FIGURE 29-25

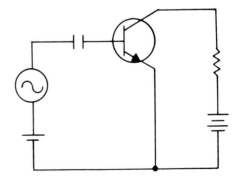

FIGURE 29-22

A. B.

RECTIFIERS

OBJECTIVES

Identify three basic rectifier circuits by their appearance and schematics.

Describe the theory of operation of three basic rectifier circuits.

Describe the characteristics and application of three basic rectifier circuits.

Determine the parameters of three basic rectifier circuits by measurement.

PURPOSE OF RECTIFIERS

The objective of this chapter is to introduce the theory of operation and characteristics of three basic rectifier circuits. A **rectifier circuit** changes alternating current to direct current. To do so, rectifiers use a diode. For this reason, diodes are often called rectifiers. Some rectifier circuits use only one diode, but others may use two or more. We will discuss the three most common rectifier circuits in this chapter.

One point should be noted: Rectifiers only change the current so that it does not alternate. That is, the direct current produced does flow in only one direction. However, it is not smooth like the current produced from a battery. It pulses. For a smooth current, filters are required. We will discuss filters in the next chapter.

If we need direct current, why not just use batteries? There are several reasons for not using batteries. Batteries wear out and are bulky. Producing high voltages can be difficult with batteries. Furthermore, alternating current is readily accessible. AC circuits have another distinct advantage over DC circuits: The voltage can easily be changed up and down with a transformer. For these reasons, most electronic products that use DC get it from AC with the aid of rectifiers. Of course, for portable uses, such as in automobiles or radios, batteries are still used.

HALF-WAVE RECTIFIER

The **half-wave rectifier** is the simplest of the three common rectifier circuits. It consists of only one component, a diode. However, an AC voltage source is also needed, and that voltage source is often a transformer. Of course, the circuit also includes a load, the component or components that require the DC voltage. Note, though, that the AC-to-DC conversion is performed in this circuit by the single diode.

Theory of Operation

The basic half-wave rectifier circuit is shown in Figure 30–1A. In most cases, the voltage source is line voltage of 120 volts at 60 hertz. We will not be concerned about the effects of frequency until filters are discussed in the next chapter.

The transformer in the circuit serves two purposes. First, it allows us to change the voltage to a level of our choice. In the circuit shown in Figure 30–1A, we want to step the voltage down by a factor of 6. Since the primary has 6 times the number of turns of the secondary, its voltage will be 6 times as large. Thus, a primary voltage of 120 volts will produce a secondary voltage of 20 volts. Keep in mind that these voltage values are root mean square (rms) values.

The second purpose of the transformer is for the isolation of grounds. While the source may already be grounded on one side or the other, a transformer separates that ground from the rest of the circuit. It allows us to ground either side, or neither side of the load, safely. Without the transformer, the risk of shorts is much higher.

The purpose of the diode is to allow current to flow in one direction but not the other. The direction depends on how the junction is biased. Since AC is applied to the junction of the diode in our circuit, the diode is forward-biased during one-half of the cycle and reverse-biased during the other half.

Now, consider what occurs during the first half of the applied signal's cycle, as shown at the bottom of Figure 30–1A. The 120 volts rms applied to the transformer primary produces 20 volts rms in the secondary. The

FIGURE 30-1
HALF-WAVE RECTIFIER CIRCUIT

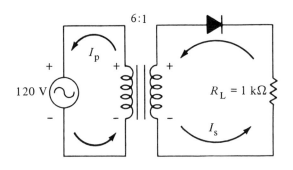

Waveshapes

A. During first 180°

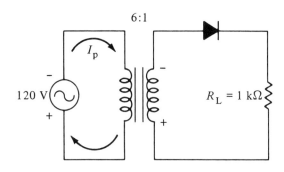

Waveshapes

B. During second 180°

important point here is the polarity. Since the sine wave is positive during the first 180°, we can say that the upper terminal of the source is positive and the lower is negative. This polarity, in turn, produces the transformer polarities noted in Figure 30–1A. Current flows through the load resistor since the diode is forward-biased.

Since the diode is not a perfect switch, when it is on, it has some resistance and, therefore, a small voltage drop. Described earlier, that drop is typically between 0.2 and 0.4 volt for a germanium diode and 0.6 to 0.8 volt for a silicon diode. Since that voltage is relatively small compared with the total secondary voltage, we will ignore it and assume that the total 20 volts rms appear across the load resistor. We convert this rms value to peak value as follows:

$$V_{rms} = 0.707 \times V_{peak}$$

$$V_{peak} = \frac{V_{rms}}{0.707} = \frac{20 \text{ V}}{0.707} = \textbf{28.3 V}$$

This peak value is the voltage across the load, V_L.

Now, we consider the second half of the cycle, as shown in Figure 30–1B. During this period, the applied voltage has the opposite polarity and produces the resulting waveshapes shown in the figure. The secondary voltage reverse biases the diode; hence, the diode does not conduct. Since no current flows through the load resistor, its voltage drop is zero for this half of the cycle. In other words, voltage appears across the load for half of the applied signal but not for the other half. For this reason, the circuit is called a half-wave rectifier.

We can convert the values to peak levels, as follows:

$$V_{in \ peak} = \frac{120 \text{ V}}{0.707} = \textbf{170 V}$$

$$V_{\text{out peak}} = \frac{V_{\text{in peak}}}{6} = \frac{170 \text{ V}}{6} = \textbf{28.3 V}$$

$$I_{\text{L peak}} = \frac{V_{\text{out peak}}}{R_{\text{L}}} = \frac{28.3 \text{ V}}{1000 \ \Omega} = \textbf{0.0283 A}$$

These peak values are then used in drawing the graphs shown in Figure 30–2, which summarize the circuit's operation.

As the figure shows, the load voltage is DC since it does not change polarity. Next, we must determine its magnitude. In this case, we are interested in an average value. And the average value of one-half cycle of a sine wave equals 0.636 times the peak value. Thus, the first half of the cycle has the following average value:

$$V_{\text{av}} \ (\tfrac{1}{2} \text{ cycle}) = 0.636 \times V_{\text{out peak}}$$
$$= 0.636 \times 28.3 \text{ V} = \textbf{18 V}$$

However, this value is only for the first half of the cycle. During the second half, the voltage is zero. To find the total DC voltage for one full cycle, we must add both halves and divide by 2, as follows:

$$V_{\text{DC}} = \frac{(18 + 0) \text{ V}}{2} = \textbf{9 V}$$

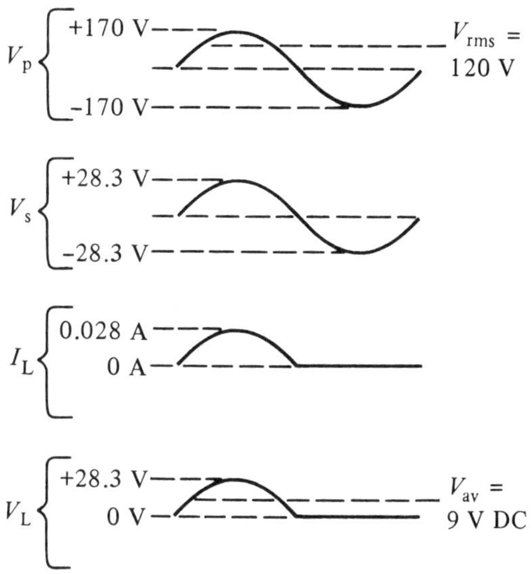

Characteristics

The characteristics described here will be used later for comparison with other rectifier circuits. One characteristic of a rectifier is called ripple. **Ripple** is the train of pulses that occurs when AC is changed to DC. Two points should be noted here. First, there is one pulse for every cycle of the input signal of ripple. So, a half-wave, 60-hertz rectifier has a 60-hertz ripple. Second, the **ripple voltage** — that is, the peak-to-peak variation of the rectifier voltage — has a peak value of 28.3 volts. Since smooth DC does not have ripple, the size and

FIGURE 30-2
HALF-WAVE RECTIFIER INPUT AND OUTPUT
VOLTAGES

frequency of the ripple is an indication of how difficult its removal will be.

A second characteristic of a rectifier is the voltage level. For future comparison, we will consider the relationship between AC into the rectifier and DC out of the rectifier. The input is the rms value at the transformer secondary, and the output is the average DC across the load. The half-wave rectifier we have been considering produces 9 volts DC from 20 volts AC. And it does so with one diode.

FULL-WAVE RECTIFIER

The most obvious limitation of the half-wave rectifier is its waste of one-half of the applied voltage. That half wave is removed because it is of the wrong polarity. The **full-wave rectifier**

FIGURE 30-3
FULL-WAVE RECTIFIER CIRCUIT

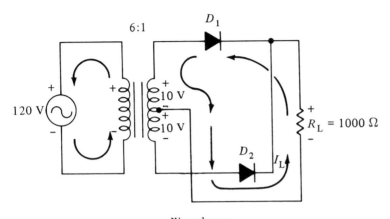

Waveshapes

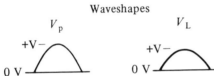

A. During first 180°

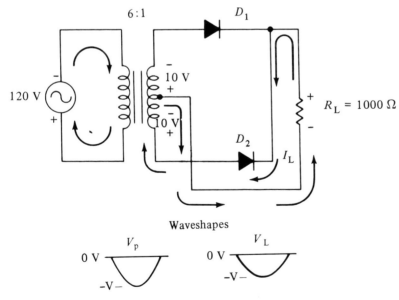

Waveshapes

B. During second 180°

takes advantage of that other half wave and makes it usable by inverting it to the same polarity as the first half. It does so with a transformer, which is a required component in this circuit.

Theory of Operation

The circuit in Figure 30–3A, a basic full-wave rectifier circuit, differs from the preceding one in two ways. The transformer now has a center tap, and a second diode has been added. Diode 1 and R_L form a series circuit with the upper portion of the secondary, and diode 2 and R_L are in series with the lower portion.

During the first half of the cycle, the upper end of the secondary is positive and the lower end is negative, as shown in Figure 30–3A. In addition, the center tap is negative as seen from the upper end and positive as seen from the lower end. The current loop for each diode only sees a force of 10 volts. If we trace the upper path, we see that D_1 is forward-biased. If we trace the lower path, we see that D_2 is reverse-biased. Current flows through D_1 and R_L during the first half of the cycle. No current flows through D_2.

The situation changes during the second half of the cycle, as shown in Figure 30–3B. Now, D_1 is not conducting because it is reverse-biased. But D_2 is forward-biased by the lower half of the secondary, and it allows current to flow up through R_L. Notice that current flows up through R_L during both halves of the applied cycle. Even though the input polarity reverses, the output polarity does not change. Thus, the top of the resistor is always positive. It is as though the negative half of the applied sine wave were inverted.

A few calculations and a summary graph will help explain the circuit's operation. The peak value of the voltage applied to the rectifier is based on 10 volts rms since only one-

half of the secondary is used at the time. Thus, the peak voltage is

$$V_{peak} = \frac{V_{rms}}{0.707} = \frac{10 \text{ V rms}}{0.707} = \textbf{14.14 V}$$

Now, we can determine V_{DC} for the first half of the cycle, as follows:

$$V_{DC} = 0.636 \times V_{peak} = 0.636 \times 14.14 \text{ V}$$
$$= \textbf{9 V}$$

The value for the second half of the cycle is the same. To find the voltage for the full cycle, we add both and divide by 2:

$$V_{DC} \text{ (full cycle)} = \frac{(9 + 9) \text{ V}}{2} = \textbf{9 V}$$

The peak current is

$$I_{L \ peak} = \frac{V_{L \ peak}}{R_L} = \frac{14.14 \text{ V}}{1000 \ \Omega} = \textbf{0.01414 A}$$

These values are used in drawing the graphs shown in Figure 30–4, which summarize the operation of the full-wave rectifier. The graphs show how one diode conducts during the first half cycle and the other diode conducts during the second half. The combination of these two currents produces a one-direction load current, which, in turn, produces a one-polarity output voltage. This circuit corrects, or rectifies, the second half wave so that it can use both halves of the input cycle, or the full wave. Therefore, it is called a full-wave rectifier.

Characteristics

The first significant difference between the full-wave rectifier and the half-wave rectifier is the ripple frequency. There are two ripple pulses for each input cycle, so the ripple fre-

FIGURE 30-4
FULL-WAVE RECTIFIER INPUT AND OUTPUT VOLTAGES

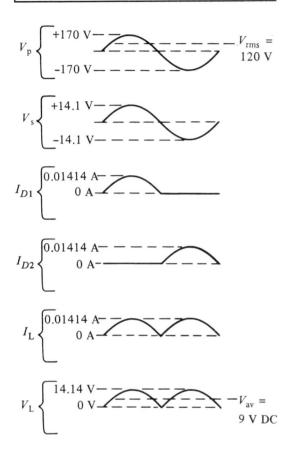

transformer was used primarily for isolation in the half-wave rectifier, it is needed here because of its center tap.

FULL-WAVE BRIDGE RECTIFIER

The advantage of the full-wave rectifier over the half-wave rectifier is the utilization of the complete cycle of the input voltage. A disadvantage is that it uses only half of the transformer at a time. Since transformers are too large and expensive to underutilize, a solution to the problem was needed. The **full-wave bridge rectifier** solved the problem. The two additional diodes of the full-wave bridge rectifier allow the full secondary of the transformer to be used all the time.

Theory of Operation

The main difference between the full-wave bridge rectifier circuit shown in Figure 30–5A and the others is that four diodes are used rather than one or two. There is also no need for a center tap on the transformer.

Once again, the voltage across the secondary has an rms value of 20 volts and a peak value of 28.3 volts. The first half cycle produces the results shown in Figure 30–5A. Electrons leaving the negative lead at the bottom of the secondary reach a junction of diodes D_2 and D_3. Since they can enter a cathode but not an anode, they enter D_3 but not D_2. As they leave D_3, they cannot enter D_4, but they can flow through R_L into D_1 and on to the positive end of the secondary. Thus, as we trace the path from each diode back to the transformer, we see that D_1 and D_3 are forward-biased and conduct. Diodes D_2 and D_4 are reverse-biased and do not conduct.

The second half of the cycle reverses the polarity, as shown in Figure 30–5B. Now,

quency here is 120 hertz. As we will see in the next chapter, this ripple makes it easier to smooth the DC. Also, the peak value of the ripple is half as large as it was before, which is another advantage.

The second characteristic to consider is the final value of DC voltage across R_L. It is the same as the half-wave rectifier output. The DC voltage is the same, even though only half of the secondary is used, since both halves of the input cycle are used. Finally, while the

FIGURE 30-5
FULL-WAVE BRIDGE RECTIFIER CIRCUIT

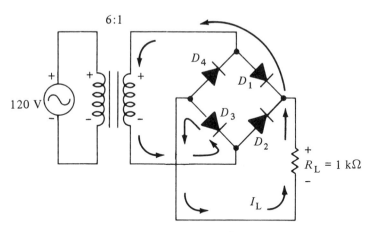

A. During first 180°

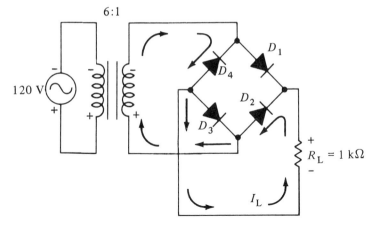

B. During second 180°

FIGURE 30-6
FULL-WAVE BRIDGE RECTIFIER INPUT AND OUTPUT VOLTAGES

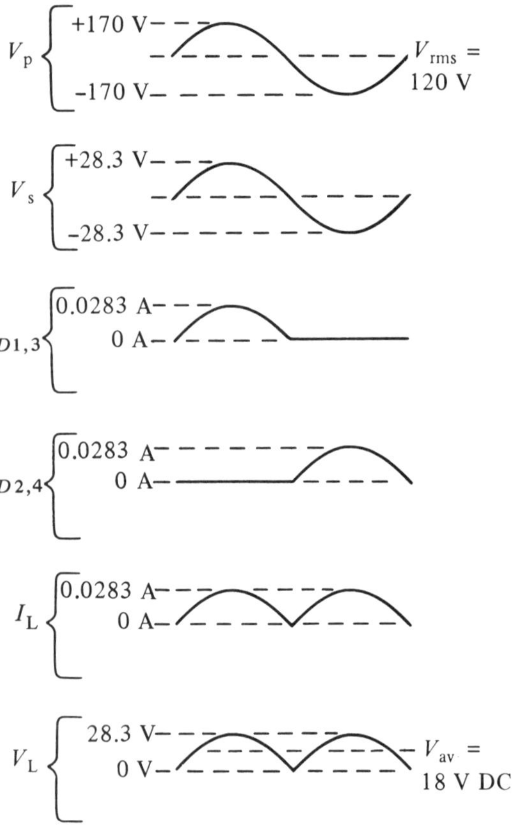

$$V_{peak} = \frac{V_{rms}}{0.707} = \frac{20 \text{ V rms}}{0.707} = \textbf{28.3 V}$$

$$V_{DC} = 0.636 \times V_{peak} = 0.636 \times 28.3 \text{ V}$$
$$= \textbf{18 V}$$

Since this circuit operates during both half cycles, this value is the DC output voltage.

The circuit has the following peak current:

$$I_{L \text{ peak}} = \frac{V_{peak}}{R_L} = \frac{28.3 \text{ V}}{1000 \text{ } \Omega} = \textbf{0.0283 A}$$

These values are used in drawing the graphs shown in Figure 30–6, which summarize the operation of the full-wave bridge rectifier. Diodes D_1 and D_3 conduct during the first half cycle, and diodes D_2 and D_4 conduct during the second half. The most significant difference between this circuit and the full-wave rectifier circuit can be seen in the peak output voltage. In this circuit, the peak output voltage is equal to the peak voltage of the transformer secondary. In the full-wave circuit, the output voltage is only one-half of that value.

We note here that the term *bridge* in full-wave bridge rectifier relates to the shape of the circuit. As you will see in future studies, a bridge is a four-branch, diamond-shaped circuit with inputs at the top and bottom and outputs at the two sides.

diodes D_2 and D_4 are forward-biased and conduct. Diodes D_1 and D_3 are reverse-biased and do not conduct. The current path begins at the upper end of the transformer secondary. Current flows down through D_4 and on to the bottom of R_L. From there, current flows through D_2 and back to the lower end of the winding. As before, current flows up through the load resistor.

The effective output can be determined by beginning with the 20 volts rms applied to the bridge, as follows:

Characteristics

The full-wave bridge rectifier circuit has the same ripple frequency as the full-wave rectifier, which is twice the applied frequency. The peak value of ripple is equal to the peak value of the voltage applied to the rectifier bridge. Since both halves of the cycle and the full secondary are used all the time, this circuit has the highest output voltage.

The three basic rectifier circuits are compared in Table 30–1. In summary, the half-

TABLE 30-1
RECTIFIER CIRCUITS

Type	V_{in} (in Volts)	Frequency (in Hertz)	$V_{out\ DC}$ (in Volts)	Ripple Frequency (in Hertz)	Ripple Peak Voltage (in Volts)
Half-wave	120	60	9	60	28.3
Full-wave	120	60	9	120	14.14
Full-wave bridge	120	60	18	120	28.3

wave rectifier is the least expensive since it requires only one diode. It is also the least efficient since it utilizes only one-half the applied voltage. The full-wave bridge requires the most diodes, but it has the highest output voltage. As we will see in the next chapter, the 120-hertz ripple frequency is also an advantage.

APPLICATIONS

Rectifier circuits are used in almost every electronic product that operates from line voltage sources. Their purpose is to convert AC to DC. The type of rectifier circuit used depends on a few factors. For instance, two questions are usually asked by designers: What quality of DC is needed, and what price can we afford to pay?

Quality here refers to the smoothness of the voltage. More ripple must be removed from circuits used in stereos, radios, and televisions than from those used in many other appliances. For that reason, full-wave rectifiers are often selected. Other, less sensitive products work just as well with a half-wave-rectified voltage.

The cost of a rectifier involves more than dollars. Space and weight are other costs

that the transformers bring with them. Thus, sometimes it is less expensive to add two more diodes and make a smaller transformer, using a bridge rather than a full-wave rectifier.

This is a good time to look at some product schematics and see what type of rectifier is used. Begin at the power cord and trace the circuit. Determine whether the product uses a half-wave, full-wave, or full-wave bridge rectifier. If the product has a power adapter, find out what is inside. Decide why a particular circuit is used by considering quality required versus costs.

EXPERIMENTS

The series of rectifier experiments outlined in this section will help you understand and compare the three basic rectifier circuits. The half-wave rectifier is considered first.

Half-Wave Rectifier

Connect the half-wave rectifier according to the schematic shown in Figure 30–7. Use a center-tapped transformer so that it can be used for all three experiments. The center tap can be ignored for this one. Make sure that the diode is rated to carry the expected current.

FIGURE 30-7
HALF-WAVE RECTIFIER

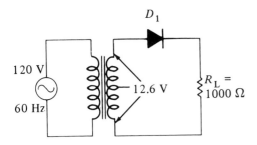

As usual, calculate the expected values before you begin. The calculations for the circuit in Figure 30–7 are as follows:

$$V_{s\ peak} = \frac{12.6\ V}{0.707} = \textbf{17.8 V}$$

$$V_{DC} = \frac{0.636 \times V_{peak}}{2} = \frac{0.636 \times 17.8\ V}{2}$$

$$= \textbf{5.7 V}$$

$$I_{peak} = \frac{V_{peak}}{R_L} = \frac{17.8\ V}{1000\ \Omega} = \textbf{0.0178 A}$$

$$I_{av} = \frac{I_{peak} \times 0.636}{2}$$

$$= \frac{0.0178\ A \times 0.636}{2} = \textbf{0.00566 A}$$

The circuit in Figure 30–7 has an expected peak output voltage of 17.8 volts. During the other half cycle, the diode will see an inverse voltage of 17.8 volts. Therefore, a diode with a 25-volt, peak inverse voltage (PIV) rating should be more than adequate. In addition, a peak current rating of 25 milliamperes and an average current rating of 10 milliamperes should also be adequate. However, a 25-milliampere average current rating will provide a greater margin of safety and will be necessary for a later experiment.

Once you obtain the parts, assemble the

circuit and have it checked. The tests you should perform are the following:

1. Measure the secondary rms voltage with a multimeter.
2. Measure the DC output voltage with a multimeter.
3. Measure the ripple voltage across R_L with the multimeter, using the output function.
4. Measure the secondary peak-to-peak voltage with an oscilloscope. Do *not* measure the primary.
5. Measure the DC output peak voltage with the oscilloscope. Carefully sketch the waveform on graph paper.

When you have completed your measurements, disable and disconnect the circuit. Then, determine the following circuit parameters:

— The DC output voltage;
— The ratio of secondary voltage to DC output, which should be the same as the following calculated value:

$$\frac{V_{DC}}{V_s} = \frac{5.7\ V}{12.6\ V} = \textbf{0.45}$$

— The ripple factor for the circuit, as calculated from the following equation:

$$\text{ripple factor} = \frac{V_{AC\ out}}{V_{DC\ out}}$$

— The ripple frequency of the circuit.

When you have these answers, you can proceed to the next experiment.

Full-Wave Rectifier

Once again, you begin with the calculations of what to expect from the circuit, which is shown in Figure 30–8. Remember that the rectifier diodes only see one-half of the transformer secondary. The calculations are as follows:

$$V_{s\ peak} = \frac{6.3\ V}{0.707} = \textbf{8.9 V}$$

$$V_{DC} = 0.636 \times V_{peak} = 0.636 \times 8.9\ V$$
$$= \textbf{5.7 V}$$

$$I_{peak} = \frac{V_{peak}}{R_L} = \frac{8.9\ V}{1000\ \Omega} = \textbf{0.0089 A}$$

$$I_{av} = I_{peak} \times 0.636$$
$$= 0.0089\ A \times 0.636 = \textbf{0.0057 A}$$

These calculations tell you what results to expect. In addition, they verify that the type of diode used in the half-wave rectifier is also satisfactory for this experiment. So, get another diode like the first one, and then connect the circuit as shown in Figure 30–8. Once it is connected, perform the following steps:

1. Measure the two secondary rms voltages with a multimeter.
2. Measure the DC output voltage with the multimeter.
3. Measure the ripple voltage across R_L, using the multimeter on the output function.
4. Measure the two, secondary, peak-to-peak output voltages with an oscilloscope. Do *not* measure the primary.
5. Measure the DC output peak voltage with the oscilloscope. Carefully sketch the waveform on graph paper.

When your data are gathered, disable and disconnect the circuit. Then, determine the following circuit parameters:

— The DC output voltage;
— The ratio of secondary voltage to DC output, which should be the same as the following calculated value:

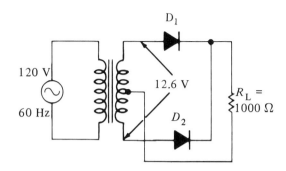

FIGURE 30-8
FULL-WAVE RECTIFIER

$$\frac{V_{DC}}{V_s} = \frac{5.7\ V}{12.6\ V} = \textbf{0.45}$$

— The ripple factor for the circuit;
— The ripple frequency of the circuit.

Once these answers are determined, you can proceed to the third experiment.

Full-Wave Bridge Rectifier

The results expected for the circuit in Figure 30–9 are as follows:

$$V_{s\ peak} = \frac{12.6\ V}{0.707} = \textbf{17.8 V}$$

$$V_{DC} = 0.636 \times V_{peak}$$
$$= 0.636 \times 17.8\ V = \textbf{11.3 V}$$

$$I_{peak} = \frac{V_{peak}}{R_L} = \frac{17.8\ V}{1000\ \Omega} = \textbf{0.0178 A}$$

$$I_{av} = I_{peak} \times 0.636$$
$$= 0.0178\ A \times 0.636 = \textbf{0.0113 A}$$

Once your calculations are complete, connect the circuit in Figure 30–9. Then, take the following measurements:

1. Measure the secondary rms voltage.

FIGURE 30-9
FULL-WAVE BRIDGE RECTIFIER

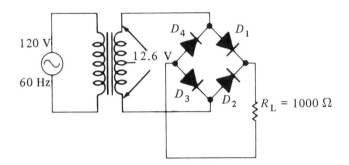

2. Measure the DC output voltage.

3. Measure the ripple voltage.

4. Measure the secondary peak-to-peak voltage.

5. Measure and sketch the output waveform across R_L.

Once again, disable and disconnect the circuit. Then, determine the following information for comparison:

— The DC output voltage;

— The ratio of secondary voltage to DC output;

— The ripple factor;

— The ripple frequency.

Lab Report

When you have completed all three experiments, begin your lab report. It should include the usual information: objective, materials and equipment lists, schematics, data, and waveforms. You should also consider the answers to the following questions:

1. Which circuit has the highest output voltage?

2. Which has the best (lowest) ripple factor?

3. Which has the best (highest) ripple frequency?

4. Which is the most efficient rectifier circuit? Which has the highest ratio of DC output to rectifier AC output?

These are the basic questions you should answer for rectifier experiments. Your conclusions should give your opinion of the best circuit overall and your reasons for your opinion. You should also comment on the following questions:

1. Did the secondary rms voltage equal the peak-to-peak value times 0.707 divided by 2?

2. Was the waveform a good sine wave? Was there waveform error?

3. Did the DC output voltage equal 0.636 times peak?

This section concludes the introduction to the rectifier portion of power supplies. Filters to remove the ripple will be described in the next chapter.

SUMMARY

A rectifier circuit is used to change AC voltages to DC voltages. The primary component used to accomplish the change is the diode. With the aid of one or more diodes, rectifier circuits remove or invert one-half of the applied sine wave to produce a direct, or nonalternating, current. The resulting DC flows only in one direction, although it is not of the quality or smoothness of the DC from a battery.

Three basic rectifiers are used in the electronics industry: half-wave, full-wave, and full-wave bridge. Almost every electronic product that operates on line voltage has one of these rectifiers in it.

In the half-wave rectifier, a transformer is usually placed between the line voltage and the remainder of the circuit. The transformer serves two purposes: It provides isolation between line and chassis grounds for safety, and it changes the voltage to the level desired. A fuse and a main power switch are usually connected in series with the primary. The half-wave rectifier has one diode connected in series between the transformer secondary and the remainder of the circuit.

The voltage across the load of a half-wave rectifier during one-half of the cycle equals the peak applied voltage times 0.636. During the second half, it equals zero, because no current flows through the diode. The diode is reverse-biased. Thus, the average DC output voltage of a half-wave rectifier equals the peak applied voltage times 0.636 divided by 2. Also, the output waveform pulses. That is, it has very large ripple of the same frequency as the applied voltage.

A full-wave rectifier requires a center-tapped transformer and a second diode. In this case, the transformer inverts the second half of the applied waveform so that it also can be utilized. Consequently, only one-half of the transformer secondary is used at a time. Interestingly enough, both halves of the cycle and of the transformer equalize, so that the full-wave and half-wave rectifiers have the same output voltage.

The ripple frequency of the full-wave rectifier is twice the applied frequency. This benefit will be of more interest in the next chapter. Another advantage of the full-wave rectifier over the half-wave is that its ripple factor is only half as large. In other words, much smoother DC is produced.

The full-wave bridge rectifer also uses a transformer, but it does not require a center tap. Furthermore, the full-wave bridge uses four diodes. With this circuit, the full transformer secondary voltage is utilized all the time, with two diodes conducting for half of the cycle and the other two conducting during the other half. Hence, the full-wave bridge rectifier has an output voltage that is twice that of the other two circuits.

The full-wave bridge rectifier has a ripple frequency similar to that of the full-wave rectifier: twice the applied frequency. Its primary advantage is that it is the most effective of the three circuits: It produces the highest DC voltage for a given AC voltage input. It also has the lowest ripple factor.

CHAPTER 30

REVIEW TERMS

full-wave bridge rectifier:
four-diode circuit that
converts AC voltage to
DC voltage

full-wave rectifier: two-diode
circuit that converts the
full AC voltage to DC
voltage

half-wave rectifier: single-
diode circuit that
converts one-half of the
AC voltage cycle to DC
voltage

rectifier circuit: circuit that
changes alternating
current to direct current

ripple: train of pulses that
occurs when AC is
changed to DC

ripple voltage: peak-to-peak
variation of the voltage
produced by a rectifier

REVIEW QUESTIONS

1. Describe the purpose and basic components of a rectifier circuit.

2. Draw the circuit of a half-wave rectifier
circuit with a 120-volt input, a 1 : 5 step-
up transformer, and a 1000-ohm load.

3. Draw the circuit of a full-wave rectifier
with a 120-volt input, a 1 : 3 step-up transformer, and a 5000-ohm load.

4. Draw the circuit of a full-wave bridge rectifier with a 3 : 1 step-down transformer,
a 120-volt input, and a 1500-ohm load.

5. Explain how current flows in the circuit
described in Question 2.

6. Explain how current flows in the circuit
described in Question 3.

7. Explain how current flows in the circuit
described in Question 4.

8. State two purposes of transformers.

9. Compare the average output voltage of the
three basic rectifiers.

10. Compare the output voltage waveforms
and ripple of the three basic rectifiers.

11. Sketch the relative input and output waveforms for the circuit described in Question
2.

12. Sketch the relative input and output waveforms for the circuit described in Question
3.

13. Sketch the relative input and output waveforms for the circuit described in Question
4.

14. Compare the ripple frequencies of the
three basic rectifier circuits.

REVIEW PROBLEMS

1. Calculate the average DC output voltage
for the circuit described in Question 2.

2. Repeat the calculation of Problem 1 with a 4 : 1 step-down transformer and a 5000-ohm load.

3. Calculate the DC output voltage for the circuit described in Question 3.

4. Repeat the calculation of Problem 3 with a 6 : 1 step-down transformer and a 470-ohm load.

5. Calculate the average DC output voltage for the circuit described in Question 4.

6. Calculate the ripple frequency and peak-to-peak ripple voltage for the circuit described in Question 2.

7. Calculate the ripple frequency and peak-to-peak ripple voltage for the circuit described in Question 3.

8. Calculate the ripple frequency and peak-to-peak ripple voltage for the circuit described in Question 4.

9. Calculate the average output voltage to be expected if one diode in the circuit described in Question 3 opened.

10. Calculate the average output voltage to be expected if one diode in the circuit described in Question 4 opened.

11. Sketch the output waveform expected for Problem 9.

12. Sketch the output waveform expected for Problem 10.

13. Calculate the expected peak-to-peak ripple and ripple frequency for the circuit described in Question 3 with one open diode.

14. Calculate the expected peak-to-peak ripple and ripple frequency for the circuit described in Question 4 with one open diode.

FILTERS

Identify three basic filter circuits by their appearance and schematics.

Describe the theory of operation of three basic filter circuits.

Describe the characteristics and application of three basic filter circuits.

Determine the parameters of three basic filters by measurement.

PURPOSE OF FILTERS

The objective of this chapter is to introduce the theory of operation and characteristics of three basic filter circuits. Simply stated, the purpose of any **filter** is to remove the undesirable elements from something. For example, air, gas, and oil filters are used to remove dirt particles from the air, gas, and oil used in automobiles. In earlier chapters, we saw how series and parallel circuits can be used to pass some frequencies and block others in a circuit. The filters described here are used in a similar way: They clean up the DC voltage produced in a rectifier circuit by removing the undesirable element, the ripple.

Rectifiers have an input of AC and an output of pulsating DC. Filters are connected to rectifiers and have an input of pulsating DC and an output of smooth DC. We will use the simple half-wave rectifier, shown in Figure 31–1, as our DC voltage source. It consists of a step-down transformer, one diode, and the load. Recall from the previous chapter that the average DC voltage of this circuit is equal to (peak voltage times 0.636) divided by 2. In this case, the peak voltage is 28.3 volts, and the average value of the DC voltage is 9 volts.

CAPACITIVE FILTER

The first circuit we will consider, shown in Figure 31–2A, is the simple **capacitive filter,** with one capacitor connected across the output of a half-wave rectifier. We will evaluate the effectiveness of this one component's ability to make the DC smoother. We will assume that the rectifier circuit is the same one used in the preceding chapter. Without the capacitor, it had a peak output of 28.3 volts and an average output of 9 volts DC. Our filter is a 10-microfarad capacitor.

Theory of Operation

As we know, when a capacitor is placed across a DC voltage source, it charges up to the level of the source. When the source is removed, the capacitor discharges through any available path. This process is exactly what occurs in the filter circuit. During the first half of the cycle, the diode is forward-biased, as shown in Figure 31–2A. Current flows through the resistor, as explained in the previous chapter. However, this time the capacitor charges up as well. By the time the input reaches its peak value at 90°, the capacitor has charged up to the full 28.3 volts.

As the 90° point is passed, the line voltage begins to drop, causing the voltage at the anode side of the diode to drop also. The capacitor is still charged to 28.3 volts, so the cathode side becomes more positive than the anode. The diode is reverse-biased and it stops conducting, as pictured in Figure 31–2B. At this point, the capacitor begins to discharge through R_L.

The discharging continues until the next pulse arrives and reaches the level that the capacitor has discharged to. Once the transformer causes the anode to be more positive than the cathode, the capacitor recharges and the cycle is repeated. This process is summarized in Figure 31–3.

There are two significant changes in the waveform because of the filter. Refer to Figure 31–3, and note that one change involves the ripple.

> **Key point:** The amount of variation, or ripple, in the signal is reduced since the voltage no longer drops back to zero between 180° and 360°.

Now, refer to Figure 31–4, and note that a second change involves the DC voltage.

FIGURE 31-1
HALF-WAVE RECTIFIER

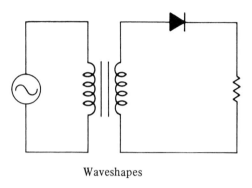

Waveshapes

Key point: The average value of the DC voltage is higher, because the voltage does not go to zero for one-half of the cycle.

Another point of significance is the rate at which the capacitor discharges. Recall that the time constant is equal to R times C. Thus, the capacitor discharges faster, or has a lower time constant, as R or C gets smaller. The ripple is larger with a smaller capacitor or resistor. Thus, a large capacitor means better filtering. Also, as the resistor gets larger, the filtering gets better. Conversely, as the load resistance gets smaller, causing greater current demand, the quality of filtering deteriorates. Therefore, the capacitive filter improves as the load current gets smaller.

Let's consider the discharge in our circuit. With an applied frequency of 60 hertz, the time for one cycle is 17 milliseconds:

$$f = \frac{1}{t}$$

FIGURE 31-2
HALF-WAVE RECTIFIER WITH A CAPACITIVE FILTER

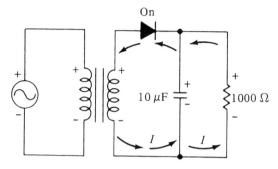

A. Current flow during first 180°

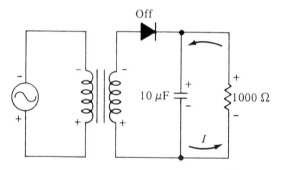

B. Current flow during second 180°

$$t = \frac{1}{f} = \frac{1}{60 \text{ Hz}} = \textbf{0.017 s}$$

The time constant for our filter circuit equals 10 milliseconds:

$$t = RC = 1 \times 10^3 \ \Omega \times 10 \times 10^{-6} \text{ F}$$
$$= 10 \times 10^{-3} \text{ s} = \textbf{0.010 s}$$

If we consider the capacitor as beginning to discharge just after 90°, we can approximate our voltage waveform on a graph, as shown in Figure 31–5. Roughly speaking, the voltage

---
FIGURE 31-3
CAPACITOR CHARGE AND DISCHARGE

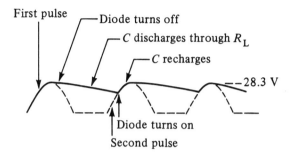

FIGURE 31-4
EFFECT OF A CAPACITOR ON THE AVERAGE DC VOLTAGE LEVEL

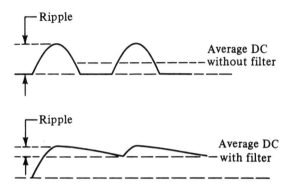

will have dropped to about 17.8 volts after 10 milliseconds. The voltage will fall a little more by the time the next pulse comes along at 17 milliseconds. This approach gives us only an estimate. More exact mathematical methods can be used, but they are beyond the approach of this text.

Ideally, the load resistor would be infinite and draw no current, and the capacitor would not discharge. It would maintain a constant charge of 28.3 volts. That assumption, of course, is not reasonable since the purpose of a power supply is to supply current. The only alternative is to increase C as the load drops and draws more current. As indicated in Figure 31–6, a large capacitor or light load (high resistance) means that the voltage will decrease only slightly between voltage source pulses. However, a heavy load (small resistance) or small filter capacitance means that large output variations will exist.

Another change is possible to improve the quality of output from this circuit. The input signal to the filter can be full-wave-rectified rather than half-wave-rectified. Let's look at the effect on just the capacitor discharge curve with the addition of the extra pulse every cycle. The two comparison curves are shown in Figure 31–7. Assume the same peak value.

FIGURE 31-5
RATE OF CAPACITOR DISCHARGE

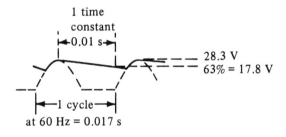

FIGURE 31-6
EFFECT OF CAPACITOR VALUE ON RIPPLE

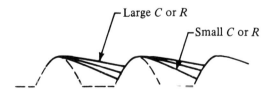

FIGURE 31-7
HALF-WAVE VERSUS FULL-WAVE INPUT
FOR A CAPACITIVE FILTER

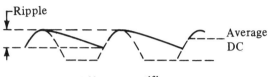

Half-wave rectifier

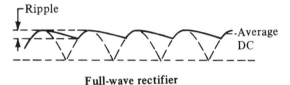

Full-wave rectifier

FIGURE 31-8
HALF-WAVE RECTIFIER WITH AN
INDUCTIVE FILTER

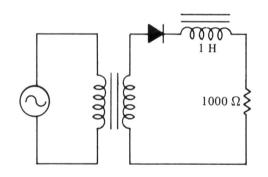

If R_L and C are not changed, the rate of discharge will be the same. However, the full-wave circuit does not discharge as much because another pulse comes along in half the time it takes a pulse to occur in the half-wave circuit. The net result is that a capacitive filter with a full-wave input has a higher average output voltage and less ripple than a filter with a half-wave input.

Characteristics

The capacitive filter consists of one or more capacitors connected across the load resistance. By charging on the rising input pulse, the capacitor is able to provide energy to the load after the pulse is gone. This result is simply the application of the definition of capacitance: opposing changes in the voltage of a circuit. The basic characteristics of the capacitive filter are as follows:

1. A capacitive filter increases the average output voltage of a rectifier circuit.
2. A capacitive filter decreases the ripple of a rectifier circuit.
3. A capacitive filter improves as the capacitance increases.
4. A capacitive filter improves as the load current demand decreases.
5. A capacitive filter's output voltage increases and its ripple decreases when the rectifier is changed from half-wave to full-wave of the same peak value.
6. A capacitive filter has a sawtooth-shaped ripple.

INDUCTIVE FILTER

Once again, we begin with the basic half-wave rectifier circuit and its load resistor. Its peak output voltage is 28.3 volts and its average DC value is 9 volts, as was shown in Figure 31–1. Our purpose in adding a filter is to improve the quality of the output signal by making it smoother. The basic **inductive filter circuit** is shown in Figure 31–8. Since inductance is the property of a circuit opposing change in the current of the circuit, the inductor is placed in series with the load.

Theory of Operation

To evaluate the performance of this filter, we place a simple series *RL* circuit of 1 henry and 1000 ohms in the rectifier circuit to see how it affects the current flow. That current flow determines the characteristics of the voltage across the load. The 60-hertz, half-wave-rectified voltage applied to the filter takes about 17 milliseconds for one cycle. As we know, current lags the voltage in this circuit by 90°, as shown in the waveforms of Figure 31–9A.

Because the inductor slows the rise of the current, the voltage across the load resistor never reaches the peak value of the applied signal. It also does not drop to zero during the second half of the cycle since the inductive field slows the fall of current as well. In other words, the inductor causes the current to level off somewhere close to the average value of the signal. And since the voltage across the load resistor is caused by the current through it, its waveshape and phase will be similar to those of the current. With these points in mind, we combine the two waveshapes of Figure 31–9A and show the relationship between the filter input and output waveshapes, as illustrated in Figure 31–9B.

The amount of ripple this filter allows depends on the time constant of the *RL* circuit. A large time constant means a slow rise and fall, which means less ripple. Recall that the time constant of an *RL* circuit is equal to *L* divided by *R*. An increase in *L* or a decrease in *R* causes a larger, or longer, time constant. While this effect slows the rise, it also slows the fall. Thus, it leads to less variation, or a smoother signal. In other words, a larger inductor means better filtering. That effect is similar to the effect of more capacitance in the preceding filter circuit.

The time constant formula also indicates that the time constant increases as the resistance gets smaller. Smaller resistance means

FIGURE 31-9
EFFECT OF AN INDUCTOR FILTER

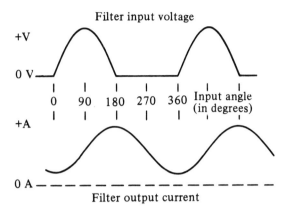

A. Output current

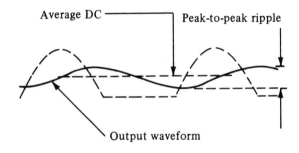

B. Output voltage

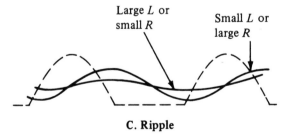

C. Ripple

more current is being drawn by the load. Hence, the inductive filter improves as the load current gets larger. This relationship is just the opposite of the one for the capacitive filter.

The relationship between inductance, resistance, and output voltage is summarized by the graph shown in Figure 31–9C. Note that the average output voltage does not change; only the ripple changes.

Another way to improve the quality of the output signal is to change the input signal from half-wave to full-wave-rectified. Two changes are immediately apparent, as illustrated in Figure 31–10. The average DC value doubles, and pulses arrive twice as fast. This second difference means that the current tries to vary twice as fast and, therefore, can vary less. Hence, the inductive filter with a full-wave input has a higher average output voltage and less ripple than a filter with a half-wave input.

Characteristics

The inductive filter consists of one or more inductors connected in series with the load resistance. By opposing the rapid change in current caused by the pulsating input, the inductor reduces the amount of change and, therefore, smooths the current. The primary characteristics of the inductive filter are as follows:

1. An inductive filter has little effect on the average output voltage of a rectifier circuit.
2. An inductive filter decreases the ripple of a rectifier circuit.
3. An inductive filter improves as the inductance increases.
4. An inductive filter improves as the load current demand increases.
5. An inductive filter's output voltage increases and its ripple decreases when the

FIGURE 31-10
HALF-WAVE VERSUS FULL-WAVE INPUT FOR AN INDUCTIVE FILTER

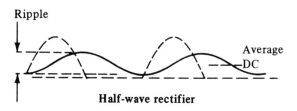

Half-wave rectifier

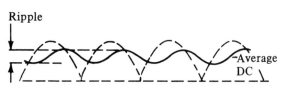

Full-wave rectifier

rectifier is changed from half-wave to full-wave of the same peak value.

6. An inductive filter has a sine wave–shaped ripple.

PI FILTER

The **pi filter** combines the characteristics of the two preceding circuits. See Figure 31–11A. The components used are capacitors in parallel and an inductor in series. If you look through the schematics of many electronic products, you will notice that the full-wave bridge rectifier with a pi filter is probably the most common rectifier-filter combination used.

Theory of Operation

We know what effect a capacitor can have and what effect an inductor can have. The pi filter circuit combines those advantages as well as

---

FIGURE 31-11
FULL-WAVE BRIDGE RECTIFIER CIRCUIT
WITH A PI FILTER

6:1

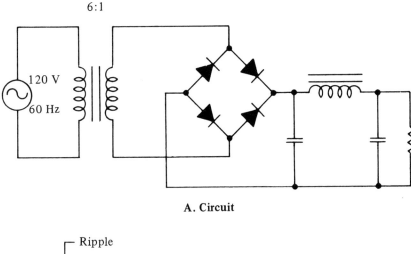

A. Circuit

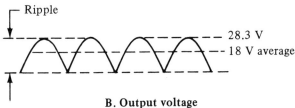

28.3 V
18 V average

B. Output voltage

the inductor's improvement with load and the capacitor's improvement with less current load. We will not analyze this circuit with the approach used in the two previous cases. Instead, we will use two methods of approximation. The full-wave-rectified signal entering the filter will be separated into two components: its average DC value and its peak-to-peak ripple value. We will evaluate how the filter responds to these two voltages.

The average DC voltage is 0.636 times 28.3 volts, or 18 volts DC. The ripple has a peak-to-peak value of 28.3 volts since it goes up to 28.3 volts and down to zero, as shown in Figure 31–11B. Although the ripple is not a sine wave, we will assume it is for con-

venience in the mathematical calculations. Otherwise, we would need more advanced mathematical methods. Furthermore, we can assume the sine wave shape since we are determining the approximate rather than exact effects of the filter.

Two more assumptions must be made: First, we assume that the capacitor is perfect with no leakage; thus, it has infinite resistance. Second, we assume that the inductor is perfect; hence, it has zero resistance. These assumptions ignore phase shift or vectors, and hence, they make our work more reasonable. We will look at the DC analysis first.

The DC equivalent circuit of the rectifier is shown connected to the filter in Figure 31–

FIGURE 31-12
DC EQUIVALENT OF A PI FILTER

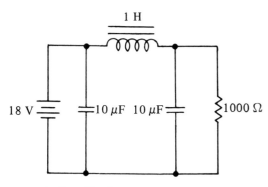

A. Rectifier DC equivalent pi filter

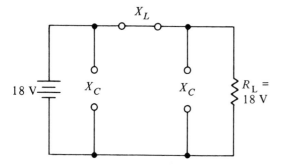

B. Equivalent circuit at 0 hertz

12A. Since the frequency of DC is zero, we can calculate the oppositions (reactance) of the inductor and the capacitor as follows:

$$X_L = 2\pi fL = 6.28 \times 0 \text{ Hz} \times 1 \text{ H}$$
$$= 0 \ \Omega$$

$$X_C = \frac{1}{2\pi fC}$$

$$= \frac{1}{6.28 \times 0 \text{ Hz} \times 10 \times 10^{-6} \text{ F}} = \infty$$

The equivalent circuit of the filter network at 0 hertz can be drawn as illustrated in Figure 31–12B. The inductance is a short, and the ca-

pacitors are open. In other words, the applied average DC voltage appears across R_L. We realize, of course, that the average could be a bit higher because of the capacitors and because of the shape of the input signal. However, this result is a good approximation.

The effects on the ripple are a bit more complex. We begin with an input signal of 28.3 volts peak to peak at 120 hertz. The circuit is illustrated in Figure 31–13A. Remember that the ripple frequency of a full-wave rectifier is twice that of the applied signal.

Once again, we can calculate the values X_L and X_C and replace those components with equivalent resistances. The calculations are as follows:

$$X_L = 2\pi fL = 6.28 \times 120 \text{ Hz} \times 1 \text{ H}$$
$$= 754 \ \Omega$$

$$X_C = \frac{1}{2\pi fC}$$

$$= \frac{1}{6.28 \times 120 \text{ Hz} \times 10 \times 10^{-6} \text{ F}}$$
$$= 133 \ \Omega$$

Our next step is to determine the amount of applied voltage that finally appears across R_L. We can begin by repositioning the components into the more familiar form of Figure 31–13B. As shown, we have a series-parallel circuit, which can be simplified with the following calculations.

The first step is to combine the parallel combination of output capacitor and load resistor, as shown in Figure 31–13C. The calculation is

$$R_L X_{C2} = \frac{R_L \times X_{C2}}{R_L + X_{C2}} = \frac{1000 \ \Omega \times 133 \ \Omega}{(1000 + 133) \ \Omega}$$
$$= 117 \ \Omega$$

Next, we add that value to the inductive reactance of the filter. This combination produces the simplified AC equivalent circuit of

FIGURE 31-13
EQUIVALENT CIRCUIT AT 120 HERTZ

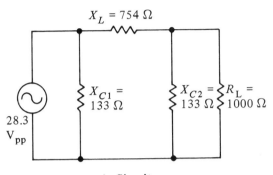

A. Circuit

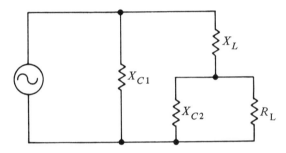

B. Components repositioned into traditional form

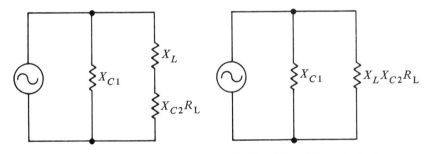

C. Load resistor and output
capacitor combined

D. Simplified AC equivalent
of pi filter

the filter, shown in Figure 31–13D. The remaining calculations are as follows:

$$R_LX_{C2}X_L = R_LX_{C2} + X_L$$

$$= 117 \ \Omega + 754 \ \Omega = \mathbf{871 \ \Omega}$$

$$I_{(RL)(XC2)(XL)} = \frac{V_{in}}{R_LX_{C2}X_L} = \frac{28.3 \ \text{V pp}}{871 \ \Omega}$$

$$= \mathbf{0.0325 \ A \ pp}$$

$$V_L = I_{(RL)(XC2)(XL)} \times R_LX_{C2}$$

$$= 0.0325 \ \text{A} \times 117 \ \Omega$$

$$= \mathbf{3.8 \ V \ pp}$$

This filter has reduced the ripple, 3.8 volts peak to peak, to about an eighth of what it was. A larger inductor or capacitor would reduce it even more.

Basically, the inductor contributes by slowing down or opposing the ripple while allowing the DC to pass. The second capacitor tends to charge to the DC level and sustain it. This capacitor gives less opposition to the ripple than the load resistor does. In this way, any ripple that has gone through the inductor can bypass the higher-opposition load. The first capacitor contributes by charging up and raising the overall DC level closer to the peak input.

Characteristics

The pi filter circuit combines the features of inductive and capacitive filters. Capacitors raise the average DC level and provide filtering under light current loads. Inductors smooth the shape of the ripple and provide filtering under heavy current loads. The combination of these features allows one component to cover the weaknesses of the other. The basic characteristics of the pi filter are as follows:

1. A pi filter tends to increase the average output voltage of a rectifier circuit.

2. A pi filter decreases the ripple of a rectifier circuit.

3. A pi filter improves as capacitance or inductance is increased.

4. A pi filter provides good filtering over a wide range of loads.

5. A pi filter has a sine wave–shaped ripple.

APPLICATIONS

Filters can be found in almost every electronic product powered by line voltage where smooth DC is required. Radios, stereos, and televisions are just a few examples. Without a good-quality filter, a 60-hertz hum would be heard from the speaker, and horizontal bars would appear on the screen of a television. Once again, the type of filter used depends on the quality of DC required balanced against a reasonable cost.

In some applications, a simple capacitor across the incoming signal is sufficient to suppress or absorb undesired pulses or noise that may also be on the line. At other times, a series inductor will be used. The component selected depends on the magnitude of current. Inductors do work better in high-current situations, but a high-current inductor can be a problem. Inductor power losses result because of the winding resistance, and these losses are proportional to the current squared.

Space is always a cost that designers must consider as well. Thus, the final design is usually a result of a compromise made by engineers considering circuit needs, available space, and reasonable cost. As we suggested in the previous chapter, look at some schematics for a television, a CB base station, or a radio. Examine the rectifier-filter circuits used. See what typical voltages and waveforms are indicated.

FIGURE 31-14
CAPACITIVE FILTER EXPERIMENT

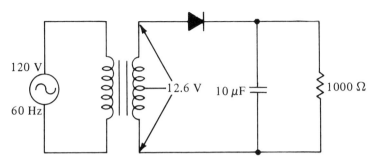

A. Circuit for half-wave capacitive filter

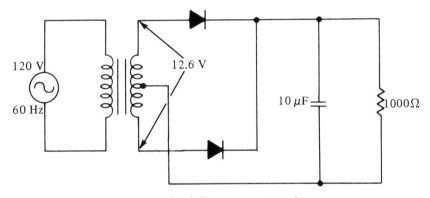

B. Circuit for full-wave capacitive filter

EXPERIMENTS

The series of experiments in this section will give you an opportunity to compare the three basic filter circuits. In each case, you will evaluate the filter as it acts on a half-wave- and a full-wave-rectified input signal.

Capacitive Filter

The first rectifier you will use is the same one described in Chapter 30. However, there is one important difference. You should select a

diode with a 500-milliampere rating to be sure that it can tolerate the current surge as the capacitor first charges.

For the half-wave rectifier experiment described in the previous chapter, we found the following expected results without a filter connected:

$$V_s = 12.6 \text{ V rms or } 17.8 \text{ V peak}$$
$$V_{DC} = 5.7 \text{ V}$$

These values can be used here, too.

Now, connect the circuit, including the filter, as shown in Figure 31–14A. Make sure the resistor is across the capacitor. In fact, *never* connect a capacitive filter without a resistor across it. Once the circuit is assembled, proceed with the experiment, as follows:

1. Measure the output DC voltage across R_L with a multimeter.
2. Measure the ripple voltage across R_L with a multimeter on the output function.
3. Measure the peak-to-peak ripple waveform across R_L with an oscilloscope. Carefully sketch the waveform on graph paper.

When these tests are completed, disable and disconnect the circuit. Then, determine the following:

— The DC output voltage of the circuit;
— The ripple factor of the circuit;
— The effect this filter had on the incoming 5.7 volts DC.

Next, connect this filter to a full-wave rectifier, as shown in Figure 31–14B. When this circuit was evaluated without a filter in Chapter 30, we determined the following values:

V_s = 6.3 V rms or 8.9 V peak

V_{DC} = 5.7 V

These values are used here, also.

Once you have the circuit connected, repeat the steps taken with the half-wave rectifier, as follows:

1. Measure the DC output voltage.
2. Measure the rms ripple voltage.
3. Measure the peak-to-peak ripple voltage and sketch the waveform.

Now, disable and disconnect the circuit. Then, determine the following:

— The DC output voltage;
— The ripple factor;
— The effect this circuit had on the incoming 5.7 volts DC.

This experiment should provide you with enough basic information to compare this filter with the other two.

Inductive Filter

The inductive filter will also be evaluated on its performance in smoothing the output of half-wave and full-wave rectifier circuits. Your first step is to connect the half-wave circuit shown in Figure 31–15A. Make sure that the inductor you choose has a current rating greater than the peak current expected in this circuit, which is

$$I_{peak} = \frac{V_{peak}}{R} = \frac{17.8 \text{ V peak}}{1000 \ \Omega} = 0.0178 \text{ A}$$

Once you have connected the circuit, repeat the measurements described for the capacitive filters. Also, determine the three answers as they apply to this circuit. Then, disable and disconnect the circuit.

Now, connect the inductive filter to a full-wave rectifier, as shown in Figure 31–15B. Repeat the three measurements for this circuit, determine the three answers, and then disconnect the circuit.

By now, you can see that you are gathering data for comparing the three basic filter circuits. It is a good idea to prepare a table similar to Table 31–1, which shows filter and rectifier types as well as input voltage and frequency data. This data can be obtained from the preceding chapter. Information relating to output voltage and ripple factors may be derived from the experiments in this section and then added to the table.

FIGURE 31-15
INDUCTIVE FILTER EXPERIMENT

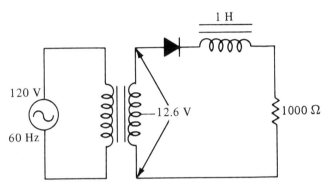

A. Circuit for half-wave inductive filter

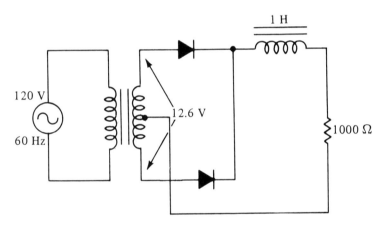

B. Circuit for full-wave inductive filter

Pi Filter

The pi filter circuit can be connected with the components already used plus another capacitor. The half-wave rectifier circuit should be connected as shown in Figure 31–16A. The tests and resulting information for this circuit are similar to those for the previous circuits. Input and output data values should be recorded in a table (see Table 31–1), waveforms sketched, and answers saved for future reference.

Next, repeat the process for the pi filter with a full-wave-rectified input. The circuit should be connected as shown in Figure 31–16B. Record your data in a table, and save your answers.

Lab Report

When the tests are completed, put all the parts away. Now, consider your report. Compare the characteristics of these three filters. For

TABLE 31-1
INPUT VALUES FOR FILTER EXPERIMENTS

Filter Type	Rectifier Type	Input (in Volts)		Frequency (in Hertz)
		DC	Ripple	
Inductive	Half-wave	5.7	17.8	60
	Full-wave	5.7	8.9	120
Capacitive	Half-wave	5.7	17.8	60
	Full-wave	5.7	8.9	120
Pi	Half-wave	5.7	17.8	60
	Full-wave	5.7	8.9	120

instance, in your report, consider the following questions:

1. How are the filters different?
2. Which combination provides the highest voltage?
3. Which combination has the least ripple?
4. Which combination has the lowest ripple factor?
5. Does the frequency difference resulting from half-wave or full-wave rectification influence ripple? Does it influence the ripple factor?

If you had repeated all these experiments with a higher and a lower load, you would have seen more differences in these circuits. However, that comparison is not necessary since we know that the capacitive filter gets worse and the inductive filter better as the load current increases.

One more point must be mentioned again: There should always be a resistor connected across the output of a filter with capacitors. Its purpose is to discharge the capacitors, in case the load does not, after the power has been turned off. This **bleeder resistor** should be high enough in value so that its demand is no greater than 10% of the normal load of the sup-

ply. Without a bleeder, filter capacitors can stay charged for days, and some unsuspecting person may get hurt. So, it is important to be especially careful when working with capacitors.

SUMMARY

Filters are passive circuits used for removing undesired content in an electric signal. Like rectifiers, they can be found in almost every electronic product that is powered by line voltage. The filters described in this chapter are used with rectifiers to remove the undesired ripple content. Generally speaking, the basic components of filters are capacitors and inductors. Various arrangements using one or more of these parts are possible.

The simplest version, a capacitive filter, has one capacitor connected across the load. It works in the following manner: When the diode is forward-biased, the diode conducts, causing current to flow through the load. Since the capacitor is in parallel with the load, it immediately charges up to the peak level of the input signal. Once the peak is passed (beyond 90°), the diode becomes reverse-biased and stops conducting. For the remainder of

FIGURE 31-16
PI FILTER EXPERIMENT

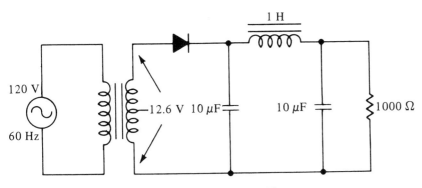

A. Circuit for half-wave pi filter

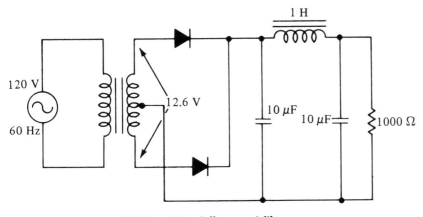

B. Circuit for full-wave pi filter

the cycle, current continues through the load R_L as a result of the capacitor discharge. The process reverses when the next pulse arrives.

If the next pulse comes along soon, as in a full-wave rectifier, the capacitor cannot discharge very much, so there is not very much ripple. A half-wave signal produces a pulse much later than a full-wave signal does, so there is more ripple. Still, it is less than the ripple would be if the capacitor were not there.

Two other factors affect the amount of rip-

ple that will exist: the size of the capacitor and the current demand. A larger capacitor means a longer time to discharge, which results in less ripple output. A large current demand means a lower R_L, which shortens the time constant and increases the ripple.

The characteristics of the capacitive circuit can be summarized as follows: A capacitive filter has a higher output voltage than other filters since the capacitor can charge up to the peak input voltage. Filtering increases when capacitance is increased but deteriorates as the load current increases. A capacitive filter must always have a bleeder resistor connected across it in order to discharge the capacitor when the circuit is de-energized.

While capacitors oppose changes in voltage, inductors oppose changes in current. The basic inductive filter consists of an inductor in series with the load. A pulsating input signal attempts to vary the load current between the peak value and zero. The inductor causes the current to level off somewhere close to the average value of the signal. The result is a current — and, therefore, a load voltage — that varies in a sine wave shape around the average value of the signal. The amount of variation depends on the time constant of the circuit.

When inductance is increased, the time constant increases, which means less variation in the output. As the load current in-creases, the apparent load resistance is less, once again producing a longer time constant and, hence, less ripple. So, while the capacitive filter deteriorates with increased load current, the inductive filter gets better.

These two components can be used together and with resistors in a variety of combinations in order to utilize their individual advantages. Probably the most common arrangement is the pi filter. In this device, a parallel input capacitor helps raise the DC average voltage while providing a low-impedance path for ripple. A series inductor provides a high-impedance obstacle for ripple yet little opposition to the DC flow. A parallel capacitor across the load helps smooth the voltage at that point while acting as a low-impedance path for any ripple in the current that got beyond the inductor.

CHAPTER 31

REVIEW TERMS

bleeder resistor: relatively high resistance connected across the output of a filter circuit for the purpose of discharging the capacitors when the power is turned off

capacitive filter: filter circuit utilizing a capacitor connected in parallel with the load

filter: circuit connected to the output of a rectifier for the purpose of reducing ripple

inductive filter: filter circuit utilizing an inductor connected in series with the load

pi filter: filter circuit utilizing both series inductance and parallel capacitance

REVIEW QUESTIONS

1. Describe the purpose of a power supply filter.

2. Describe the connection and operation of a capacitive filter.

3. What are the advantages of a capacitive filter? What are the disadvantages?

4. Describe the connection and operation of an inductive filter.

5. What are the advantages of an inductive filter? What are the disadvantages?

6. Describe the connection and operation of a pi filter.

7. What are the advantages of a pi filter?

8. Describe the connection and purpose of a bleeder resistor.

9. What are the dangers of operating a power supply without a bleeder resistor?

10. State the characteristics of an inductive filter.

11. State the characteristics of a capacitive filter.

12. State the characteristics of a pi filter.

13. Which rectifier circuit provides voltage easiest to filter? Why?

14. Which rectifier circuit provides voltage most difficult to filter? Why?

REVIEW PROBLEMS

1. Draw the circuit of an inductive filter connected to the output of a half-wave rectifier with a transformer input.

2. Calculate the approximate DC and AC output voltages for the circuit described in Problem 1 if the input voltage is 120 volts rms, the transformer is 1 : 3 step up, the inductor is 1.5 henrys, the load is 10 kilohms, and the input frequency is 60 hertz.

3. Sketch the expected output waveform for the circuit described in Problem 2.

4. Draw the circuit of a capacitive filter connected to the output of a half-wave rectifier with a transformer input.

5. Sketch the expected output waveform for the circuit described in Problem 4.

6. Draw the circuit of a pi filter connected to the output of a half-wave rectifier with a transformer input.

7. Repeat Problems 1 and 2, but use a full-wave rectifier circuit.

8. Repeat Problems 1 and 2, but use a full-wave bridge rectifier circuit.

9. Sketch the relative waveshapes of the output voltages for Problems 7 and 8, and compare them with the results of Problem 3.

10. Repeat Problem 4, but use a full-wave rectifier circuit.

11. Repeat Problem 4, but use a full-wave bridge rectifier circuit.

12. Sketch the relative waveshapes of the output voltages for Problems 10 and 11 and compare them with the results of Problem 5.

13. Repeat Problem 6, but use a full-wave rectifier circuit.

14. Repeat Problem 6, but use a full-wave bridge rectifier circuit.

15. Sketch the relative waveshapes of the output voltages for Problems 13 and 14, and compare them with the results of Problem 5.

32

ELECTRONIC EQUIPMENT EVALUATION

State the purposes of equipment evaluation.

Describe the procedures for equipment testing.

Describe the process of evaluating test results.

Describe the process for correcting defects.

PURPOSES OF EVALUATING EQUIPMENT

The objective of this chapter is to summarize the purposes and methods of evaluating electronic equipment. This chapter may be the most important one in the text. In it, you will learn how to apply the information presented in previous chapters. Much of the electronic theory presented earlier can be kept in notebooks, memorized, or retained in other ways. Now, you are at a point where you must understand how components work in relation to each other. You must be able to conduct tests, evaluate data, analyze symptoms, and conclude possible causes. Remembering is no longer enough. You must **understand,** think, and then act. The results of your actions will be what is important.

This chapter describes the procedures a technician follows in the process of evaluating the performance of electronic equipment. For example, in the evaluation process, symptoms are observed and compared with what is expected. But before a technician looks for symptoms in electronic equipment, he or she must know the purpose of the evaluation. In the electronics industry, equipment is typically evaluated for one of four reasons: to locate defects, to determine equipment characteristics, to provide quality control, or to perform preventive maintenance. We will examine each of these purposes in the subsections that follow.

Locating Defects

Quite often, electronic equipment is evaluated because someone believes that it is defective. For instance, a customer returns a computer because it will not operate. Another customer has distortion on a mobile telephone. These customers conclude that, since the operation of the device is not acceptable, the product is defective. In cases like these, the defect may or may not be within the product. Most technicians obtain more information from the customer before examining the equipment.

Sometimes, the device has smoke coming from it, making the decision easier. Obviously, there is a defect. Locating the defect, however, involves more than finding a burned part. The difficult task is determining the cause of the malfunction. In other words, the defect may, in fact, be defects, symptoms and cause. Your task, then, is to locate the cause, not just replace the burnt-out component. Otherwise, the new part will also burn.

The process of locating defects requires knowledge, skill, and imagination. You must understand what the circuit is supposed to do. You must be able to properly use test equipment to determine what is happening.

Key point: You must understand the meaning of your measurement results.

Remember that troubleshooting is simply a matter of applying what you have learned.

Skilled troubleshooters are valuable to a company because they save money for the company. That is, there are reasonable limits to what a customer can be charged for the repair of an electronic product. So the faster you can locate defects, the more valuable you are.

Determination of Characteristics

Technicians who work with design engineers often must determine the characteristics of a particular circuit. This task involves conducting tests to determine voltage, current, or resistance values as well as waveforms. While this task may seem simple, you should think and plan before proceeding. For example, you should know what parameters you are going to measure and what they are likely to be. The designer will usually provide you with them. Otherwise, you must calculate them. Your task also includes the selection of the equipment to be used. Again, there are questions you must answer. Which meter should be used? Is the meter's sensitivity high enough? Is it's frequency response adequate?

Determining the characteristics of a circuit should be a simple task. You merely take some measurements and write them down. This task is not difficult to do correctly. However, it is also not difficult to do incorrectly. For instance, the meter may be defective. It may be the wrong one for the job. Or you may make an error in reading the indicated values. A key to success in measurement is to always ask yourself, Do these numbers seem reasonable?

Quality Control

Most manufacturers have quality control departments and procedures to ensure that finished products are within predetermined standards. **Quality control** is a process in which components and products are inspected at various stages of manufacture for the early detection of defects. A defective component is put aside and is not installed in equipment. Subassemblies are tested before they are installed in large products. The point is to find defects as early in the manufacturing process as possible to reduce the time and cost of locating and correcting them later.

Quality control methods vary from one manufacturer to another. One reason for the variation is the volume of their manufacturing systems. Thus, a company that makes 1000 items per day cannot check every one at every step along the way. So samples are taken and checked at various points. If the samples are acceptable, the items pass and the sampling process continues. When samples indicate defects at one point in production, the sampling

percentage increases. Typically, all products are given some final inspection before they are shipped.

Large products such as computer systems undergo dozens of quality control tests as they progress through the manufacturing process. ICs are tested before they are installed. Circuit boards are tested when they are finished. Thus, by the time a computer system is connected, every part is known to be working. Quality control tests at several stages increase the likelihood that, when subsections are interconnected, the system will work.

Preventive Maintenance

Products used in their normal application can be checked on a regular basis to reduce the likelihood of future defects. Front-end alignment and wheel balancing are preventive maintenance steps taken to prolong the life of automobile tires. Replacing multimeter batteries on a regular basis prevents malfunction as well as damage from defective batteries. Regular equipment calibration is another form of preventive maintenance. Thus, we see that **preventive maintenance** is a scheduled process of inspection and correction conducted to ensure proper operation of a product.

Broadcasters have a wide range of tests that must be conducted for preventive maintenance. Certain parameters such as power and frequency must be measured several times per day, for example. Although these measurements may be taken and logged automatically, they are certified as completed and correct by a licensed technician. The technician guarantees the accuracy of the results, not the broadcaster. These tests ensure that the signals transmitted are in compliance with the law. Broadcasters are also required to conduct proof-of-performance tests a number of times each year. These tests consist of evaluating the signal from microphone to transmitter to ensure audio signal quality. Table 32–1 shows

TABLE 32-1
TYPICAL FORMAT FOR DOCUMENTING FREQUENCY RESPONSE AND DISTORTION OF THE MAIN TRANSMITTER

	50% Modulation					
Frequency (Hz)	50	100	400	1000	5000	7500
Reference	−66.1	−66.1	−66.1	−66.1	−66.1	−66.1
Attenuation	−63.6	−65.4	−66.1	−66.2	−66.2	−66.8
Response	−2.5	−0.7	0	+0.1	+0.1	+0.7
Distortion	2.0	1.65	1.45	1.45	1.0	1.0

a typical format used for documenting this process.

All electronic equipment involved in the broadcasting process is evaluated regularly, and records of the tests are kept. The point here is to locate any potential problems before they become defects. For this reason, broadcasting systems rarely fail. When they do, it is often because of some factor, such as an electrical storm, beyond the technician's control.

PROCEDURES FOR EQUIPMENT EVALUATION

The process of evaluating electronic circuits and systems is best conducted in a planned and organized manner. Thus, you need to know what circuits you will evaluate, what instruments you will use, what procedures you will follow, and what results you should expect. Information about expected results comes from specifications.

Obtaining Specifications

There is no way you can find what is wrong if you do not first know what is right. Test measurements have little significance unless you have the **circuit specifications** — that is, a description of voltages, waveforms, and other behaviors to be expected from a circuit or product during normal operation. There are two methods for obtaining specifications: You can get them from someone else, or you can determine them yourself. The second method requires that you fully understand the operation of the circuit and then spend considerable time drawing a schematic and calculating the values. Although you should eventually be able to determine specifications, it is an ineffective and time-consuming method. Hence, it should be avoided if at all possible.

Manufacturer's or designer's specifications should be your first choice since those individuals know how the circuit is intended to operate. They also quite often specify which locations are the best to examine, which parameters need to be measured, what equipment should be used, and what range of values can reasonably be expected. All you need to do is take measurements and compare results.

When specifications from the manufacturer or designer are not available, you may find other sources. For instance, supply houses often sell service manuals that provide the same information. Service manuals are often more valuable than manufacturer's specifications since they are prepared specifically for technicians who will be testing the circuits for known defects. Quite often, the manuals include schematics, parts lists, alignment procedures, and other troubleshooting information you may need to know. Charts like the one presented in Figure 32–1 can ease the troubleshooting process by describing probable causes and corrective actions for certain conditions. Pictorials such as the one shown in Figure 32–2 show the location of internal diagnostic indicators that, like the chart in Figure 32–1, can simplify the troubleshooting process.

In summary, the first step in evaluating equipment is to obtain specifications. If printed specifications are not available, you must calculate them. Only after you have obtained specifications can you evaluate a circuit. Here is the important point to remember.

Key point: A circuit is working properly only when it is working to specification.

Analyzing Symptoms

The cause of many malfunctions can quite often be determined by analyzing the symptoms before a product is tested. Generally, tests are conducted to confirm a technician's opinion. You can learn to analyze symptoms for many electronic products. We will consider a television as a typical example. Suppose a customer brings you a television that does not work. You are asked to fix it. The first thing you should do is look for the symptoms. Is the sound working? Is it good quality? Is the picture quality good? Does the television receive all channels?

Once you have found the symptoms, you should consider two routes to the problem. What sections work properly? What sections do not work properly? This step helps you narrow the range of defective circuits.

Suppose there is no sound or picture and the picture tube does not have a raster (does not light up). You know, then, that there must be a problem with a circuit upon which all of these functions depend. It may be the power

cord, the fuse, the low-voltage power supply, or some other circuit. However, if there is no sound or picture but there is a raster, the power cord, fuse, switch, and power supply must be working. Without all of them, there would be no raster. Now, you can determine whether there is noise from the speaker that changes as the volume control is changed, even though there is no audio (that is, no broadcast sound).

To follow this process, you must understand the function of each section, the symptoms of normal operation, and how they depend upon each other. Step by step, you can check each circuit and determine whether or not it is working according to the type of output you obtain.

Sometimes, the wrong cause is given to a problem. For example, a person may complain that a brand new television has a poor-quality picture. But the problem may not be with the television at all — that is, it may be working within its advertised specifications. Perhaps it is not a sensitive model and is not receiving a strong signal. In this case, a new antenna may be needed, or the antenna location may need to be changed. Although the symptoms indicate a poor picture, the problem here is a poor signal. So in equipment evaluation, you must ask the following question: Is the product doing what its specifications say it should be doing? If it is but the results are not satisfactory, look elsewhere for the cause. It is the cause of the problem that must be corrected.

Determining Test Procedures

As noted in the preceding sections, you should not begin any tests until you have considered the symptoms of the problem and know just what steps you will follow in your test. You must also know what results you should get under normal conditions — that is, the specifications. Sometimes, this information is pre-

FIGURE 32-1
TROUBLESHOOTING CHART FROM A
SERVICE MANUAL

Condition	Cause	Corrective Action
POWER PROBLEM 1. No lights with POWER switch ON.	a. Facility AC power problem. b. Power cord unplugged. c. Open fuse (F1).	a. Turn on facility AC power and ensure voltage is correct. b. Plug cord in proper AC outlet. c. Replace Fuse (F1). See Fuse Replacement section in this manual. If fuse blows again call Instron for assistance.
2. STOP button not lit or dimly lit, or blank EXTENSION readout.	a. AC input power low.	a. Set line voltage select tab to match facility AC input power. See Installation section in Operation Manual. b. If power set up is correct go to Power Problem 3.
3. No indicators lit except power on switch.	a. Open circuit breaker on control board. b. Defective control assembly.	a. Reset both circuit breakers on control board. Use diagnostic procedure D1. b. If circuit breakers trip again replace control assembly. Use repair procedure R1 or call Instron for asssistance.

sented in the form shown in Figure 32–3. This table lists what voltage can be expected at each pin of a given connector and what the color of the wire should be. The troubleshooter's task is to measure the actual voltages and compare the results with the specifications.

Specifications may also indicate the circumstances under which you should get certain results. For example, manufacturers will often note that a specified value was measured with a 20,000-ohms-per-volt multimeter when the television was on and receiving a program. Ideally, the schematic you follow will indicate expected values, the equipment used for testing, and the circumstances under which these values were obtained. For instance, the schematic in Figure 32–4 shows voltage waveshapes that can be expected at key points in a circuit. If this type of information is not available, you must plan your test accordingly.

Knowing that you have a DC voltage to measure may not be all the information you need before you begin your tests. For example, you must also consider what type of meter to use. A VOM is often inadequate for this measurement, since a low-sensitivity meter

FIGURE 32-2
PICTORIAL FROM A SERVICE MANUAL
SHOWING THE LOCATION OF
DIAGNOSTIC LAMPS

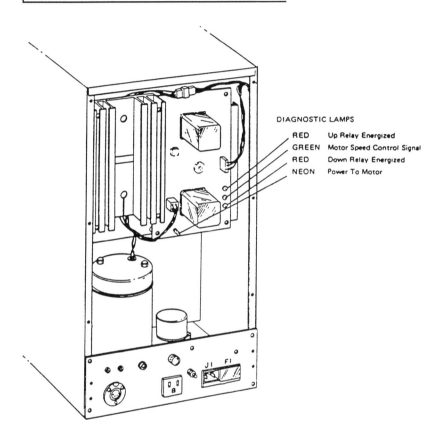

DIAGNOSTIC LAMPS

RED   Up Relay Energized
GREEN   Motor Speed Control Signal
RED   Down Relay Energized
NEON   Power To Motor

can load down a high-impedance circuit, causing incorrect results. If you do not know the characteristics of the circuit and the equipment is not specified, it is wise to avoid this potential problem by using an electronic multimeter. An AC circuit presents the same problem plus many others. For instance, you must know whether the frequency response of the meter you intend to use is high enough for the frequency being measured. In other words, the adequacy of your methods is as important a concern as the magnitude of the value to be measured.

So in planning your test procedures, you must consider the following questions: What part of the circuit and which parameters must be measured? Should an oscilloscope or a multimeter be used? What is the frequency in this section of the equipment? What is the voltage level? Is this a high-impedance or a low-impedance point? The answers to these questions will help you choose the test equip-

FIGURE 32-3
TABLE FROM A SERVICE MANUAL
SHOWING CONNECTOR PIN VOLTAGES

Power Supply Connections	Wire Color Code	Prog. Unit Pin No.	Function
Power Supply Chassis			
J6-1	Green	D	OV1 Power Ground
J6-2	Orange	K	OV2 Logic Ground
J6-3	White-Green	W	+15V Unregulated
J6-4	Tan	P	OV3 Signal Ground
J6-5	White-Brown	Z	−15V Unregulated
J8	White-Yellow	X	+5 Unregulated
J9-1	Red	C	+24V
J9-2	Grey	L	+28V
J9-4	White-Red	V	−12V
J9-5	White-Blue	Y	+12V
Chassis Gnd.	Green	R	OV4 Chassis Ground
Regulator Module A461-2			
YEL	Yellow	43K	+15V Unregulated
RED	Red	J	+15V Regulated
GRN	Green	M	OV3 Signal Ground
WHT	White	43E	−15V Unregulated
BLK	Black	T	−15V Regulated

ment, which is a major step in determining your test procedure.

Another issue to consider is how to conduct the tests without causing injury to yourself or the circuit. Plan on making no disconnections. Have a schematic and the test equipment ready before you begin. Keep paper and a pencil handy for taking notes. Inspect the device before you plug it in to look for obvious defects. Find out just where the area of interest is so that you do not have to hunt through a live circuit. Once you have located the specific problem area, you can energize the circuit.

Conducting Tests

There is one more question to answer before you make any measurements: What is the reference or common point? Most voltage measurements use some common or ground point as a reference. It may be a metal chassis, or it may be some other point. The schematic is your best source for determining where to connect the common lead of the meter or oscilloscope. Connecting it to the wrong place will cause incorrect results, malfunctioning of the product being tested, and possible damage to the product.

Once this connection is made, you can proceed to measure and record voltages, observe and sketch waveforms, and de-energize the circuit and make resistance measurements. All of these steps are done according to your pretest results in an orderly manner. Your next step is to compare your results with the specifications, which will be done with the power off and the product set aside.

FIGURE 32-4
SCHEMATIC FROM A SERVICE MANUAL
SHOWING EXPECTED RESULTS

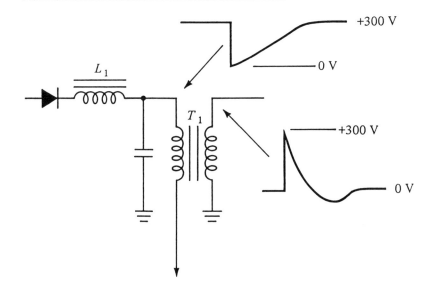

EVALUATION OF TEST RESULTS

Gathering data is only the first step in the troubleshooting-repair process. The next step is to determine what the results mean. In other words, is the circuit working properly? If it is not, are there any clues about what the problem might be?

In this step, the knowledge of theory and the ability to analyze information become very important. For example, you must compare your results with the specifications. Then, on the basis of your comparison, you must decide what is causing the problem.

Comparing Results with Specifications

The comparison step is conducted with a pencil. That is, your test results are recorded on the schematic adjacent to the specified values. Mostly, these results will be voltages. This recording step gives you an opportunity to make sure that you have information about all parts of the circuit in question and provides an easy way for you to compare what is with what should be. In your comparison, you will not be looking just for differences — you will be looking for *significant* differences.

Determining Significant Differences

A **significant difference** is defined as a difference between measured and specified values that is beyond what is considered to be normal or acceptable. Do your test results show any significant differences from the specifications? This is one of the more difficult questions you will have to answer when evaluating the performance of a circuit. An understanding of basic electrical theory such as Ohm's law, series circuits, and parallel circuits is often enough to answer the question. What is required here is thought and common sense. Let's consider some examples.

Suppose a voltage is much lower than specified. You compare it with others around it. If they are also low, then you may conclude that the source is low. There is no defect in this section of the circuit. So one good method of evaluating a result is to compare it with results from components in the test area. They may all be a bit low or a bit high. If one is further from the specification than the others, it should be tested again.

At what point does a difference become significant? Unfortunately, an exact answer is not easy to give. Results that are 50% more or less than specification generally indicate a defect; those with a 5% difference generally do not. Since many components have a 10% tolerance, they can normally cause that much variation. When the difference exceeds 10%, it may be significant.

There are times, however, when a 10% variation is not acceptable. For example, suppose the circuit you are working on has a power supply with a voltage regulator, which is a circuit that maintains a constant output even though the input varies. Your circuit may specify a DC voltage of 18 volts $\pm$ 0.5 even if the line voltage varies between 105 and 130 volts. For this specification, a DC voltage outside the range 17.5 to 18.5 volts indicates a defect in the circuit. Normally, though, you can accept a power supply variation that is proportional to the line variation. That is, if the input is 10% high, the output can be 10% high.

Finding the Cause

Once you have concluded that your test results give a significant difference, you must return to basic electrical theory. Causes can usually be determined by the application of Ohm's law and series and parallel circuit rules. Suppose a voltage is 0 volts when it should be something else. What are the possible causes? The source may be 0 volts. There may be an open in the series circuit, which means that there is no current flow and no voltage drop. Or there may be a short across the circuit. At this point, a resistance measurement is needed. The short will be obvious in a resistance measurement.

Suppose a voltage is much higher than it should be. Once again, basic rules can be applied. A voltage may be very high because the source is. Assume that you checked the source and it was normal. Now, you must look for other causes. A high voltage in a series circuit indicates that a voltage drop is out of proportion. Perhaps an opposition is too high or is open. Let's consider this problem for the transistor circuit shown in Figure 32–5.

Assume that the symptoms are a collector voltage that is too high and an emitter voltage that is too low. What can cause a low drop across R_E? A short resulting in little current flow can cause a low voltage drop. Notice also that there is no drop across R_C. What can cause that problem? Once again, if the voltage drop is zero, then no current is flowing. The conclusion here is that the transistor is not conducting.

Now, you must determine why the transistor is not conducting. First, you check the collector voltage. Your measurements indicate that there is a collector voltage. Next, you check the base voltage. Your measurements also show a base voltage. Furthermore, the difference between the emitter and base voltages is the expected difference of a few tenths of a volt that exists when that junction conducts. Since you have all the necessary conditions for flow of collector current but no collector current is flowing, you conclude that the junction must be open.

Your intention at this point is to localize the problem to a specific component or con-

FIGURE 32-5
TRANSISTOR VOLTAGES

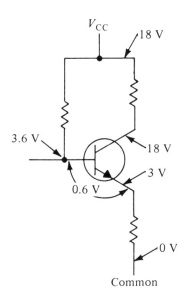

V_{CC}

18 V

3.6 V

18 V

0.6 V

3 V

0 V

Common

nection. Voltage tests can almost always help you here. Their advantage is that they can be done without any disruption to the circuit. Once you have found the faulty component, you may want to make a few more tests to be sure. Resistance checks are good techniques in this situation. You may also want to remove one lead to isolate the component from the others in the parallel section. The final result of all of these tests is a positive location of the defect.

Planning Repairs

Sometimes, a problem can be corrected with only a little effort. For instance, a printed circuit board may have a small crack that can be repaired with a touch of solder. Other times, more drastic steps are required. However, you should do only the work required to solve the problem. Remember that, as a troubleshooter,

you want to save time and expenses. In addition, the more you work on or handle a circuit, the more likely you are to introduce additional problems. Two rules apply here.

> **Rules:** Leave things alone unless you have no choice. If a component is not broken, do not fix it.

Do not quickly jump into the wholesale replacement of components. Think about the problem. Can it be corrected with cleaning or resoldering? Can it be repaired in the circuit? If it must be removed, do you have a replacement, or can you get one? Before you take any steps, know just what you are going to do, how you will do it, where the new parts will come from, and how you will install them. When you are sure you are ready, proceed with the repairs.

DEFECT CORRECTIONS

The major part of the troubleshooting-repair process is over once the defect has been identified. This does not mean, however, that there is nothing left to do or that what must be done is trivial. Several issues must still be resolved, such as who will do the repair work, where the repair will be done, what procedures will be followed. But the most important issue here is the quality of the repair work.

> **Key point:** The quality of the repair work must be equal to the workmanship of the original product.

Deciding Who Performs Corrections

The decision of who corrects the defect may depend upon the nature of the product and its defect. A product that is under warranty must be returned to the supplier; you should not work on it at all. Thus, if the item has a 12-month guarantee and it is only 4 months old, do not open it. You can consider symptoms and diagnose the problem, but do not otherwise work on the product. Once you work on it, the supplier will no longer be responsible.

If you are employed in the test department of a manufacturing operation, you probably will only locate defects, tag the product, and then set it aside. Your job is to find defects; other personnel have the task of replacing parts. In other situations, the replacement jobs may require fine skill and dexterity so that only highly skilled personnel can do the work. Or the replacement work may simply be routine, in which case semiskilled technicians will do the repairs. Thus, as a test technician, you may be required to find defects but not correct them.

If you are a service person on the road or if you have a shop of your own, you will probably have to locate and correct defects yourself. But if you feel that you cannot repair a product, you should take it to someone else. However, by the time you have your own shop or are out on service calls, you probably will be fully prepared for making repairs. Your procedures will be routine and will be similar to the ones described in the next section.

Procedures to Be Followed

Your intention is to repair or replace the defective part so that the product is once again operating normally. You should not introduce any new problems, such as scratches or external damage. So protect the equipment from all external damage. Use proper tools, because improper tools also cause damage. If you have to replace a part, make sure that you remove the correct one and that you have the correct replacement available. Do not remove a part before you have the replacement unless you need it as a sample or a trade-in.

Make a sketch of the part location and lead connections before you remove it. Do not rely on memory. Quite often, a manufacturer provides layout drawings and detailed instructions for the replacement of parts. Your best approach is to follow these drawings if you have them. Otherwise, make your own sketch. Take your time, and think before you act.

Replacing Parts

When you have located the defective part and obtained its replacement, you are ready to replace it. Proceed carefully. Some parts can be easily removed, but others may be damaged in the process.

Disassembling a circuit can be a tedious job. The first step is to disconnect electrical connections. The second step is to disconnect mechanical ones. Electrical connections can be opened with an iron and desoldering. Remember that you are removing a component from a finished circuit. It was probably not installed in the order in which you are removing it. So things may be in your way that were not in the way of the assembler. The danger that you must avoid is allowing your soldering iron to get too near other components. Take a good look at the component and its surrounding area before you bring the iron close.

Once the desoldering is complete, you can open the mechanical connections of the leads. This task must also be done gently since adjacent components can easily be broken. Remember that if you break good components, you have added work and cost that cannot reasonably be charged to the customer.

When the mechanical connections of the leads are separated, the component can be removed. For this job, you may simply be able to pick it up, or you may need to remove some hardware. If hardware is to be removed, be on the alert for accompanying washers. Often, a technician will turn a screw with one hand and hold the nut with the other, forgetting about the washers, which may fall and disappear in a maze of circuitry. When the power is reapplied, another defect shows up in the product.

Once the part is removed, it can be set aside. If it is defective, mark it as such before it gets mixed in with the good parts. You may wish to save it to show the customer. Generally, parts can be disposed of by putting them in the trash. Some parts, though, such as CRTs, have special disposal procedures that should be followed for safety reasons.

The installation of replacement parts is the next step. If you sketched the location of the original part and you were careful in its removal, you should have no difficulty now. Mechanically secure the part, mechanically connect the leads, and solder the leads if required. Your goal here is make sure that your workmanship is identical to that of the original assembler.

Post-Repair Inspection

When your repair work is completed, two conditions should be satisfied. The product should work according to specifications, and it should look the same way it did when you received it. Generally, a repaired product is operated long enough to ensure that it works and that the defect has been found. Although the device may work immediately, it could fail a few minutes or a few hours later. To avoid the cost and embarrassment of being recalled by the customer, give the equipment time to prove its condition.

Before your work is finished, you will also want to make sure that the product looks good. Secure loose parts, and clean the outside. The product should be as important to you as it is to the customer. Finally, be sure that the product meets the manufacturer's specifications and yours.

Repair work will also include some form of documentation. Where lab reports were used before, service reports are used here. A service report lists the name of the customer, product identification, and the nature of the defect. This information is given to you before you begin. When you are finished, list the cost of materials, the time you spent, and the total cost for the repair work. Once more, accurate records are essential. The final result is a satisfied customer and a product operating according to specification.

SUMMARY

Technicians evaluate equipment for a variety of reasons. The equipment user may be dissatisfied with the performance and believe a defect exists. A designer may want to know just how a circuit is functioning. Or a manufacturer may wish to obtain performance data. For example, tests are conducted throughout the manufacturing process to ensure product quality control. In this way, defects can be found and corrected early in the process, avoiding complicated and costly repairs later. Finally, products are tested prior to shipment to ensure that the product operates as specified.

Preventive maintenance tests of finished products are conducted on a scheduled basis as a way of avoiding downtime caused by failures. In this way, variations from normal operation can be detected before a product breaks down. So for one reason or another, the performance of electronic products must be evaluated.

Evaluation of a circuit begins by knowing just what the circuit should do. Manufacturer specifications or other published data is your best source for this information. In the absence of published specifications, you must draw a schematic and calculate all values. This approach is both time-consuming and subject to errors. However, there will be times when you have no choice but to use this method.

Once you know what should be, you can determine what is. The first step is to look for symptoms. What functions of the circuit are performing as specified and which ones are not? What is the interrelationship between these sections? Do they share a common power supply? By collecting symptoms of performance, you can often localize a defect to one specific area of a product. At this point, you can think about looking inside the product.

This next phase begins with your decision of the tests to conduct, the instruments to use, and the required specifications. What sensitivity is required? What frequency response is needed? What are the safety precautions? When these questions are answered, you can select the test equipment, energize the circuit, and take measurements. Measure and record voltages around the section of concern. Observe and sketch the waveforms. Then, de-energize and disconnect the circuit.

The analysis of results involves comparing measured values with specified values. Your major task is to find any differences that are significant. The level of significance varies, although some general rules can be followed. In the absence of specified limits of variation, a difference of 5% is acceptable; a difference of 50% is not. A normal resistor can vary as much as 10% or more. But quite often, the variation is more drastic than that. For example, a voltage may rise to the power supply level or drop to zero. In either case, you have found the location of the defect. The next question is, What is the cause?

Finding the cause of a defect is often just a matter of applying Ohm's law or other basic circuit rules. A voltage that is too low or too high may be the result of a simple short or open. So a clear understanding of these effects is invaluable. Recall that an open in a series circuit causes the total applied voltage to appear across the open and 0 volts to appear elsewhere. A series short causes other voltages in the series circuit to rise while the drop across the short goes to zero.

Once you have found the cause of a defect, you can correct it. Before a part is replaced, though, you need to consider who is to do it. The manufacturer may be responsible. Within a manufacturing organization, another department may be responsible. If replacement is your task, proceed with caution. Desolder connections, disconnect leads, and remove the part. In these replacement procedures, be sure that you do not cause external damage or additional internal damage to the product.

Once the part is removed, the new one should be installed. The quality of your work here should be similar to that of the original assembler. The old part should be tagged and saved or safely disposed of. Repaired equipment should be operated for a reasonable period of time to ensure that the defect is corrected. A repaired product should also be cleaned and loose parts tightened.

Now, the paper work can be completed. Service reports list the owner of the product, the product's name, and its suspected defect. The final part of this report indicates what was done, the items that were replaced, the adjustments that were made, the time spent, and the cost of this work. When you reach this point, you are sure that the product is working according to its specifications.

CHAPTER 32

REVIEW TERMS

circuit specifications:
description of voltages, resistances, waveforms, and other behaviors to be expected from a circuit or product during normal operation

preventive maintenance:
scheduled process of inspection and correction conducted to ensure proper operation of a product

quality control: process in which components and products are inspected at various stages of manufacture for the early detection of defects

significant difference:
difference between measured and specified values that is beyond what is considered normal or acceptable

REVIEW QUESTIONS

1. Name three reasons for evaluating the performance of a circuit.

2. Explain the difference between symptoms and cause.

3. Explain why a technician must not only identify defective components but also find the cause of the defect.

4. Name two methods you can use to determine how a circuit should operate.

5. Of the two methods you named in Question 4, which is the best? Why?

6. Describe what is meant by quality control.

7. Explain why your test methods are as critical as your results.

8. Describe how you can perform some troubleshooting without using test equipment and without operating the malfunctioning product.

9. Describe what is meant by common.

10. Name four kinds of information that can be found on a manufacturer's schematic.

11. Explain why voltage measurement is a preferred method of troubleshooting.

12. Explain why a technician should know what type of instrument was used to measure specified values.

13. What is an important difference between the voltage indication of an oscilloscope and that of a meter?

14. Why is meter sensitivity a consideration in circuit testing?

15. Explain how you can decide whether the difference between an indicated voltage and its specified value is significant.

16. Describe the procedures you use to locate an open circuit.

17. Describe the procedures you use to locate a short circuit.

18. What is the primary characteristic of acceptable repair work?

19. What type of evaluation is performed after a product has been repaired?

20. Explain how knowledge, skill, attitude and experience are part of the evaluation-repair process.

EVALUATION OF A POWER SUPPLY

Determine, through measurement, the normal output characteristics of a power supply.

Determine the regulation percentage and ripple percentage of a power supply.

Describe the symptoms of some common defects in a power supply.

CIRCUIT TO BE EVALUATED

The objective of this chapter is to apply your electronics knowledge and skill to the evaluation of a DC power supply. The exercise described in this section will give you an opportunity to experience the evaluation of a circuit design. Components will be selected according to the design specifications. Then, the circuit will be assembled. Measurements will be made to determine how the circuit is actually performing. The first step is to identify the circuit and describe what it is expected to do.

Circuit Description

The circuit you will work with is a low-voltage DC power supply. As Figure 33–1 shows, it is a full-wave bridge rectifier with a pi filter. Its output is intended to be slightly above 12 volts without a load connected and slightly below 12 volts with a full load. The load will be simulated by a resistor (R_L) that you will add. The resistor value should be such that it will demand the specified full load of this circuit, which is 150 milliamperes. The bleeder resistor (R_B) will always be connected.

The measurements you will make are DC output voltages with no load and with a full load as well as ripple output under no load and full load. This data will be used to determine the regulation and ripple percentages of the circuit. Regulation will be described in more detail later in this chapter. The percentage of regulation indicates how constant the output remains as the load current varies. You have already been introduced to ripple (see Chapter 31).

When these voltage tests are finished, you will perform a few more tests in order to become familiar with the symptoms caused by some common defects. After you evaluate this data, you should be familiar with the normal and abnormal operation of a full-wave bridge rectifier with a pi section filter.

Selecting Components

A good place to begin selecting components is at the load. Resistor R_L is intended to draw 150 milliamperes at 12 volts, so its resistance and power values can be calculated as follows:

$$R_L = \frac{V_L}{I_L} = \frac{12 \text{ V}}{0.150 \text{ A}} = \textbf{80 } \mathbf{\Omega}$$

$$P_L = VI = 12 \text{ V} \times 0.150 \text{ A} = \textbf{1.8 W}$$

Since this resistance value is not standard, you will need to produce it by combining standard values in parallel. Four 330-ohm resistors in parallel will be close enough.

The power rating of 1.8 watts does not take ripple into consideration. Therefore, we should consider the worst possible case. The worst case will occur if the peak voltage occurs and causes a peak current. For this case, we begin with the secondary voltage of the transformer and assume no losses between it and the load. The calculations are

$$V_{peak} = \frac{V_{rms}}{0.707} = \frac{12.6 \text{ V}}{0.707} = \textbf{17.8 V}$$

$$I_{peak} = \frac{V_{peak}}{R_L} = \frac{17.8 \text{ V}}{80 \text{ }\Omega} = \textbf{0.222 A}$$

$$P_{peak} = V_{peak} \times I_{peak}$$
$$= 17.8 \text{ V} \times 0.222 \text{ A} = \textbf{3.95 W}$$

It is wise to double an estimated value when you are selecting an actual power rating. For this reason, an 8-watt power rating is advised. Thus, your load resistance can be four 330-ohm, 2-watt resistors in parallel.

The next component to be selected is the bleeder resistor. It should have a current flow of approximately a tenth of the supply's maximum load current. The calculations are

FIGURE 33-1
POWER SUPPLY CIRCUIT

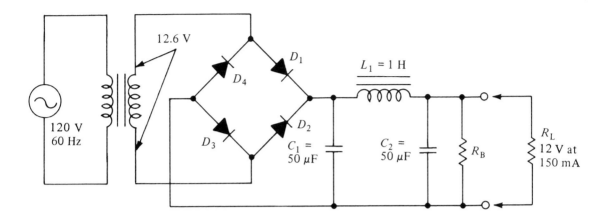

$$I_B = 0.1 \times I_L = 0.1 \times 0.150 \text{ A}$$

$$= \mathbf{0.015 \text{ A}}$$

$$R_B = \frac{V_L}{I_B} = \frac{12 \text{ V}}{0.015 \text{ A}} = \mathbf{800 \ \Omega}$$

Since 800 ohms is the minimum value for a bleeder resistor, a 1000-ohm resistor is a reasonable choice. Once again, you can determine the peak power it will dissipate by calculating peak current:

$$I_{\text{peak}} = \frac{V_{\text{peak}}}{R_B} = \frac{17.8 \text{ V}}{1000 \ \Omega} = \mathbf{0.0178 \text{ A}}$$

$$P_{\text{peak}} = V_{\text{peak}} \times I_{\text{peak}}$$

$$= 17.8 \text{ V} \times 0.0178 \text{ A} = \mathbf{0.316 \text{ W}}$$

For a margin of safety, the bleeder can be a 1000-ohm, 1-watt resistor.

Filter capacitors are selected next, and 50 microfarads have been specified for each. The concern here is their voltage rating. Filter capacitors generally have an average or working voltage rating (VDCW), and they can be safely operated at that value or well below it. Since 15 and 25 volts are standard values for capac-

itors, a reasonable margin is provided if you select 50-microfarad, 25-VDCW capacitors.

The next component to be selected is the inductor or filter choke, and it has a specified value of 1 henry. The combined load and bleeder currents are about 165 milliamperes. But current may also flow through C_2, and the inductor current rating must be higher than 165 milliamperes. In addition, a higher-current inductor has larger wire and lower DC resistance, which will reduce DC losses. The disadvantage of this higher current rating is that it requires a physically larger component. However, since this test is only an experiment, a large component is not a problem. Therefore, a 1-henry inductor rated between 300 and 500 milliamperes will be satisfactory.

The four diodes should be able to safely tolerate two parameters: maximum current (I_{max}) and peak inverse voltage (PIV). Although the normal average current in this circuit approaches 200 milliamperes, a much larger surge occurs when the circuit is first energized and C_1 begins to charge. Remember that an uncharged capacitor provides very little opposition. This surge current can be dif-

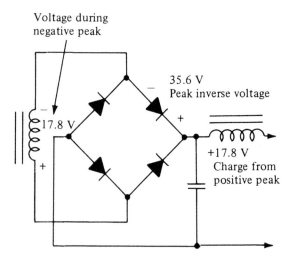

Voltage during negative peak

17.8 V

35.6 V
Peak inverse voltage

+17.8 V
Charge from positive peak

former rated at 0.5, 1.0, or 1.2 amperes will be satisfactory.

Fabricating the Circuit

Since this test is an experiment, the assembly should not be permanent. Some disconnections will be made in the process of taking measurements. However, the possibility of shock, shorts, and other dangers does arise when components are electrically connected and mechanically insecure. For this reason, some form of secure breadboarding should be used. As always, have the circuit checked before you energize it.

Before you begin your measurements, you should prepare a table for your data. Table 33–1 is an example of the type of table you need. Note that, in order for sample calculations to be presented in this chapter, hypothetical measurements will be used. These values may not be the same as the results you get, but the process should be the same.

DC OUTPUT VOLTAGE

Once your circuit is assembled, inspected, and energized, you can begin your measurements. These measurements will indicate just how the circuit performs. Since the circuit is expected to provide voltage to some other product, it must be evaluated with and without a load.

No-Load Output

Because you will be testing the circuit first under no-load conditions, R_L should not be connected. However, R_B should always be connected. For this measurement, you can use a multimeter. It does not need to have a high sensitivity. Even a 1000-ohm-per-volt meter on the 25-volt range will place 25,000 ohms

ficult to determine, but a reasonable margin can be estimated. Diodes with an I_{max} of 1 ampere should be adequate.

Calculation of peak inverse voltage is done next. Imagine that capacitor C_1 charges up to the first positive peak. It reaches and can hold a level of 17.8 volts. The transformer voltage changes, and within 180°, the upper winding of the secondary has reached its negative peak of -17.8 volts. Thus, the diode is momentarily reverse-biased with 35.6 volts, as indicated in Figure 33–2. For a reasonable operating margin, the diodes should have an I_{max} of 1 ampere and a PIV of 50 volts.

The last component to be selected is the transformer, and it has a specified secondary voltage of 12.6 volts. We have already estimated an average power supply current of approximately 200 milliamperes, and most transformers with the specified voltage far exceed that capacity. Therefore, a 12.6-volt trans-

TABLE 33-1
TEST RESULTS

Condition	Output Voltage		
	DC	AC Peak to Peak	AC rms
No load			
Full load			
D_1 open			
C_1 open			
C_2 open			

across the 1000-ohm resistor. Since the meter represents a resistance 25 times larger than the load it is connected in parallel with, it will have little loading effect. Loading error should not be a problem.

The measurement you are to make is DC output voltage. It is likely to be greater than 12 volts. Once you measure it, you should record the value in the table and de-energize the circuit.

Full-Load Output

Your first step in this full-load test is to connect the load resistance (R_L) across the output terminals. Now, you can energize the circuit and measure and record the DC output voltage. It will be lower than the unloaded voltage for several reasons. The increased current flow owing to the change from bleeder current only to bleeder current plus load current causes a larger voltage drop across the DC resistance of the filter choke. In addition, the filter capacitors discharge a greater amount between pulses because of the increased load.

Once this measurement is made, you can de-energize the circuit and determine the regulation percentage of this power supply.

Regulation Percentage

The regulation percentage describes how constant the output voltage of a power supply remains from no load to a full load. A low percentage indicates a more constant output voltage. Assume that this supply has an output voltage of 12.7 volts with no load and 9.7 volts with a full load. The regulation percentage is calculated as follows:

$$\text{regulation \%} = \frac{V_{NL} - V_{FL}}{V_{FL}} \times 100\%$$

$$= \frac{12.7 \text{ V} - 9.7 \text{ V}}{9.7 \text{ V}} \times 100\%$$

$$= \mathbf{30.9\%}$$

This value is too high. A more efficient power supply will have less of a voltage change under a full load. For example, the full-load voltage should drop to only 12.2 volts. For this situation, the regulation percentage is

$$\text{regulation \%} = \frac{V_{NL} - V_{FL}}{V_{FL}} \times 100\%$$

$$= \frac{12.7 \text{ V} - 12.2 \text{ V}}{12.2 \text{ V}} \times 100\%$$

$$= \mathbf{4\%}$$

An ideal power supply would have no change in output voltage for all specified loads, which would lead to a regulation of 0%.

One way to compensate for loading effects is to use an adjustable power supply. An adjustable power supply can be obtained in a variety of ways. An adjustable transformer or an autotransformer can be added at the input and turned to zero before the supply is energized. Then, the load can be connected with a voltmeter across it. Next, the power supply is energized and adjusted to the desired load voltage. Other, more complex regulation circuits are also used; you will learn about them in your next course.

Effects of a Bleeder Resistor

To determine the effects of the bleeder resistor, disconnect R_L, connect the voltmeter across R_B, and energize and de-energize the circuit. Notice how long it takes C_2 to discharge. With the 1000-ohm resistor R_B across the capacitor C_2, the time for discharge should be about 5 time constants. It should discharge quickly, as indicated by the following calculation:

$$\text{discharge time} = 5t = 5 \times RC$$
$$= 5 \times 1 \times 10^3 \ \Omega \times 50$$
$$\times 10^{-6} \ F$$
$$= \mathbf{0.25 \ s}$$

Now, de-energize the circuit and change R_B from 1000 ohms to 100,000 ohms. This larger value is equivalent to having a poor bleeder. Once R_B is in place, connect a voltmeter across it. Energize and de-energize the circuit. Now, notice how long capacitor discharge takes. It is probably quite slow, as the next calculation shows:

$$\text{discharge time} = 5RC$$
$$= 5 \times 0.1 \times 10^6 \ \Omega \times 50$$
$$\times 10^{-6} \ F$$
$$= \mathbf{25 \ s}$$

Suppose there was no bleeder resistor across the output and no load was connected. How long would discharge take? Hours? Days? In some high-quality circuits, a discharge time of days is possible. For this reason, always observe the following safety precaution.

Safety tip: Never operate a power supply without a bleeder resistor.

When the capacitor has fully discharged, replace the 100,000-ohm resistor with the original 1000-ohm one.

RIPPLE OUTPUT VOLTAGE

Once the DC voltage measurements are complete, the ripple can be measured. Ripple voltage and ripple percentage provide an indication of the quality or smoothness of a DC voltage. Like DC voltage, ripple must be measured under both load and no-load conditions.

No-Load Output

Ripple is best-measured with an oscilloscope. The results you obtain will be peak-to-peak values, which you will later convert to rms values. Connect an oscilloscope across the power supply terminals, and energize the circuit. This first measurement is across R_B with R_L removed. Once the value is measured, you can de-energize the circuit.

Full-Load Output

Connect R_L across R_B and the oscilloscope. Energize the circuit, measure and record the peak-to-peak ripple voltage, and de-energize the circuit. The full-load ripple should be larger and have a different shape than the no-load ripple. With no load, the capacitor maintains its charge longer, and the inductor has little effect. When a substantial load is added, the capacitor will discharge more, but the inductor will filter more. The ripple waveshape should change from a sawtooth to a larger sine wave.

Ripple Percentage

Another way to describe the output quality of a power supply is with its ripple percentage. Ripple percentage indicates how much ripple you get from the DC voltage. A low ripple percentage indicates a high-quality filter. You can determine the ripple percentage for this circuit with the data in your table. Begin by converting the peak-to-peak values to rms values. Let's assume that the no-load ripple was 0.026 volt rms. The calculation is

$$\text{ripple } \% = \frac{V_{\text{rms}}}{V_{\text{DC}}} \times 100\%$$
$$= \frac{0.026 \text{ V}}{12.7 \text{ V}} \times 100\% = \mathbf{0.2\%}$$

When the load was added, the ripple voltage increased and the DC voltage decreased. Assume that the DC voltage dropped to 10.5 volts, and the ripple increased to 1.2 volts. Once again, determine the ripple percentage:

$$\text{ripple } \% = \frac{V_{\text{rms}}}{V_{\text{DC}}} \times 100\%$$
$$= \frac{1.2 \text{ V}}{10.5 \text{ V}} \times 100\% = \mathbf{11.4\%}$$

This result is not very desirable for most applications.

Ripple percentage, like regulation, can be improved with added circuitry. For example, power supplies can be obtained having both of these specifications as a fraction of 1%. Whenever a power supply is needed, these four specifications should be known:

1. Output voltage,
2. Rated load,
3. Regulation percentage,
4. Ripple percentage.

SYMPTOMS OF COMMON DEFECTS

The ability to recognize the symptoms of common defects is a valuable asset when you are troubleshooting electronic circuits. Other than an open fuse or circuit breaker, an open diode is the most common defect in power supplies. Shorted, open, and leaking capacitors are also common. The symptoms produced by these defects are usually recognizable to a skilled technician.

Open Diode

Power supply overload or a shorted filter capacitor can often cause excessive diode current. This excess current, in turn, can cause a diode to open. Figure 33–3 shows the circuit that will result. Notice that the open branch causes the circuit to become a half-wave rectifier. Conduction can only occur during half of the cycle.

This defect causes some very obvious symptoms visible at the load. Recall that a half-wave rectifier produces only half the voltage that a full-wave bridge does, using the same transformer. Therefore, this symptom should indicate to you that a diode may be open. Another symptom is the change in ripple, another difference you should recall about half-wave rectifiers. The defect should cause the ripple frequency to change from 120 to 60 hertz.

As part of this study, you should remove one diode and energize the circuit. Compare the output characteristics of the open-diode circuit with the original normal circuit. Do this test under full load. Measure the DC and ripple levels, and notice the ripple waveshape. Have they changed?

Now, measure the DC voltage across the three good diodes and across the location where the defective one should be. Notice the difference. The voltage across the three good diodes will be somewhat similar, but the voltage across the open diode will probably be much larger. As mentioned in earlier chapters, the voltage across an open will be much larger than expected and will probably be equal to the source voltage. When you are finished ex-

FIGURE 33-3
CIRCUIT WITH AN OPEN DIODE

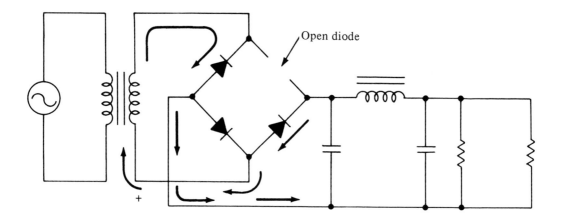

Open diode

amining this defect, de-energize and correct the circuit.

Open Input Capacitor

One way to consider the open-capacitor problem is to imagine what the input capacitor does. First, it charges to the peak value of the input signal. Therefore, opening or removing this capacitor should cause the output voltage to drop. It will not drop as much as it would with an open diode, however. The other contribution of this capacitor is the filtering of ripple. Again, its removal will cause the ripple to increase.

You can simulate this defect by removing the capacitor, as indicated in Figure 33–4. Energize the circuit, and make some measurements. Use the full-load condition for comparison purposes. Once again, measure and record the DC output voltage, ripple voltage, and ripple frequency. Note the parameters that have changed and how much they have changed.

You should have noticed two characteristics. An open input capacitor will reduce the average value of the DC voltage and increase the value of the ripple. If these symptoms were detected, the capacitor can be removed and checked with a capacitor checker. Or a temporary replacement can be connected in parallel with the defective capacitor to see whether the conditions improve. When you have finished taking your measurements, disable and correct the circuit.

Open Output Capacitor

The capacitor at the output side of the filter has an important role in removing ripple. Its effect on the DC voltage level may not be as noticeable as that of the input capacitor. The easiest way to determine its effect is to remove it and compare the results with those obtained when it was in the circuit.

For this test, remove the capacitor, as indicated in Figure 33–5. Energize the circuit, make your measurements, and de-energize the

FIGURE 33-4
CIRCUIT WITH AN OPEN INPUT CAPACITOR

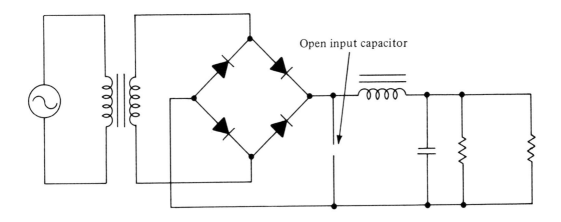

Open input capacitor

FIGURE 33-5
CIRCUIT WITH AN OPEN OUTPUT
CAPACITOR

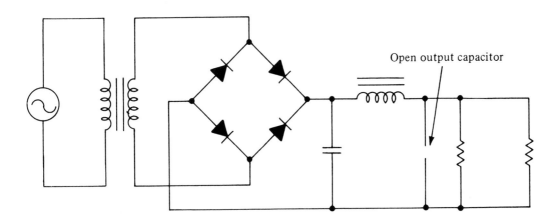

Open output capacitor

FIGURE 33-6
CIRCUIT WITH A SHORTED LOAD

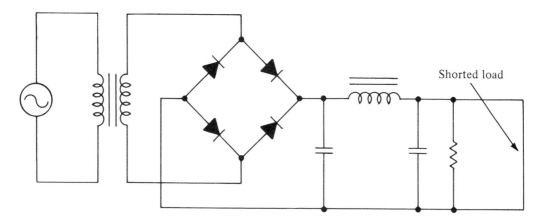

Shorted load

circuit. You are now finished with the measurement part of these experiments, so you can set the assembly aside.

Consider your results. Notice that the ripple is again higher than expected. Also, the open output capacitor has less effect on the DC output voltage because it is lower than it should be but not as low as it was when the input capacitor was open. In other words, both capacitors have an effect on ripple and DC average, but the input capacitor has the greatest effect.

Shorts

The effects of shorts are for discussion only since experiments with shorts cause damage. Consider first a short in the load, as shown in Figure 33–6. The immediate effect of this short would be an increase in output current, because the only opposition is the DC resistance of the filter choke. Since that resistance is quite low, excessive current would flow. The excess flow would stop quickly — as soon as the I_{max} rating of some component is reached. That component is likely to be a diode.

A shorted filter capacitor (Figure 33–7) would have a similar effect on the circuit. Excess current would flow through the diodes. Although capacitors do sometimes short, a more common problem is capacitors that leak. That is, their dielectric begins to fail, and they allow more and more electrons to pass from one plate to the other. This problem does not cause an immediate short. Rather, it allows the rectifier current to increase gradually above its limit. Leaky capacitors can be detected by removing and testing them with a leakage tester.

Fuses are a normal part of most power supplies. The output fuse in Figure 33–8 prevents damage caused by an overload. The input fuse protects the equipment from internal shorts.

Fuses are an inexpensive investment and are easy to replace. If fuses are not used, the damage and expense can be considerable. A short in an unfused power supply leads to downtime, troubleshooting, and expenses for parts and labor. Thus, fuses are a symbol of an important point.

Key point: It is much easier to avoid damage than to repair it.

FIGURE 33-7
CIRCUIT WITH A SHORTED CAPACITOR

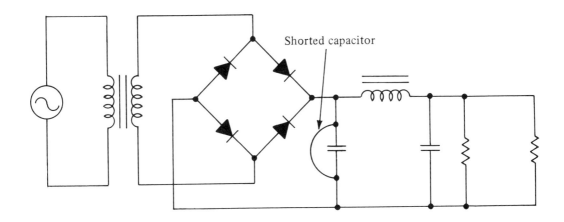

Shorted capacitor

SUMMARY

The selection of components for evaluating a power supply is accomplished by using specifications or calculations. If specific values are not provided, you must calculate them. Calculations are done by proceeding from the power source to the load and back again. Once approximate values are determined, a margin should be added. The magnitude of the margin varies, but a useful guideline is to use diodes with an I_{max} twice that needed and to use resistors with double the estimated power.

To assemble an experimental circuit, you connect it so that you can take it apart easily. At the same time, it should not fall apart with handling. However, do not attach the components so securely that they must be damaged to be disconnected. Likewise, make sure that wires do not touch each other when you try to make other connections. You should use one of the many breadboarding systems available today.

Power supplies are tested for three basic characteristics: output voltage, regulation percentage, and ripple percentage. The output voltage is the average DC voltage that appears across the output terminals.

Regulation percentage gives an additional description of the output voltage value. Regulation percentage describes how well a power supply maintains a constant output voltage from no-load to full-load conditions. The lower this number is, the better-regulated the supply is.

Ripple percentage is another specification that further describes the output voltage. Just as regulation indicates stability, ripple percentage indicates quality. It is equal to the rms value of ripple divided by the average value of the DC voltage times 100%. A low ripple percentage is an indication of a smooth output voltage.

Some common defects that can occur in a power supply are open diodes and shorts. An open diode in a full-wave bridge rectifier will reduce the DC output voltage by 50%, increase the ripple value, and decrease the ripple frequency from 120 to 60 hertz. An open diode can be found by measuring the voltage drop across each diode in the bridge. Open components have significantly higher voltage drops.

FIGURE 33-8
FUSE LOCATIONS

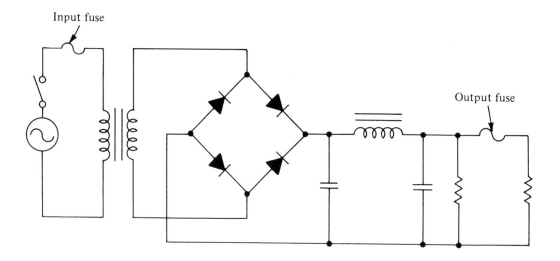

Input fuse

Output fuse

An open input capacitor will cause a small reduction in DC output voltage and an increase in the ripple magnitude. If the output capacitor opens, it will cause an increase in ripple magnitude and a change in the DC voltage level. The effect of an open output capacitor on DC voltage may not be as noticeable as the effect of an open input capacitor. Questionable capacitors can be removed and tested for capacitance and leakage with a capacitor checker.

Shorts cause other problems. The most frequent result of a short is an open diode or a component in series with it. A short in an unfused power supply will cause excessive current to flow, which, in turn, will open a diode. Replacing the open diode will not correct the problem. If the short is not also corrected, an open diode will recur. Once again, curing symptoms is not sufficient; you must also correct the cause of the problem. The best way to test a circuit suspected of having shorts is to take measurements with the device removed from the power source.

REVIEW QUESTIONS

1. Describe no-load output voltage.

2. Describe full-load output voltage.

3. What is regulation percentage?

4. What is ripple percentage?

5. Name four measurements you should make to determine the operating characteristics of a power supply.

6. List two calculations you can do by using the data you named in Question 5.

7. Explain why a bleeder resistor is essential.

8. Describe some possible dangers of working with power supplies.

9. What output symptoms will be displayed by an open diode in a half-wave rectifier circuit? In a full-wave circuit? In a full-wave bridge?

10. What symptoms will be displayed by an open filter capacitor? By a shorted filter capacitor?

11. What symptoms will be displayed by an open filter choke?

12. What steps should you follow to troubleshoot the defects described in Question 9?

13. What steps should you follow to troubleshoot the defects described in Question 10?

14. What steps should you follow to troubleshoot the defect described in Question 11?

15–18. Consider the circuit in Figure 33–9, and suppose that it has been measured on

FIGURE 33-9

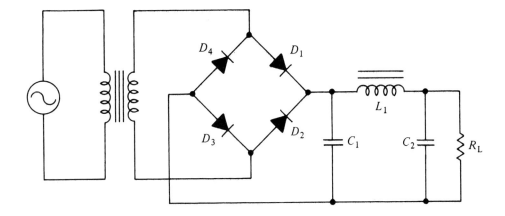

four different occasions. The results are presented in Table 33–2. For each set of measured values in Table 33–2, determine whether or not the results indicate accept-

able operation. If the operation is not acceptable, state why it is not; and give your opinion of the probable defect.

TABLE 33-2
POWER SUPPLY VOLTAGES

Data	Component					
	C_1		L_1		$C_2 - R_L$	
Voltage	DC	AC	DC	AC	DC	AC
Specified value	350	2	10	1.5	340	0.5
Question 15 measured value	175	1	5	0.75	170	0.25
Question 16 measured value	350	1	350	1	0	0
Question 17 measured value	350	2	10	1	340	1
Question 18 measured value	350	25	9	18	320	6

APPENDIX TABLES

TABLE A-1
RESISTIVITY OF MATERIALS

Material	Resistivity (in Ohm-meters)
Silver	1.47×10^{-8}
Copper	1.72×10^{-8}
Gold	2.44×10^{-8}
Aluminum	2.62×10^{-8}
Iron	9.71×10^{-8}
Germanium	4.5×10^{-1}
Silicon	8.5×10^{2}
Bakelite	2×10^{6}
Glass	1×10^{10}
PVC	1×10^{14}
Mica	1×10^{15}
Quartz	1×10^{17}

TABLE A-2
APPROXIMATE CAPACITIES OF WIRES (VARIES WITH MATERIAL AND TEMPERATURE)

Wire Size	Current (in Amperes)
10	30
12	20
14	15
16	10
18	7
20	4
22	2
24	1
26	0.7
28	0.4

TABLE A-3
COPPER WIRE TABLE

Gage Number	Diameter (in Mils)	Square Inches	Ohms per 1000 Feet 25°C (77°F)	Ohms per 1000 Feet 65°C (149°F)
0000	460.0	0.166	0.0500	0.0577
000	410.0	.132	.0630	.0727
00	365.0	.105	.0795	.0917
0	325.0	.0829	.100	.116
1	289.0	.0657	.126	.146
2	258.0	.0521	.159	.184
3	229.0	.413	.201	.232
4	204.0	.0328	.253	.292
5	182.0	.0260	.319	.369
6	162.0	.0206	.403	.465
7	144.0	.0164	.508	.586
8	128.0	.0130	.641	.739
9	114.0	.0103	.808	.932
10	102.0	.00815	1.02	1.18
11	91.0	.00647	1.28	1.48
12	81.0	.00513	1.62	1.87
13	72.0	.00407	2.04	2.36
14	64.0	.00323	2.58	2.97
15	57.0	.00256	3.25	3.75
16	51.0	.00203	4.09	4.73
17	45.0	.00161	5.16	5.96
18	40.0	.00128	6.51	7.51
19	36.0	.00101	8.21	9.48
20	32.0	.000802	10.4	11.9
21	28.5	.000636	13.1	15.1
22	25.3	.000505	16.5	19.0
23	22.6	.000400	20.8	24.0
24	20.1	.000317	26.2	30.2
25	17.9	.000252	33.0	38.1
26	15.9	.000200	41.6	48.0
27	14.2	.000158	52.5	60.6
28	12.6	.000126	66.2	76.4
29	11.3	.0000995	83.4	96.3
30	10.0	.0000789	105.0	121.0
31	8.9	.0000626	133.0	153.0
32	8.0	.0000496	167.0	193.0
33	7.1	.0000394	211.0	243.0
34	6.3	.0000312	266.0	307.0
35	5.6	.0000248	335.0	387.0
36	5.0	.0000196	423.0	488.0
37	4.5	.0000156	533.0	616.0
38	4.0	.0000123	673.0	776.0
39	3.5	.0000098	848.0	979.0
40	3.1	.0000078	1,070.0	1,230.0

TABLE A-4
RESISTOR COLOR CODE

Color	Number	Multiplier	Tolerance
Black	0	1	
Brown	1	10	1%
Red	2	100	2%
Orange	3	1,000	
Yellow	4	10,000	
Green	5	100,000	0.5%
Blue	6	1,000,000	0.25%
Violet	7	10,000,000	0.10%
Gray	8		0.05%
White	9		
Silver		0.01	10%
Gold		0.1	5%

First digit
Second digit
Multiplier
Tolerance

TABLE A-5
CAPACITOR COLOR CODE

Color	First Digit (B)	Second Digit (C)	Multiplier (D)	Tolerance (E) More than 10 pF (in Percent)	Tolerance (E) Less than 10 pF (in pF)	Temperature Coefficient (A)
Black	0	0	1.0	±20	±2.0	0
Brown	1	1	10	±1		−30
Red	2	2	100	±2		−80
Orange	3	3	1,000			−150
Yellow	4	4	10,000			−220
Green	5	5		±5	±0.5	−330
Blue	6	6				−470
Violet	7	7				−750
Gray	8	8	0.01		±0.25	+30
White	9	9	0.1	±10	±1.0	+120 to −750 (EIA)
						+500 to −330 (JAN)
Silver						+100 (JAN)
Gold						Bypass or coupling (EIA)

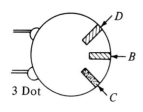

3 Dot

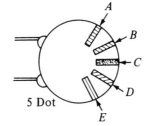

5 Dot

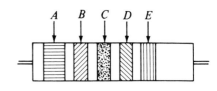

TABLE A-6
TRIGONOMETRIC TABLE

Angle	Sine	Cosine	Tangent	Angle	Sine	Cosine	Tangent	Angle	Sine	Cosine	Tangent
1°	.0175	.9998	.0175	31°	.5150	.8572	.6009	61°	.8746	.4848	1.8040
2°	.0349	.9994	.0349	32°	.5299	.8480	.6249	62°	.8829	.4695	1.8807
3°	.0523	.9986	.0524	33°	.5446	.8387	.6494	63°	.8910	.4540	1.9626
4°	.0698	.9976	.0699	34°	.5592	.8290	.6745	64°	.8988	.4384	2.0503
5°	.0872	.9962	.0875	35°	.5736	.8192	.7002	65°	.9063	.4226	2.1445
6°	.1045	.9945	.1051	36°	.5878	.8090	.7265	66°	.9135	.4067	2.2460
7°	.1219	.9925	.1228	37°	.6018	.7986	.7536	67°	.9205	.3907	2.3559
8°	.1392	.9903	.1405	38°	.6157	.7880	.7813	68°	.9272	.3746	2.4751
9°	.1564	.9877	.1584	39°	.6293	.7771	.8098	69°	.9336	.3584	2.6051
10°	.1736	.9848	.1763	40°	.6428	.7660	.8391	70°	.9397	.3420	2.7475
11°	.1908	.9816	.1944	41°	.6561	.7547	.8693	71°	.9455	.3256	2.9042
12°	.2079	.9781	.2126	42°	.6691	.7431	.9004	72°	.9511	.3090	3.0777
13°	.2250	.9744	.2309	43°	.6820	.7314	.9325	73°	.9563	.2924	3.2709
14°	.2419	.9703	.2493	44°	.6947	.7193	.9657	74°	.9613	.2756	3.4874
15°	.2588	.9659	.2679	45°	.7071	.7071	1.0000	75°	.9659	.2588	3.7321
16°	.2756	.9613	.2867	46°	.7193	.6947	1.0355	76°	.9703	.2419	4.0108
17°	.2924	.9563	.3057	47°	.7314	.6820	1.0724	77°	.9744	.2250	4.3315
18°	.3090	.9511	.3249	48°	.7431	.6691	1.1106	78°	.9781	.2079	4.7046
19°	.3256	.9455	.3443	49°	.7547	.6561	1.1504	79°	.9816	.1908	5.1446
20°	.3420	.9397	.3640	50°	.7660	.6428	1.1918	80°	.9848	.1736	5.6713
21°	.3584	.9336	.3839	51°	.7771	.6293	1.2349	81°	.9877	.1564	6.3138
22°	.3746	.9272	.4040	52°	.7880	.6157	1.2799	82°	.9903	.1392	7.1154
23°	.3907	.9205	.4245	53°	.7986	.6018	1.3270	83°	.9925	.1219	8.1443
24°	.4067	.9135	.4452	54°	.8090	.5878	1.3764	84°	.9945	.1045	9.5144
25°	.4226	.9063	.4663	55°	.8192	.5736	1.4281	85°	.9962	.0872	11.4301
26°	.4384	.8988	.4877	56°	.8290	.5592	1.4826	86°	.9976	.0698	14.3007
27°	.4540	.8910	.5095	57°	.8387	.5446	1.5399	87°	.9986	.0523	19.0811
28°	.4695	.8829	.5317	58°	.8480	.5299	1.6003	88°	.9994	.0349	28.6363
29°	.4848	.8746	.5543	59°	.8572	.5150	1.6643	89°	.9998	.0175	57.2900
30°	.5000	.8660	.5774	60°	.8660	.5000	1.7321	90°	1.0000	.0000	

Source: Reprinted courtesy of U.S. Air Force.

ANSWERS

SELECTED REVIEW PROBLEMS

CHAPTER 6
1. $100,000\ \Omega \pm 10\%$ **3.** $39\ \Omega \pm 10\%$ **5.** $490,000\ \Omega,\ 0.33\ \Omega$
7. $90,000\ \Omega,\ 110,000\ \Omega$ **9.** $35.1\ \Omega,\ 42.9\ \Omega$
11. $68,000\ \Omega,\ \pm 10\%,\ 61,200\ \Omega,\ 74,800\ \Omega$
     $2,700\ \Omega,\ \pm 5\%,\ \ 2,565\ \Omega,\ \ 2,835\ \Omega$
13. $472,000\ \Omega$ **15.** $\pm 2\%,\ 462,560\ \Omega,\ 481,440\ \Omega$

CHAPTER 7
1. $0.54\ \text{V},\ 270\ \text{V}$ **3.** $1.5\ \text{mA},\ 0.03\ \text{A}$ **5.** $1.6\ \text{V}$
7. $0.563\ \text{V},\ 0.588\ \text{V}$ **9.** $0.571\ \text{V},\ 0.579\ \text{V}$
11. 11.1% **13.** 8.3% **15.** $2\ \text{M}\Omega,\ 500\ \text{k}\Omega$

CHAPTER 8
1. $6\ \text{mA},\ 54\ \text{mW}$ **2.** Increase to $216\ \text{mW}$
5. $7.52\ \text{V}$ **7.** $93.8\ \text{mW}$
10. $R_T = 8\ \text{k}\Omega,\ I_T = 5\ \text{mA},\ V_1 = 11\ \text{V},\ V_2 = 16.5\ \text{V},\ V_3 = 5\ \text{V},\ V_4 = 7.5\ \text{V},\ P_1 = 55\ \text{mW},\ P_2 = 82.5\ \text{mW},$
$P_3 = 25\ \text{mW},\ P_4 = 37.5\ \text{mW},\ P_T = 200\ \text{mW}$
11. $R_T = 8\ \text{k}\Omega,\ I_T = 7.5\ \text{mA},\ V_1 = 16.5\ \text{V},\ V_2 = 24.75\ \text{V},\ V_3 = 7.5\ \text{V},\ V_4 = 11.25\ \text{V},\ P_1 = 124\ \text{mW},\ P_2 = $
$186\ \text{mW},\ P_3 = 56\ \text{mW},\ P_4 = 84\ \text{mW},\ P_T = 450\ \text{mW}$
12. $R_T = 7\ \text{k}\Omega,\ I_T = 5.7\ \text{mA},\ V_1 = 12.57\ \text{V},\ V_2 = 18.86\ \text{V},\ V_3 = 0\ \text{V},\ V_4 = 8.57\ \text{V},\ P_1 = 71.8\ \text{mW},\ P_2 = $
$107.8\ \text{mW},\ P_3 = 0\ \text{W},\ P_4 = 49\ \text{mW},\ P_T = 228.6\ \text{mW}$

CHAPTER 9
1. R_2 changed to $790\ \Omega$ **3.** R_2 opened
5. R_3 opened   **7.** Open lamp
9. $18.2\ \text{mA},\ 22.2\ \text{mA},\ 19\text{mA},\ 21.1\ \text{mA}$
11. $22.7\ \text{mA}$ ($44.9\ \text{V}$ and $1980\ \Omega$)

CHAPTER 12
1. 0.5 A-turns, 64 A-turns
3. 20 A-turns/m, 2.56 kA-turns/m, 10 A-turns/m, 1.28 kA-turns/m
5. $125\ \mu\text{V}$ **7.** $4\ \text{H}$ **9.** $4.6\ \text{H},\ 3.4\ \text{H}$
11. $4.55\ \text{ms}$

13. t = 4.55, 9.1, 13.6, 18.2, 22.8 ms; I_L = 22.9, 31.6, 34.5, 35.6, 36 mA; V_R = 7.6, 10.4, 11.4, 11.7, 11.9 V; V_L = 4.4, 1.6, 0.6, 0.3, 0.1 V

CHAPTER 13

1. 442 pF   **3.** 0.029 µF   **5.** 2045 pF

7. 0.005 µF, 0.02 µF

9. 7.5 V (0.01 µF), 2.5 V (0.03 µF)

11. 1.5 ms

13. t = 1.5, 3.0, 4.5, 6.0, 7.5 ms; I = 4.4, 1.6, 0.6, 0.2, 0.1 mA; V_C = 7.6, 10.4, 11.4, 11.8, 11.9 V; V_R = 4.4, 1.6, 0.6, 0.2, 0.1 V

CHAPTER 14

1. 24 Ω   **3.** 66 Ω   **5.** 97 Ω

7. I_1 = 30 mA, I_2 = 20 mA, I_T = 50 mA, R_T = 360 Ω, P_1 = 0.54 W, P_2 = 0.36 W, P_T = 0.9 W

9. 200 kΩ: I_1 = I_2 = I_3 = I_4 = 240 µA, I_T = 960 µA, P_1 = P_2 + P_3 = P_4 = 11.52 mW, P_T = 46.1 mW

1.2 MΩ: I_1 = I_2 = I_3 = I_4 = 40 µA, I_T = 160 µA, P_1 = P_2 = P_3 = P_4 = 1.92 mW, P_T = 7.68 mW

11. 200 kΩ: I_1 = I_2 = I_3 = 180 µA, I_T = 540 µA, P_1 = P_2 = P_3 = 6.48 mW, P_T = 19.4 m

1.2 MΩ: I_1 = I_2 = I_3 = 30 µA, I_T = 90 µA, P_1 = P_2 = P_3 = 1.08 mW, P_T = 3.24 mW

13. I_1 = 60 mA, I_2 = 40 mA, I_T = 100 mA, P_1 = 2.16 W, P_2 = 1.44 W, P_T = 3.6 W

14. I_1 = 15 mA, I_2 = 10 mA, I_T = 25 mA, P_1 = 135 mW, P_2 = 90 mW, P_T = 225 mW

CHAPTER 15

5. R_3 open   **7.** One open resistor

9. R_1 open   **11.** R_2 and R_3 open

13. 21.8 mA, 26.7 mA, 22.9 mA, 25.3 mA

15. 21.4 mA (11.76 V and 550 Ω), 27.2 mA (12.24 V and 450 Ω)

CHAPTER 16

1. 72 Ω   **3.** 30 Ω   **5.** 1333 Ω   **7.** 2.732 MΩ

9. R_T = 3.6 kΩ, I_T = I_1 = 3.33 mA, I_2 = 1.11 mA, I_3 = 2.22 mA, V_1 = 8 V, V_2 = V_3 = 4 V, P_1 = 26.6 mW, P_2 = 4.44 mW, P_3 = 8.8 mW, P_T = 40 mW

12. R_T = 635 kΩ, I_T = 14.2 µA, I_1 = 6 µA, I_2 = I_3 = 8.18 µA, V_1 = 9 V, V_2 = 3.84 V, V_3 = 5.15 V, P_1 = 54 µW, P_2 = 31.4 µW, P_3 = 42.1 µW, P_T = 128 µW

CHAPTER 17

1. R_1 open   **3.** R_3 open   **5.** R_2 open

7. R_2 and/or R_3 shorted   **9.** V_1 = 7.1 V, V_2 = 7.9 V

11. V_1 = 12 V, V_3 = 2 V, V_4 = 10 V

13. R_T increased from 7.92 to 8.75 Ω; I_T decreased from 1.5 to 1.37 A

CHAPTER 18

1. V_T = 50 V, I_T = 0.08 A, I_1 = 0.02 A, I_2 = 0.05 A, I_3 = 0.01 A, P_1 = 1 W, P_2 = 2.5 W, P_3 = 0.5 W, P_T = 4 W

4. I_1 = 0.1 A, I_2 = 0.3 A, I_3 = 0.4 A, V_1 = 1.8 V, V_2 = 10.8 V, V_3 = 7.2 V

6. I_1 = I_5 = 9 mA, I_3 = 23.3 mA, I_2 = I_4 = 14.3 mA, V_1 = 1.98 V, V_2 = 2.57 V, V_3 = 5.57 V, V_4 = 2.85 V, V_5 = 1.44 V

8. $I_1 = 295$ mA, $I_2 = 117$ mA, $I_3 = 412$ mA, $V_1 = 4.72$ V, $V_2 = 2.34$ V, $V_3 = 4.94$ V, $V_{int\ A} = 2.36$ V, $V_{int\ B} = 0.7$ V
10. $I_1 = I_5 = 22.6$ mA, $I_2 = I_4 = 11.5$ mA, $I_3 = 34.7$ mA, $V_1 = 5.4$ V, $V_2 = 3.8$ V, $V_3 = 7.6$ V, $V_4 = 6.4$ V, $V^5 = 10.6$ V, $V_{int\ A} = 0.3$ V, $V_{int\ B} = 0.1$ V
12. $I_5 = 16.7$ mA, $V_5 = 0.833$ V

CHAPTER 19

1. 80 V   **2.** 0.2 mV   **5.** $\times 1000$ (1 mV), $\times 0.1$ (10 V)
7. $+0.15$ V   **9.** -25 μV

CHAPTER 20

1. 77.8 V   **3.** 50 V, 0 V, -100 V   **5.** 800 kHz
7. 16.7 ms   **9.** 5 mA   **11.** 0.8 V pp, 0.283 V rms
13. 100 V pp, 35.4 V rms

CHAPTER 21

3. 15 V   **5.** 5:1   **7.** 25 kΩ **9.** 180 Hz, 300 Hz
11. 6.25 kHz   **13.** 90 Hz   **15.** 5 A, 120 W

CHAPTER 22

1. 8 kΩ **3.** 9.85 kΩ, 49.85 kΩ, 99.85 kΩ, 499 kΩ
5. 10.2 Ω **7.** 1 kΩ/V
10. with meter $= 6.67$ V, without meter $= 13.33$ V   **11.** 33.3%

CHAPTER 24

1. 565.5 Ω, 1131 Ω **3.** 13.6 mA pp
5. $Z = 10,150$ Ω, $I = 4.73$ mA, $V_R = 32.2$ V, $V_L = 35.6$ V, $\theta = 48°$, $P = 152$ mW
9. $I_L = 717$ μA, $I_R = 529$ μA, $I_T = 891$ μA, $Z = 40.4$ kΩ, $\theta = 53.6°$, $P = 19$ mW
13. 39°

CHAPTER 25

1. 1769 Ω, 885 Ω **3.** 2.1 mA pp
5. $Z = 7129$ Ω, $I = 561$ μA, $V_C = 3$ V, $V_R = 2.64$ V, $\theta = 48.7°$, $P = 1.48$ mW
9. $I_C = 746$ μA, $I_R = 851$ μA, $I_T = 1.13$ mA, $Z = 3540$ Ω, $\theta = 41.2°$, $P = 3.4$ mW

CHAPTER 26

1. 59,972 Hz
5. $Z = 6.8$ kΩ, $I = 3.53$ mA, $V_L = V_C = 199$ V, $V_R = 24$ V, $\theta = 0°$
7. 8.3, 7226 Hz
8. Q increased to 83, BW decreased to 722.6 Hz
11. Upper $f = 63,585$ Hz, $X_L = 59,897$ Ω, $X_C = 53,283$ Ω, $Z = 9486$ Ω, $I = 2.53$ mA, $V_R = 17.2$ V, $V_L = 151.5$ V, $V_C = 134.8$ V, $\theta = 44.2°$
Lower $f = 56,359$ Hz, $X_L = 53,090$ Ω, $X_C = 60,114$ Ω, $Z = 9777$ Ω, $I = 2.45$ mA, $V_R = 16.7$ V, $V_L = 130.3$ V, $V_C = 147.6$ V, $\theta = 46°$
13. 0.701, 42.6 mW, 41.4 mW

CHAPTER 27
1. 59,972 Hz
4. Decrease f_r to 51,937 Hz
6. $I_L = 422 \ \mu A$, $I_C = 425 \ \mu A$, $I_R = 160 \ \mu A$, $Z = 150 \ k\Omega$, $\theta = 0°$
8. 8.3, 7226 Hz
10. Upper $f = 63,585$ Hz, $I_L = 398 \ \mu A$, $I_C = 450 \ \mu A$, $I_R = 160 \ \mu A$, $Z = 143 \ k\Omega$, $\theta = 18°$
Lower $f = 56,359$ Hz, $I_L = 448 \ \mu A$, $I_C = 399 \ \mu A$, $I_R = 160 \ \mu A$, $Z = 143 \ k\Omega$, $\theta = 17°$
12. Upper power factor $= 0.951$, $P = 3.8$ mW; lower power factor $= 0.956$, $P = 3.8$ mW

CHAPTER 28
1. Signal, power   **4.** 1N1614, 1N2069   **6.** Good, open

CHAPTER 29
2. Signal, power   **6.** Good   **7.** pnp

CHAPTER 30
1. 270 V DC   **2.** 13.5 V DC   **3.** 161.9 V DC
6. 60 Hz, 848 V pp   **9.** 81 V DC
13. 255 V pp, 60 Hz

CHAPTER 31
2. 161.9 V DC, 170 V rms   **7.** 161.9 V DC, 80.9 V rms
8. 323.7 V DC, 161.7 V rms

INDEX

RESISTORS

INDUCTORS

CAPACITORS

TRANSFORMERS

CONNECTIONS

Connected

- Not Connected